Bioenergy from Wood

Managing Forest Ecosystems

Volume 26

Series Editors:

Klaus von Gadow

Georg-August-University,
Göttingen, Germany

Timo Pukkala

University of Joensuu,
Joensuu, Finland

and

Margarida Tomé

Instituto Superior de Agronomía,
Lisbon, Portugal

Aims & Scope:

Well-managed forests and woodlands are a renewable resource, producing essential raw material with minimum waste and energy use. Rich in habitat and species diversity, forests may contribute to increased ecosystem stability. They can absorb the effects of unwanted deposition and other disturbances and protect neighbouring ecosystems by maintaining stable nutrient and energy cycles and by preventing soil degradation and erosion. They provide much-needed recreation and their continued existence contributes to stabilizing rural communities.

Forests are managed for timber production and species, habitat and process conservation. A subtle shift from *multiple-use management to ecosystems management* is being observed and the new ecological perspective of *multi-functional forest management* is based on the principles of ecosystem diversity, stability and elasticity, and the dynamic equilibrium of primary and secondary production.

Making full use of new technology is one of the challenges facing forest management today. Resource information must be obtained with a limited budget. This requires better timing of resource assessment activities and improved use of multiple data sources. Sound ecosystems management, like any other management activity, relies on effective forecasting and operational control.

The aim of the book series ***Managing Forest Ecosystems*** is to present state-of-the-art research results relating to the practice of forest management. Contributions are solicited from prominent authors. Each reference book, monograph or proceedings volume will be focused to deal with a specific context. Typical issues of the series are: resource assessment techniques, evaluating sustainability for even-aged and uneven-aged forests, multi-objective management, predicting forest development, optimizing forest management, biodiversity management and monitoring, risk assessment and economic analysis.

For further volumes:
http://www.springer.com/series/6247

Thomas Seifert
Editor

Bioenergy from Wood

Sustainable Production in the Tropics

Springer

Editor
Thomas Seifert
Department of Forest and Wood Science
Stellenbosch University
Matieland, Western Cape
South Africa

ISSN 1568-1319 ISSN 1568-1319 (electronic)
ISBN 978-94-007-7447-6 ISBN 978-94-007-7448-3 (eBook)
DOI 10.1007/978-94-007-7448-3
Springer Dordrecht Heidelberg New York London

Library of Congress Control Number: 2013956124

Printed on acid-free paper

Springer is part of Springer Science+Business Media (www.springer.com)

Preface

This book was developed to provide a guideline for biomass production, procurement and energy production for scientists, practitioners, and decision makers who are interested in a value-chain perspective of bioenergy production from wood in tropical and sub-tropical countries. It was written as a collaborative effort of several specialists on the topic, with a core group at the Department of Forest and Wood Science of Stellenbosch University in South Africa. In integrating all the authors' expertise in a multi-disciplinary context, the goal was to address an identified gap of knowledge on dealing with sustainable wood-based bioenergy concepts in the tropical regions of the Southern Hemisphere.

The decision to cover the whole value-chain of bioenergy production from wood was made to provide a holistic view on bioenergy production and pinpoint issues that might be relevant for a truly sustainable implementation. In particular also socio-economic side effects of bioenergy production and effects on local and global environment were addressed. In the development of such a book it is always a balancing act to provide the right degree of detail and scientific depth. Due to size limitations of this book, such a rather wide scope inevitably led to sections that had to be formulated concise and more detail had to be acquired with help of the given references. However, we hope that the chosen approach will provide accessibility to a wider readership and at the same time open up the way for enough in-depth information.

Writing this multi-disciplinary book on bioenergy was a learning experience for all of us authors, and our hope is that this book provides valuable information on the many complex aspects of sustainability that are involved in bioenergy production from wood in the tropics in all its many aspects from the tree to the energy.

Professor for Forest Growth and Yield Science
Department of Forest and Wood Science
University of Stellenbosch, Stellenbosch, South Africa

Thomas Seifert
The editor

Acknowledgements

We would like to dedicate this book to our late colleagues Anthonie van Laar († 2009) and Kobus Theron († 2011), two extraordinary scientists, who pioneered biomass research in South Africa and contributed to this book in the early stages with their helpful comments. Sadly they did not have the opportunity to see this book finished, what would have surely pleased them since it took their work a step further.

The authors would further collectively like to acknowledge the contribution of several persons that provided helpful comments on chapters in this book or contributed illustrations:

Dr. Wale Aboyade
Environmental & Process Systems Engineering Research Group
Chemical Engineering Department, University of Cape Town

Prof. Harro von Blottnitz
Associate Professor at the Department of Chemical Engineering
University of Cape Town, South Africa

Dr. Michel Brienzo
National Institute of Metrology, Quality and Technology, Brazil

Dr. Willie Claassen
CEO EC Biomass, South Africa

Dr. Lutz Fehrmann
Abteilung Waldinventur und Fernerkundung
Fakultät für Forstwissenschaften und Waldökologie
Georg-August-Universität Göttingen, Germany

Dr. Johannes Gediga
PE International
Stuttgart, Germany

Tarquinio Magalhaes
Faculty of Agronomy and Forestry
Eduardo Mondlane University
Maputo, Mozambique

Dr. Matthias Schmidt
Abteilung Waldwachstum
Norwestdeutsche Forstliche Versuchsanstalt
Göttingen, Germany

Prof. Almeida A. Sitoe
Associate Professor, Silviculture and Ecology of Tropical Forests
Faculty of Agronomy and Forestry
Eduardo Mondlane University
Maputo, Mozambique

Jonathan Timberlake
Editor – Flora Zambesiaca
Herbarium, Library, Art & Archives, Royal Botanic Gardens Kew, England

Dr. Wessel J. Vermeulen
SANParks Science Services (Knysna)
South African National Parks (SANParks), South Africa

John de Wet
Facility Management (Sustainability)
University of Stellenbosch

Russell de la Porte, Esther van Vuuren and Hannes Wiese for their language editing work and Elisabete Machado from Springer Publishing for her professional work with us and her patience. We would also like to thank Prof. Klaus von Gadow for his encouragement to publish this book.

Contents

Chapter 1
Biomass from Wood in the Tropics

Thomas Seifert, Pierre Ackerman, Paxie W. Chirwa, Clemens von Doderer, Ben du Toit, Johann Görgens, Cori Ham, Anton Kunneke, and Martina Meincken

1.1 Woody Biomass – An Antiquated or a Modern Source of Energy?

Bioenergy production from wood is one of the oldest forms of energy and it was for a long time considered a primitive energy source in many industrialised countries. However, it is currently experiencing an increase in attention worldwide. Considering its importance and history, it is astonishing that the widespread cognizance of wood as an important energy source in modern times is a recent phenomenon. It has been mainly driven by the pressure of diminishing fossil fuel resources in industrialised countries, as well as the wish to become more independent from nuclear power and its risks in some developed countries. In addition, amongst other renewable energy sources, bioenergy was identified as an alternative to

T. Seifert (✉)
Department of Forest and Wood Science, University of Stellenbosch, Private Bag X1, 7602 Stellenbosch, South Africa
e-mail: seifert@sun.ac.za

P. Ackerman • B. du Toit • C. Ham • A. Kunneke • M. Meincken
Department of Forest and Wood Science, Stellenbosch University, Private Bag X1, 7602 Matieland, South Africa

P.W. Chirwa
Forest Science Postgraduate Programme, University of Pretoria, 5-15 Plant Sciences Complex, Pretoria 0028, South Africa
e-mail: paxie.chirwa@up.ac.za

C. von Doderer
Department for Agricultural Economics, Stellenbosch University, Private Bag X1, 7602 Matieland, South Africa

J. Görgens
Department of Process Engineering, Stellenbosch University, Private Bag X1, 7602 Matieland, South Africa

T. Seifert (ed.), *Bioenergy from Wood: Sustainable Production in the Tropics*, Managing Forest Ecosystems 26, DOI 10.1007/978-94-007-7448-3_1,

fossil fuels, which could also help to prevent furthering an anthropogenic climate change by attempting to reduce greenhouse gas emissions. Currently, two different development routes are recognizable, which appear to go in opposite directions and result in competition for land resources in tropical countries. One route is driven by developing countries and the other by developed countries. Globally, wood is the most important source of renewable energy and is used to produce more energy than all other renewable energy sources combined (ren21 2013; FAO 2011). According to (FAO 2011), the global annual woodfuel consumption, which comprises fuelwood, charcoal and other wood based energy sources, sums up to 1.87 billion m^3. Of this amount, 13 % are consumed in the tropical and sub-tropical regions of America and the Caribbean region, 30 % in Africa and 30 % in Asia and the Pacific region. In total, almost three quarters of the global woodfuel consumption occur in tropical countries.

1.1.1 Wood Based Energy in the Developing Countries

The strong contribution to woodfuel consumption by tropical and subtropical countries is not surprising since they account for a majority of the two billion people worldwide who are dependent primarily on firewood for cooking and heating (Mathews et al. 2000). The correlations of population growth, wood fuel demand and deforestation are well known (Allen and Barnes 1985; Barnes 1990). As an example for this process the Southern African Development Community (SADC) region in sub-Sahara Africa should be given, where the dependency on biofuel and charcoal is bigger than in most other regions of the world (Hall et al. 1994). Figure 1.1 illustrates the correlation between population growth and deforestation in the SADC countries. SADC has one of the fastest growing populations in the world and as such faces the challenge of increasing food and fuelwood demands (Barnes 1990; Hall et al. 1994). This has inevitably resulted in large forested areas being cleared or sequentially degraded. Non-sustainable fuelwood use is the second biggest cause of deforestation in that region (FANR 2011). The relatively limited access and high cost of electricity and fossil fuels in rural areas, where over 70 % of the population reside, and in urban areas worsens the situation (FANR 2011).

Due to a lack of methodological knowledge and financial means, biomass conversion to energy in the tropics comprises mainly of low-tech fuelwood use and charcoal production. With industrial growth in many tropical countries, there is a growing movement from traditional firewood use to coal and other fossil based energy sources. This change is furthered by the fact that the natural ecosystems are often not capable of sustaining the supply of fuelwood for the growing population, and thus demand. These challenges and trends are similar to those previously experienced in developed countries during the industrialisation process two centuries ago.

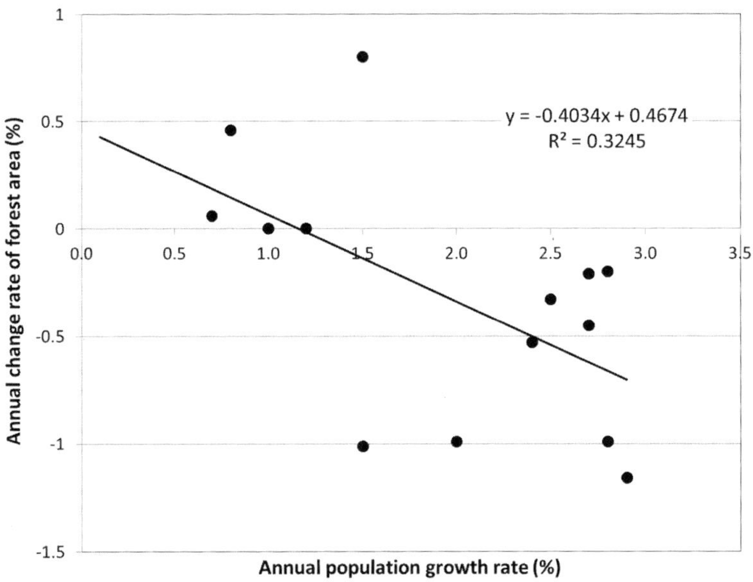

Fig. 1.1 Correlation between annual population growth rate (2010) and percentage forest loss in the SADC countries (excluding Zimbabwe) from 2005 to 2010, based on data from FAO (2010)

1.1.2 Wood Based Energy in the Industrialised Countries

Energy development in many industrialised countries shows a reverse trend, turning away from fossil fuels and nuclear power towards regenerative energy sources. International commitments towards clean development mechanisms, such as the Kyoto protocol, increasing fossil prices due to a dwindling supply, and the general aversion to nuclear energy in many developed countries, lead to a dynamic development of alternative energy sources in the past two decades. The European Union, which may serve here as an example for the industrialised nations, has defined the goal to increase renewable energy in their energy portfolio from the current value for the year 2011 of 13 % (Eurostat 2013) to 20 % in the year 2020 and is planning to move beyond that mark (European Commission 2010). About 70 % of the current European renewable energy production was from biomass in 2011 and about 70 % of that biomass was based on wood and wood residues (Eurostat 2013). In the last decade the amount of wood based biomass used for energy has increased by more than 50 % (Fig. 1.2). In the same time frame (2002–2011), net imports of biofuels have increased by a factor of ten, indicating that the import of solid and liquid biofuel is part of the strategy to transform the energy portfolio in the European Union. Due to restricted access to land and high production costs, industrialised

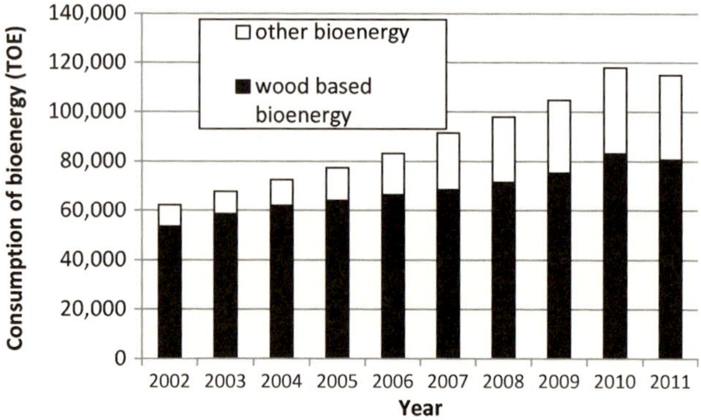

Fig. 1.2 Consumption of bioenergy and wood based bioenergy in the EU countries (EU27), measured in Thousand Oil Equivalents (TOE)

nations increasingly try to source their biofuels from tropical countries, preferably in the form of easily transportable fuels such as oils, fatty acid distillates, bioethanol or solid pellets, all of which can facilitate co-firing in power plants or be blended in fuel for motor vehicles.

However, to embark on commercial bioenergy projects for the mitigation of climate change might be a two-edged sword. Non sustainable practice, in particular with palm oil production, have raised concerns about the suitability of biofuel production for mitigating climate change since it may degrade existing natural resources and may further increase climate change due to deforestation and the deterioration of natural ecosystems, especially in tropical countries (Wicke et al. 2008; Butler et al. 2009). Many of these concerns are valid, particularly in the tropics, where conversion from one land-use to another is quite common, where it is often loosely regulated and controlled and land tenure is frequently unsolved. Conversely, if bioenergy is produced sustainably, it offers the potential to provide an energy resource, sequester carbon and at the same time alleviate poverty in many developing countries in the tropics. In this context it is important to see both aspects of current bioenergy use. The traditional low-tech fuelwood aspect for everyday cooking and heating that is prevalent in most developing countries, and the high-tech approach of many industrialised countries, which will also be realised mainly in the developing countries of the tropical and sub-tropical areas. Some tropical and sub-tropical countries have embarked on commercial bioenergy production, driven partly by local demands and by the increasing resource needs of the industrial countries of the Northern Hemisphere (Mathews et al. 2000; Wright 2006).

1.2 The Key Concept of Sustainable Production of Bioenergy

Sustainability of bioenergy production from wood in the tropics is mainly endangered by the 'gold-rush fever' phenomenon on the bioenergy market, which sometimes fosters developing projects in tropical countries, without a clear concept of sustainable resource supply. This applies to both traditional and commercial bioenergy production. Tropical countries face the challenge of designing holistic concepts for a sustainable implementation of bioenergy use that is adapted to local conditions. These concepts must embrace all three aspects of sustainability and carefully balance bioenergy production with all other socio-economic and ecologic demands (Fig. 1.3).

It entails that all levels of sustainability have to be met, starting with sustained economic feasibility, long-term beneficial impact on society and the avoidance of negative impacts on the environment. This challenge is best met with an integrated land-use management system, where different land-use forms and eco-system services such as food-production, fibre production, biodiversity conservation, water provision, job creation and biomass production are balanced (Fig. 1.4). A bias of land-use towards a singular objective of biomass production does not meet the sustainability criteria. Unfortunately, decision making support tools to balance land-use portfolios are rare and still have to be developed or adapted to tropical conditions (Fürst et al. 2013; Seifert et al. in press).

Frequently, basic knowledge on the implementation of sustainable systems for biomass production and conversion to energy is also not readily available in many tropical and sub-tropical countries. The vast majority of the current literature on biomass production originates from countries in the temperate and boreal zone of the Northern Hemisphere and due to differences in climate zones; it may not be directly applicable to countries in the tropics and sub-tropics.

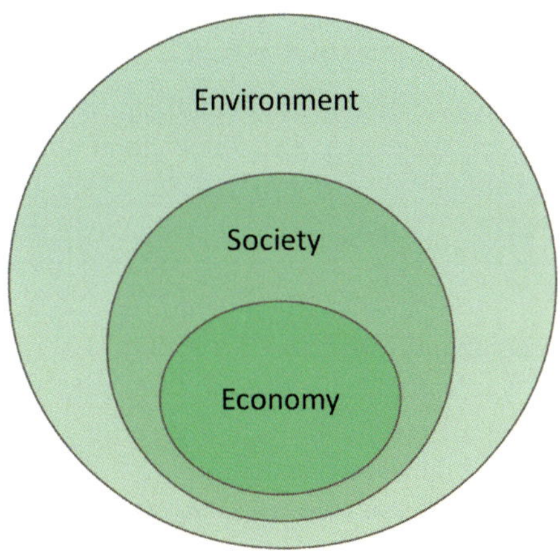

Fig. 1.3 The three spheres of sustainability

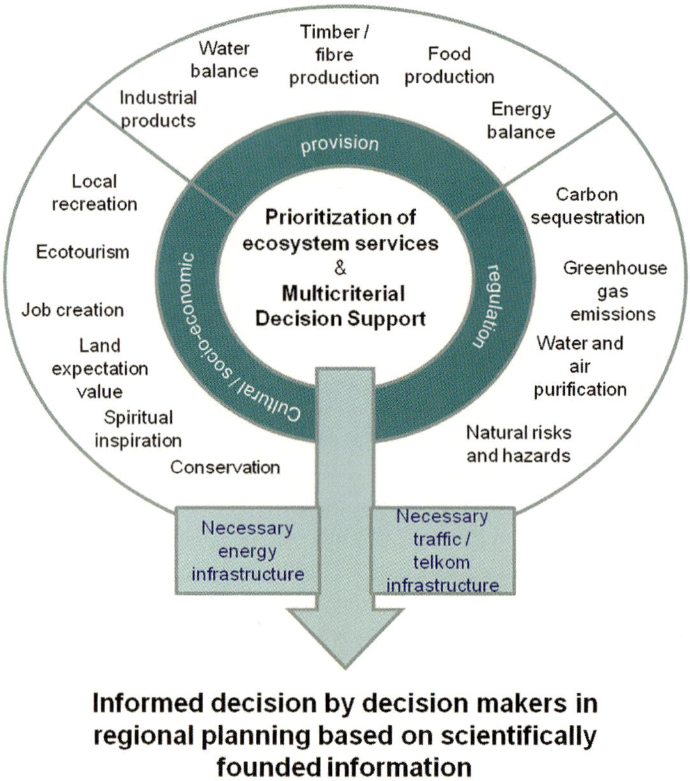

Fig. 1.4 Balancing different ecosystem services, and bioenergy production as one of them, is the major challenge for a sustainable land-use management

1.3 Managing the Value Chains

A supply chain management approach is applicable to biomass to bioenergy conversion processes. The value chain approach allows sustainable linkages between individual solutions in the value chain such as the availability of resources in time and space and transport and conversion techniques to a sustainable concept (Fig. 1.5). As a result the value chain approach provides the basis for the comprehensive analysis of the economic and ecologic consequences of bioenergy production from wood.

The value chain illustrated in Fig. 1.5 shows the bioenergy concept of supply chain management. In this case the chain is initiated with the assessment of the current and future availability of a resource. It goes on to include sound sustainable resource management to maintain production, harvesting and logistics planning and ends with characterisation of biomass and processing. Undelaying the value chain are aspects concerning economic sustainability, socio-economic considerations and environmental impact assessments at both local and global levels (Fig. 1.6).

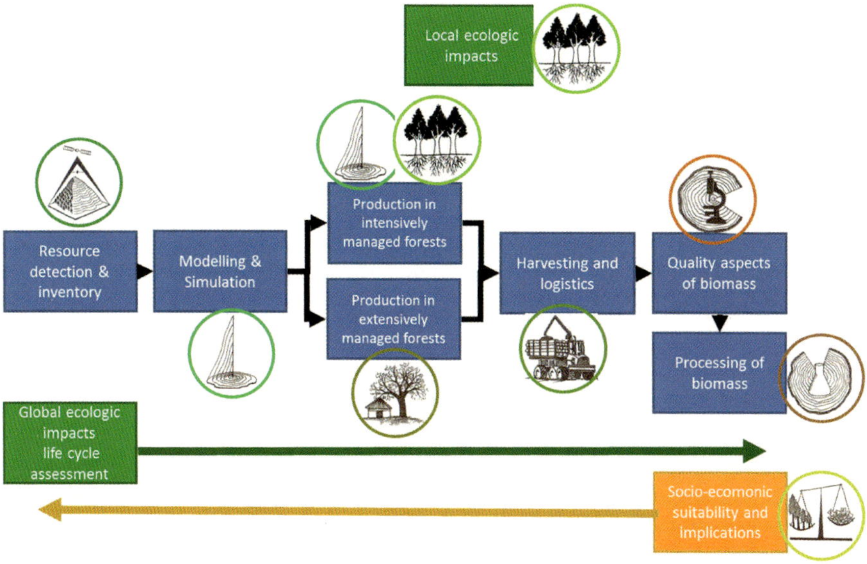

Fig. 1.5 Value chain approach for sustainable biomass production systems

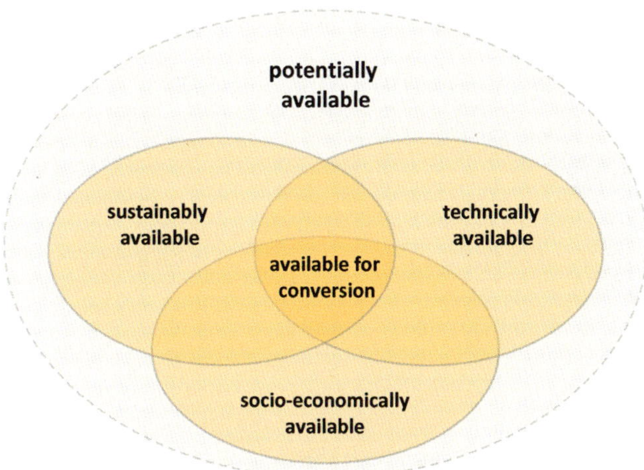

Fig. 1.6 Different constraints for the availability of biomass

In cases where the total supply chain is not taken into account with its technical, environmental and socio-economic constraints in the planning of bioenergy projects the available resource is frequently overestimated. Another challenge to the system is the choice of the conversion processes to match the biomass supply in particular to ensure efficient utilisation of biomass for an end-purpose in mind. Most conversion plants entail large capital investment and stationary biomass is sensitive to transport costs. As such spatial resource localisation planning is a priority at the entry point to any bioenergy project to ensure sustained fuel supply.

The choice of the best method to convert the available biomass and the scale of the operation are often decisive for the success of a project and depend on available biomass. This biomass may be different from region to region as regards quantity and quality. In addition forest based biomass may only be one part of a larger initiative to feed a bioenergy plant. On the other hand the scale of the biomass processing plant is also relevant and must be matched to the potentially available biomass. Another factor to take into account is current and future market segmentation. This concerns more traditional markets for biomass in terms of pulp and paper or particle board, which may inhibit local wood markets.

Ensuring sustainable use of woody biomass for bioenergy strongly relies on a value-chain approach to supply solutions. It is clear from the above that any bioenergy project must involve a multidisciplinary approach from remote sensing and growth and yield modelling to ascertain the extent of the resource; silviculture in terms of biological production; forest engineering and logistics to make the biomass in question available and delivered; environmental aspects and maintenance of biodiversity; socio-economics to consider communities and the financial viability and process engineering for the development of the envisaged products, and finally all relevant step analysed through a life cycle assessment of the system.

1.4 The Scope and Structure of This Book

This book is intended to close the short comings highlighted above. It is fully embracing the value-chain approach for sustainable production of bioenergy from wood. Furthermore it is meant to address all relevant aspects of biomass production and conversion along the production chain, with a particular focus on tropical and sub-tropical countries of the Southern Hemisphere.

The authors admit to bias towards Southern Africa, from where most of the case studies were taken but assume that the examples can be applicable to many other countries in the tropical and sub-tropical zone and in the Southern Hemisphere. The intention of this book is to present the state of the art in bioenergy production from a technical perspective, domestic and global consequences for the environment, economic feasibility and the socio-economic implications. All of these points address typical challenges of a bioenergy production system in tropical countries, considering both intensively managed plantation forestry and small growers and community forests.

This approach addresses all essential aspects of the value added chain of bioenergy production. The book is written for scientists that are involved or want to become involved with research on bioenergy as well as for forest practitioners and forest managers who are looking for an up to date compendium on the topic.

The book may also serve as a concise introduction into bioenergy production for stakeholders and decision makers that have to create the framework for sustainable production of bioenergy from forests and woodlands.

Chapter 2 introduces biomass inventory concepts for the localisation of woody biomass using terrestrial and remote sensing techniques. It is closely aligned to Chap. 3 which is dedicated to modelling and simulation of biomass. In Chaps. 4 and 5 silvicultural management aspects of biomass production for bioenergy in natural woodlands and commercial plantations are discussed. Chapter 6 introduces relevant topics of biomass harvesting, transport and logistics and provides the interface between biological production and biomass processing. The latter is dealt with in Chap. 7, and provides an overview on conversion techniques for woody biomass and their application range. Chapter 8 provides the reader with information on biomass quality testing, which is a prerequisite for establishing optimum conversion techniques. In Chap. 9 an analysis of socio-economic impacts of biomass production is presented, and provides the linkage to society as a major stakeholder when it comes to the implementation of bioenergy value chains. Chapter 10 introduces constraints to the implementation of biomass production systems resulting from potential impacts on water, soil fertility and biodiversity. Chapter 11 finally provides an overview on more global impacts of bioenergy production with a life cycle assessment based on a case study.

References

Allen JC, Barnes DF (1985) The causes of deforestation in developing countries. Ann Assoc Am Geogr 75(2):163–184

Barnes DF (1990) Population growth, wood fuels, and resource problems. In: George TF Population growth and reproduction in Sub-Saharan Africa. Worldbank, Washington DC, 44–59

Butler RA, Koh LP, Ghazoul J (2009) REDD in the red: palm oil could undermine carbon payment schemes. Conserv Lett 2:67–73

European Commission (2010) Analysis of options to move beyond 20% greenhouse gas emission reductions and assessing the risk of carbon leakage. Communication from the Commission to the European Parliament, the Council, the European Economic and Social Committee of the Regions, Brussels, 26 May 2010, COM265 online: http://eur-lex.europa.eu/LexUriServ/LexUriServ.do?uri=COM:2010:0265:FIN:en:PDF

Eurostat (2013) Statistical data base of the European Union. Online: http://epp.eurostat.ec.europa.eu/

FANR (2011) SADC Food Agriculture and Natural resources webpage. Online: http://www.sadc.int/fanr/naturalresources/forestry/management.php

FAO (2010) Global forest resources assessment 2010: main report. Food and Agriculture Organization of the United Nations, Rome

FAO (2011) World forestry report. Food and Agriculture Organization of the United Nations, Rome

Fürst C, Frank S, Witt A, Koschke L, Makeschin FJ (2013) Assessment of the effects of forest land use strategies on the provision of ecosystem services at regional scale. Environ Manage 127:96–116

Hall DO, Mao YS, Karenzi PC, Kgathi DL, Sekhwela MBM, Tietema T, Mpotokwane MA, Diphaha JBS, Maya SR (1994) Biomass energy and coal in Africa. ZEN Books, London, 196pp
Mathews E, Payne R, Rohweder M, Murray S (2000) Forest ecosystems – a pilot study of global ecosystems. World Resources Institute, Washington DC. Online: www.wri.org/wr2000
ren21 (2013) Renewables 2013. Global Status report. Paris, ren21 Secretariat
Seifert T, Kunneke A, Ackerman A, Baptista C, Ham C. The design and use of forest Decision Support Systems in South Africa. In: Borges JG, Trasobares A, Nordstrom E, Garcia-Gonzalo J, Hujala T, Eriksson L (eds) Computing based tools for supporting forest management. The experience and the expertise world-wide Book Chapter. Opoce, pp 364–375 (in press)
Wicke B, Dornburg V, Junginger M, Faaij A (2008) Different palm oil production systems for energy purposes and their greenhouse gas implications. Biomass Bioenergy 32(12):1322–1337
Wright L (2006) Worldwide commercial development of bioenergy with a focus on energy crop-based projects. Biomass Bioenergy 30:706–714

Chapter 2
Localisation of Biomass Potentials

Anton Kunneke, Jan van Aardt, Wesley Roberts, and Thomas Seifert

2.1 Introduction

The aim of this chapter is to provide an overview of methods of estimating woody biomass from inventory information.

In understanding any resource, the extent (spatial localisation) as well as the amount of resource should be estimated. Inventory is the term used in forestry practice for assessing the timber resource. The result of an inventory should establish at least three values. The first is the estimated mass or volume of resource per unit area, for a given time period, the second value is the total area of the resource, and the third is an error value (accuracy and precision) associated with the estimate. In a heterogeneous resource a subdivision of the resource in classes is also necessary. Common subclasses could be vegetation types (e.g., 0–30 % tree cover, 30–50 % tree cover, broadleaved forest and savannah, etc.) or qualitative age classes (e.g., young, semi mature, mature). For instance, naturally regenerated forests, as opposed to commercial plantations, contain all age classes in a single stand at the same time. An inventory in those forests will therefore provide a range of diameter classes, and should rather concentrate on size than age.

Conventional forest inventory typically attempts to establish the volume of utilisable wood or bio-energy in the forest stand. The minimum information necessary for bio-energy use is a map with the available biomass and a value in

A. Kunneke (✉) • T. Seifert
Department of Forest and Wood Science, Stellenbosch University, Stellenbosch, South Africa
e-mail: ak3@sun.ac.za

J. van Aardt
Centre for Imaging Science, Digital Imaging and Remote Sensing Group,
Rochester Institute of Technology, Rochester, NY, USA

W. Roberts
BioCarbon Partners, Cape Town, South Africa

T. Seifert (ed.), *Bioenergy from Wood: Sustainable Production in the Tropics*,
Managing Forest Ecosystems 26, DOI 10.1007/978-94-007-7448-3_2,

kg m^{-2} for each section (stand or subdivision) on the map, since the localisation of the biomass will guide all further steps of harvesting (Chap. 6). Biofuel can be described by characteristics such as energy content (MJ kg^{-1}) and heating value (MWh kg^{-1}). Furthermore, it might be desirable to know the percentage of components of biomass, e.g., leaves, twigs, branches, bark and stem wood, since the comprehensive biomass composition determines the ash content, a critical factor in bio-energy conversion (Chaps. 7 and 9) and is a major determinant for the export of nutrients and thus stand sustainability (Chap. 10). Models that estimate biomass composition, based on inventory data, are thus relevant additions to the typical inventory information (Chap. 3). Biomass resource assessment or inventory is in essence then no different from conventional inventory in that the quantity and quality of biomass needs to be estimated. It differs in that the mass and not the volume has to be estimated and that the quality parameters differ. This chapter will concentrate on procedures to estimate the mass and quality parameters needed, while briefly hinting at the associated errors associated with these.

Three main approaches could be followed to provide biomass assessments:

- Sample plot-based methods using terrestrial light detection and ranging (LiDAR) scans (TLS) or standard methods, as with conventional timber inventory, related to diameter and height measurements in sample plots. The sample plots are assumed to be representative of the area in the inventory and models are used to predict the component fractions of the stand in the inventory.
- Aerial surveys based on LiDAR and photogrammetry estimates of tree parameters (number, height, crown size, etc.) at the stand level, applied to models to predict total biomass and biomass components of the stand.
- Space or aerial survey methods of estimating total aboveground biomass directly. This is possible with SAR and LiDAR methods, as well as optical infrared sensing in combination with models. The total aboveground mass is a single value, which requires additional modelling to predict a breakdown of component fractions for the stand.

This chapter will describe the methods used in measuring parameters, while the upscaling and modelling of component fractions will be dealt with in Chap. 3.

2.2 Terrestrial Inventory Methods

Methods that require a person or team of people physically working on the ground measuring physical quantities are described as terrestrial inventory. A wide array of methods from, traditional tape measurements and clinometers, to sophisticated electronic laser equipment is currently available for forest assessment. Examples of such work can be found in Seidel et al. (2012).

Inventory requires a proper sampling plan, which includes the aim or result of the measurement, error tolerance, sampling procedure, available budget, trained people and equipment. The required result will determine the error margin, which will dictate the intensity of the sample for each method selected. The method selected

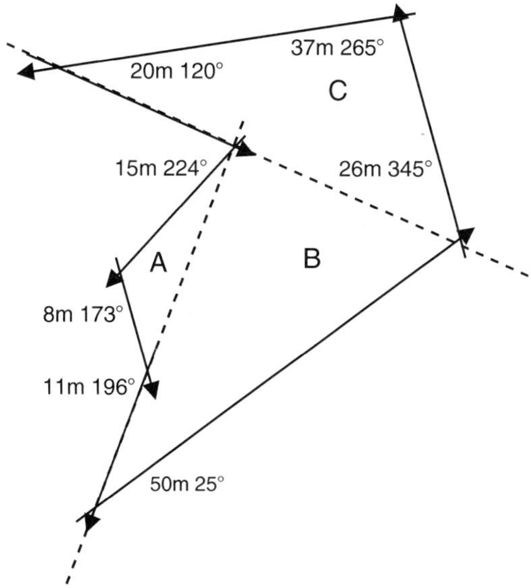

Fig. 2.1 Estimation of area graphically

will determine the equipment selected, which will indicate the size and level of training of the person or team. Finally, the available budget will have an influence on the detail of outcome, as well as the intensity of the inventory and the error of the result.

This section will focus on an overview of methods used with references to texts that provide detailed discussions on each of the methods.

2.2.1 Estimation of the Area or Extent

Maps and the electronic successors of maps, namely Geographic Information Systems (GIS), have been used for ages to describe the lay of the land. Planimetry on maps and area calculation in GIS are used to estimate extent of an area of interest, for example a stand of trees. If no map is available, a chain and compass (Fig. 2.1) or Global Navigation Satellite System (GNSS) receiver can be used to determine the boundary of an area of interest. Most modern GNSS receivers include a function to estimate the area described by a set of points (coordinates).

Chain distances and compass directions can be entered in a GIS or drawn on scale, which allow for areas to be estimated. Dimensions can be drawn on a scaled diagram and the area of the resulting polygon determined by subdividing it in a series of regular figures, for example triangles (Fig. 2.1 A, B, C). Calculating area from coordinates can be done via many methods. One such formula is provided below:

$$\text{Area} = \frac{1}{2}\{(Y_1X_2 + Y_2X_3 \cdots + Y_nX_1) - (X_1Y_2 + X_2Y_3 \cdots + X_nY_1)\} \quad (2.1)$$

where

X_l is the x coordinate of vertex 1
Y_n is the y coordinate of the nth vertex.

Area is a by-product of the mapping procedure (GIS) or from image processing, e.g., classification, and is discussed again in the section on Remote Sensing methods.

2.2.2 Sampling Techniques

Natural resource inventory is an expensive and time-consuming activity and full enumerations, where every tree in the population under investigation is counted and measured, are only done in exceptional cases. A discipline of science evolved around the methods to extract a representative subpopulation for measurement, as a cost effective replacement for full inventories. In essence a sub-sample of trees is selected from the total population and all of the parameters described by the inventory procedure are measured. The mean of the parameters measured for the sub-population is then accepted, with an error margin, as the mean for the population. A discussion on the selection of representative samples follows.

2.2.3 Selection of Representative Trees for a Stand

The selection process of representative trees for a stand has to consider the number of trees that should be sampled and how they should be selected from the population.

The actual number of sample trees is determined by the level of precision required and the variability of the resource, with the principle that model bias should be entirely avoided as stated by van Laar and Theron (2004). There is always a trade-off between accuracy and efficiency of the sampling procedure. Higher accuracy results in a higher cost and time investment. Improving the standard error of the mean by 50 % requires a four times higher sampling number, since the standard error of the mean is inversely proportional to the square root of the sample size (Eq. 2.2). This well-known fact and the effort involved in biomass studies are a strong motivation for choosing a sampling number close to the theoretical minimum number, which is still representative of the population.

$$S_y = \frac{s}{\sqrt{n}} \tag{2.2}$$

where

S_y is the standard error of the estimate of the arithmetic mean of a population
s is the estimate of the standard deviation of the population
n is the number of samples

The standard error of the mean, S_y, is then multiplied by the appropriate z-score, obtained from the Normal-(z)-Distribution, e.g., 1.96 for a 95 % confidence level, to obtain the margin of error (*ME*). The margin of error defines the confidence interval, when subtracted from or added to the estimate, $\hat{E}$.

The minimum sampling percentage for simple random sampling from large populations depends on the variance in the population and the accepted error probability of the outcome, defined by the chosen confidence limits (Eq. 2.3).

$$N = \left[\frac{z \cdot s}{E}\right]^2 \quad (2.3)$$

where

N is the minimum number of samples to be representative of the population
E is the margin of error, the maximum accepted difference between estimated ($\hat{x}$) and true mean (μ) of the population.
z is the z-score obtained from a Normal-(z)-Distribution
s is the estimate of the standard deviation of the population

Usually the true variance is unknown, thus the sample variance has to replace the population variance when calculating the t-statistic or the z-score, as indicated in Eq. 2.4 (van Laar an Akça 2007).

As stated by van Laar and Theron (2004) "There is no universally accepted criterion for the size of the acceptable margin of error, which is a function of the population variance, sample size and the statistical risk of exceeding the margin of error." The margin of error for volume calculations in forest inventory data should preferably not exceed 3–6 % of the mean (Smith et al. 2004; Bundeswaldinventur 2008 cited by Waggoner 2009). For biomass and carbon inventories sometimes higher error margins, ranging from 10–15 %, are accepted for the estimation (van Laar and Theron 2004; Hollinger 2008).

However, the minimum sampling number of trees has to be determined iteratively. Stauffer (1982) suggested an iterative procedure and expressed the margin of error as a percentage of the mean, which requires the substitution of the standard deviation by the coefficient of variation, according to Eq. 2.4. An iterative determination of the sample size with increasing n, converges after a few iterations (for an example see van Laar and Akça 2007, p. 258).

$$n = \left[\frac{t_{\alpha,n-1} \cdot s\%}{E\%}\right]^2 \quad (2.4)$$

where

n is the number of samples t_α is the t-statistic at n–1 degrees of freedom
$s\%$ is the coefficient of variance and $E\%$ the relative margin of error.

If variance information on biomass is not readily available, which might be typical, other methods can be applied like regression-based sampling. This method

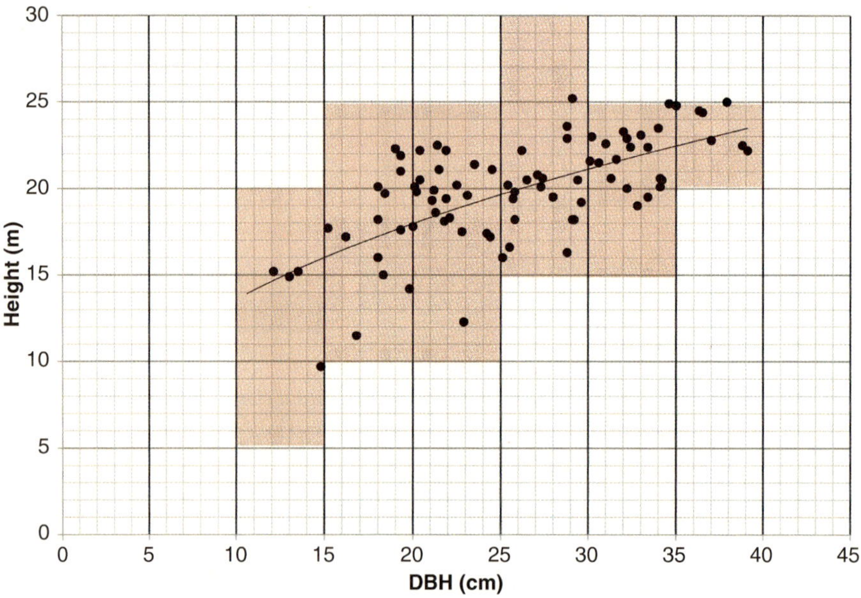

Fig. 2.2 Diameter-height plot based on inventory data of a 14 year old *Pinus radiata* stand in the Western Cape, South Africa. Marked are 15 DBH-height classes for biomass sampling

is based on the use of an auxiliary variable, which is easier and cheaper to measure and which is ideally linearly correlated to biomass. For example, tree basal area (BA) lends itself as an auxiliary variable for biomass sampling, since it is highly correlated to tree biomass and easy to determine from stem diameter measurements, which can be conducted in high numbers. The process is then a two-phase sampling procedure, where first the population mean of the BA is estimated and then a sub-sample is drawn for biomass sampling. However, Cunia (1990) raised doubts about the unbiased nature of the sample if anything other than diameter at breast height (DBH) or height were used as auxiliary variables.

Once the minimum tree number for sampling has been determined, the selection of trees which represent the stand is a substantial task. It requires prior knowledge about the representation of the independent variables that will be used for modelling. Tree height and DBH were chosen as independent variables in this example. A DBH-height-plot produced from a preceding terrestrial inventory of the stand or the region where the function should be applied, is a good help to choose the sample trees. A way to obtain robust regression estimates, which can be applied in similar stands, is to divide the classes of DBH and height and sample in each class in a stratified sampling procedure, so that the selected sample trees cover the full range of DBH-height relations. This can be done visually, based on equidistant classes as indicated in Fig. 2.2, or alternatively, in a bivariate binning procedure via a classification based on quantiles for the DBH and the height. If too few DBH-height pairs are available, a selection based on DBH-classes presents a feasible alternative.

Van Laar and Akça (2007) refer to Attiwil and Ovington (1968), who compared different sampling methods for the estimation of biomass from an area and found that the regression-based method ($N = 20$), performed better than the estimation of the biomass from the mean DBH tree ($N = 5$) sampling of four trees in five DBH classes ($N = 20$) and regression-derived mean DBH tree multiplied by the tree number. The other sampling methods produced biased results in the range of −8.5, −3.1, and −11.6 %. These results and the criticism of Cunia (1990) on importance sampling may serve as valid arguments to base tree selection on a regression approach that covers the spectrum of DBH-height relations in the population. This should also result in more generic models that are easier to adapt to other sites later on.

There are two different approaches to estimate biomass for forest areas. Both include a forest inventory component based on a sampling procedure. The first approach is based on dry mass/area relations, where upscaling to the stand level is done per sample plot first and then the area of sample plots is used as a ratio estimator to upscale biomass from the total sampled area to the total stand area. The second approach is based on single tree equations (Chap. 3) and relies on stem frequencies in diameter classes, as produced in the inventory. For each mean tree per diameter class, the biomass is upscaled with a ratio estimator relating the sampled tree numbers to the population tree numbers.

The following example illustrates the difference of the application of an established allometric biomass prediction model for *Pinus radiata* (van Laar and van Lill 1978) to estimate the total biomass based on different selections of trees from the population. The assumption that the model produces unbiased estimates for the population should fit the purpose here. The model is applied to estimate dry mass for a pine stand of one hectare in size, which was fully mapped for DBH and tree heights (Ackerman et al. 2013) to illustrate the influence of tree selection on the outcome of a model prediction. The application of an allometric equation to all individual trees serves as a reference. This is tested against the application of the biomass model, based on the diameter of the quadratic mean diameter tree (dq) and based on average dq trees in seven diameter classes. The diameter distribution of the stand is shown in Fig. 2.3.

The application of the functions for all 318 individual trees of that stand resulted in an estimated total stand dry mass of 424.99 Mg ha^{-1} (tons). The estimate based on the mean quadratic diameter was clearly negatively biased at 422.87 Mg ha^{-1} (−2.12 Mg ha^{-1}), while the diameter class-based approach produced a nearly unbiased dry mass estimate of 424.88 Mg ha^{-1} (−0.11 Mg ha^{-1}). It is obvious that the estimate of biomass for a full stand based on the dq does not take into account the nonlinear relation between diameter and biomass, and thus underestimates the contribution of the bigger diameter trees in the stand, while the estimate with diameter classes mitigated that effect to an acceptable degree. This effect can be expected to be much stronger for a more skewed diameter distribution.

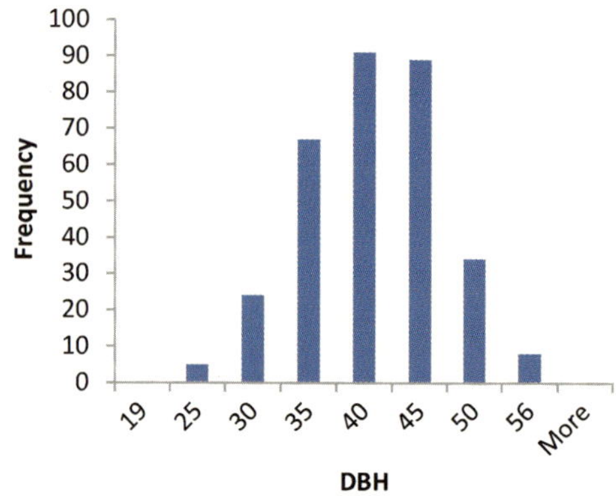

Fig. 2.3 Diameter distribution of a pine plantation stand, used as a basis for an evaluation of different biomass modelling approaches

A multitude of inventory techniques developed for volume estimation in forests can be easily adapted for biomass inventories. The same variables that are used for biomass estimation, namely DBH and height, are typically used to estimate volume. The interested reader might therefore consider some standard volumes on forest inventory, such as Kangas and Maltamo (2006), van Laar and Akça (2007), Mandallaz (2008), for detailed information.

2.3 Remote Sensing Methods Using LiDAR or Photogrammetry to Estimate Tree or Stand Parameters

Remote sensing is based on the principle of measuring reflected electromagnetic energy from a target under investigation. The source of the energy can either be energy (photons) from the sun or an artificial source, like a laser pulse in the case of a LiDAR instrument. However, the energy measured by the sensor interacted with the target, as well as having travelled through layers of the atmosphere and sensor components. Pre-processing, or the process of transforming the measured energy to levels as it were just after being emitted or reflected from/off the target, therefore is required before data are used. This process ensures compatibility between different sensors and even different times, and includes efforts related to atmospheric compensation, e.g., conversion from digital numbers (DN) to radiance (W m^{-2}) to reflectance (unitless). Feature extraction then measures the object under investigation (target), which could be a pixel, object, single tree crown, stand of trees, etc. The process is validated by comparing the extracted result with measured data from a field survey.

Inventory based on remote sensing data nearly always involves a multi-phase approach with a terrestrial component, used for calibration. In regional assessments, assumptions about the terrestrial component are made and the remote sensing

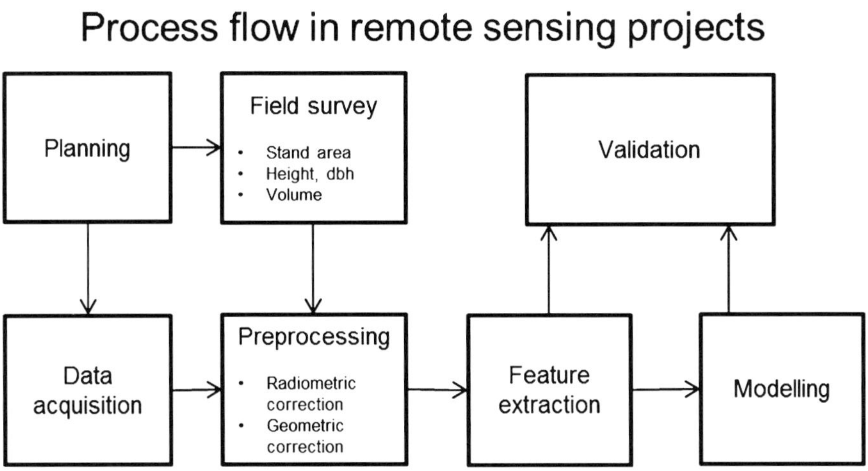

Fig. 2.4 Process flow followed in the use of remote sensing for biomass inventory

information is interpreted in accordance with measured point samples. Two phase inventories seem to be the most common option, in which detail terrestrial information is collected at a few sample locations, which is then used as training and verification of the image processing process (Fig. 2.4). A decision should be made during the planning phase on the appropriate sensor and time of acquisition, as well as the sampling strategy for field survey or ground truthing. In short, all remote sensing approaches or remote sensing inventory models should be calibrated and validated using a sample of field-measured values, especially when a model is used across different regions, site conditions, or species.

2.3.1 Aerial Photo and Satellite Image-Based Estimates of Stand Parameters

Aerial photography has been the basis of resource assessments since the start of human flight, but more intensely since the Second World War. Space photography, on the other hand, has been in use since the 1970s. Orthocorrected photographs, i.e., where the effects of terrain or topographical variations have been removed, can be used as a planimetric source of information, as all 3D distortions are removed. It is the use of photogrammetry that is of particular interest in biomass estimates. Not only can areas of the resource be mapped, but a height value can be estimated for standing biomass via stereo-photogrammetry.

Stereo-photogrammetry provides a measure of parallax, which can be interpreted as the positional difference between the base and the top of an object, due to changes in viewing geometry. By collecting points on the ground surface (terrain) and the

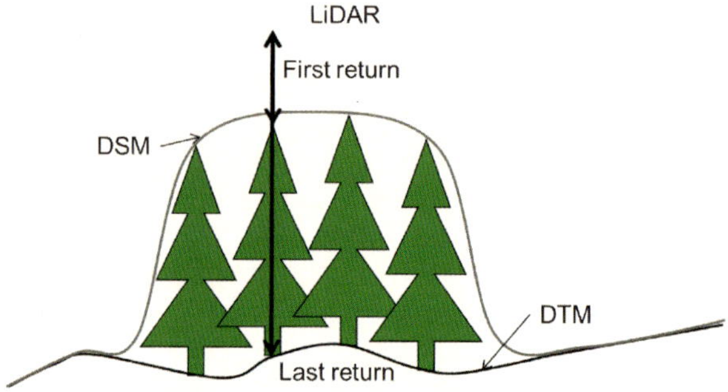

Fig. 2.5 Illustration of a Digital Terrain Model (*DTM*) and Digital Surface Model (*DSM*) and LiDAR terms, as referred to in this text

top of all objects on the terrain (surface) and performing an interpolation on these points, digital terrain models (DTM) and digital surface models (DSM) are products of the digital photogrammetry process.

Subtraction of a DTM from a DSM results in the height of an object above the terrain at each chosen point, as shown in Fig. 2.5. The digital orthophoto, which is also a product of the digital photogrammetry process, also provides a basis of measuring crown diameters. Although trained analysts can estimate individual tree sizes, which are useful for landscape-level analyses, this process must be automated to increase analyst efficiency. Automatic extraction of crown area, DBH, and biomass at the tree level is a relatively new type of image-analysis capability and associated techniques, which rely on delineating a tree crown from background soil and vegetation, and adjacent trees. A GIS layer of polygons, where the polygon area represents the crown area, can then be produced (Maltamo et al. 2003). Crown diameters are related to stem diameters for a given species and site, which means that both height and DBH could be estimated from the products of photogrammetry, with accuracy depending on the scale or resolution of the photography and the natural variability of the variables in the population.

Images that are suitable as input to the digital photogrammetry process range from satellite products to unmanned aerial vehicle (UAV) photography, which is becoming increasingly popular. For instance, individual tree canopies were mapped from 0.1 m aerial imagery to estimate carbon in a tropical forest in Belize, while canopy structure attributes can be estimated from images with resolution better than 4 m (Chambers et al. 2007; Greenberg et al. 2005). Figure 2.6 shows a graphical representation of the interaction between object size, in this case a tree canopy, and spatial resolution (pixel size) of some common satellite sensors. It is important to note, however, that (i) there is a trade-off between spatial resolution and temporal resolution (revisit time), where sensors with high spatial resolutions, e.g., IKONOS can take years to revisit the same spot on Earth; (ii) higher spatial resolutions generally imply lower spectral resolutions (broader bands); and (iii) inventory models that are developed at one spatial resolution, e.g., high spatial resolution IKONOS imagery, should ideally be coupled to and scaled with lower spatial resolution sensors, such as the Landsat suite of sensors. This approach effectively enables a tree-to-stand-to-landscape type inventory approach.

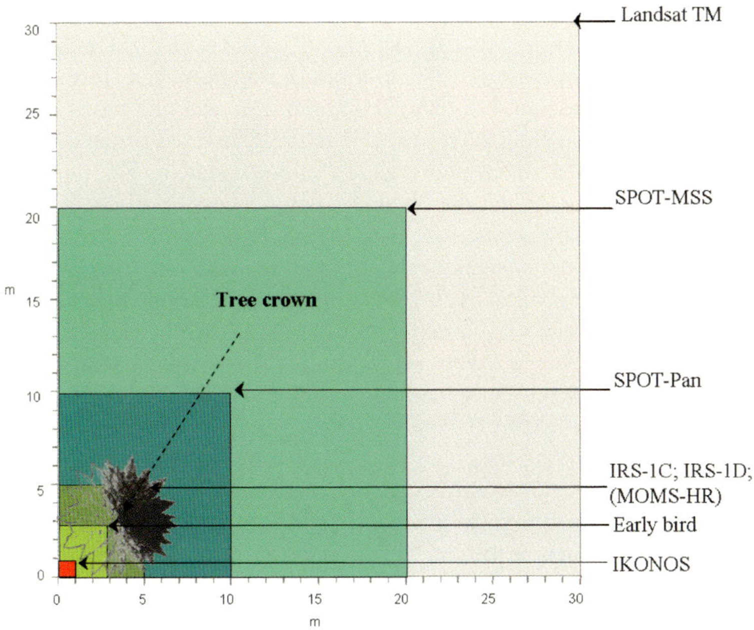

Fig. 2.6 Illustration of spatial resolution of different satellite sensors in relation to tree crown (adapted from Kätsch 1999)

2.3.2 LiDAR Estimates of Stand Parameters

Airborne LiDAR sensors are operated using an aircraft-mounted scanning laser altimeter, which emits short burst laser pulses. The round trip travel time between the emitted laser pulse, its interaction with a target, and its return to the sensor is measured, thus enabling a range (distance) measurement between the aircraft and the surface. Precise geographic coordinates and attitude (pitch, roll, yaw) of the aircraft are captured using a Differential Global Positioning System (DGPS) and an inertial measurement unit (IMU), respectively (Wehr and Lohr 1999). The exact location and pointing direction of the laser instrument are thus known. Each LiDAR return is represented by a (x;y;z) coordinate, where z refers to height or range; a large collection of such (x;y;z) returns is referred to as a "point cloud". Resulting LiDAR height point clouds can be used to create detailed three-dimensional models of the area of interest, along with estimates of the vertical distribution of sub-canopy strata (Haala and Brenner 1999; Zimble et al. 2003). For example, several studies have shown that the height estimates derived from LiDAR height point clouds are as accurate and sometimes more accurate than traditional methods (Holmgren 2004). In fact, the 3D nature of LiDAR height data makes this technology especially suitable for assessing forest volume and biomass. Such LiDAR-based forest measurements are not only useful to forest inventory and canopy structure modelling (Lefsky et al. 2002a, b; Næsset 2002; Popescu et al. 2002, 2004), but also to estimation of forest fuel loads (Riaño et al. 2003; Seielstad and Queen 2003), and extraction of digital elevation models (DEMs) (Popescu et al. 2002; Hodgson et al. 2003); all of these aspects are essential to forest management and site mapping.

LiDAR volume modeling has focused on both tree- and plot-based approaches (Nilsson 1996; Popescu et al. 2004), as well as stand-level assessments (van Aardt et al. 2006). These are based on either concrete surface modeling, i.e., top-of-canopy and ground, or LiDAR height distributional approaches, based on extraction of height distributional parameters (e.g., median, mode, percentiles) from all LiDAR returns within spatial units, such as grid cells or segments (stands) (Means et al. 2000; Næsset 2002; van Aardt et al. 2006, 2008). Whichever approach is selected, either tree-, plot-, or stand-level, *or* the use of surface height models vs. LiDAR distributions, the user should always keep in mind that the scale and geography will dictate the approach. For instance, homogenous even-aged stands are well-suited to LiDAR surface modeling (Fig. 2.5), while mixed stands with uneven ages are often best assessed by exploiting the complexity of LiDAR distributions throughout the canopy layer. Some specific examples are discussed next.

The use of locally varying window sizes, applied to height grids, is one popular approach for tree crown detection, height estimation, and determining average LiDAR values within a predefined area. Popescu et al. (2002) assessed both static and variable window size approaches for determining plot-level tree height. The authors reported that the maximum height within a predefined area returned an R^2 of between 0.85 and 0.90, while an R^2 of 0.85 was obtained for plot-level tree height when the local varying window size filters were applied. The local filter algorithm produced improved results in the case of co-dominant trees when compared to the maximum height approach. An alternative method includes the use of quartile-based LiDAR height metrics (Magnussen and Boudewyn 1998; Andersen et al. 2005; Næsset and Gobakken 2005), which makes use of distributional measures of LiDAR point height as opposed to absolute LiDAR height values (Holmgren et al. 2003; van Aardt et al. 2006, 2008). Segmentation and regression modelling (Coops et al. 2004; van Aardt et al. 2008) and statistical analyses using field enumerated data have also been successfully employed (Nilsson 1996). van Aardt et al. (2008), for instance, showed that LiDAR distributions can be used to (i) segment a complex coniferous, deciduous, and mixed stand environment into homogenous structural stands, (ii) assess volume and biomass for these stands, and even (iii) map taxonomic types (coniferous vs. deciduous), for a wall-to-wall forest inventory by forest type. However, an estimate is of little use if not accompanied by an associated precision.

Most LiDAR-based forest inventory studies report high R^2 values, except in highly variable or complex forest environments. And while most studies report root mean square error (RMSE) values that approach field inventory efforts, e.g., in the region of 10 % of the mean estimate, LiDAR is by no means a panacea. The calibration-validation approach of using a small sample of locally measured variables when transporting LiDAR models across regions or species, remains an absolute requirement. Many efforts have been made to assess LiDAR accuracy and precision when it comes to forest inventory, and while most such studies report these values, a couple of efforts or implications deserve special mention:

Duncanson et al. (2010) used an airborne LiDAR dataset in combination with forest inventory data to explore the relationship between their model error and canopy height, aboveground biomass (AGB), stand age, canopy rugosity, mean diameter at breast height (DBH), canopy cover, terrain slope, and dominant species type. The authors found that fusion of LiDAR and space borne imagery exhibited high associated error when applied to areas with greater than 200 Mg ha^{-1} of AGB; it therefore is recommended that practitioners always evaluate bias and precision using a field-based sub-sample. Generally speaking, however, more than 150 studies exist that have shown the usefulness of LiDAR sensing when it comes to forest volume or biomass estimation. The question thus becomes one of not necessarily proving the utility of the approach, but rather of constraining the estimation errors. Over and above biomass, it should be noted that there is one variable where a typical, consistent bias is often present, namely tree height. Tree height underestimation is primarily related to sensing characteristics, namely LiDAR point density and at-target beam diameter, dictated by sensor beam divergence (Baltsavias 1999). Current small-footprint discrete return LiDAR sensors have at-target beam diameters of approximately 0.5 m. These small footprints are less likely to interact with the top of the canopy, especially when the survey point density is less than the average crown size, which is the case with crown apexes of many conifer trees (Clark et al. 2004). Another reason for height underestimation is overestimations in the DTM, or digital elevation model (DEM). In old-growth high forests, where a thick understorey is usually present, it is difficult to differentiate between ground and non-ground LiDAR returns. As such, many DTM's overestimate ground height; this results in underestimated tree height (Maltamo et al. 2004). One example of this is a universal LiDAR canopy height indicator that has been developed by Hopkinson et al. (2006). The authors predicted plot-level canopy height of various vegetation types using the standard deviation of de-trended (topographically normalised) first and second return LiDAR points (Fig. 2.5). The method returned a correlation coefficient of 0.80 when compared to field enumerated heights. However, the authors noted that when the survey was conducted over homogenous vegetation types, the local maximum LiDAR metric returned improved results.

LiDAR sensors are widely regarded as the future of forest inventory, but their application in operational environments remains limited. This is largely due to the costs associated with a LiDAR survey, the documented underestimation of tree heights (Gaveau and Hill 2003; Suarez et al. 2005), and the computational requirements of the LiDAR data processing and analysis. In addition, the application of LiDAR can be impaired by heavily undulating terrain, e.g., in mountainous areas, because of shading effects of the terrain. However, this problem applies to most airborne and satellite-based remote sensing techniques. Even given all of these caveats, LiDAR is bound to play an increasingly important role in forest inventory. It is evident that an increase in markets and vendors will drive costs down, while improved algorithms are increasingly able to address accuracy-precision concerns. These observations are borne out by the importance and investment that large international forest companies, e.g., Sappi, Mondi, Weyerhaeuser, Georgia-Pacific,

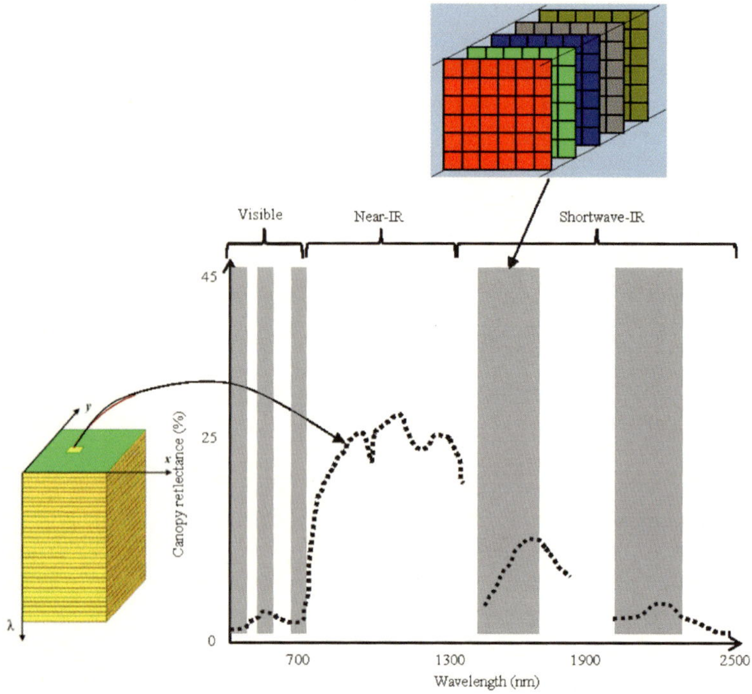

Fig. 2.7 Hyperspectral reflectance signature compared against multispectral bands e.g. Landsat

etc., are ascribing to LIDAR. This technology arguably will become more adopted, even if forest practitioners will never be able to completely circumvent field measurements for model validation purposes.

2.3.3 Species Identification and Classification Using Hyperspectral Sensing

Another remote sensing option, related to the assessment of biomass, is the combined use of different wavelengths of light to differentiate tree species in stands, i.e., enabling biomass-per-species maps. This is relevant because biomass functions for total biomass and biomass components are usually species-specific (Chap. 3). So knowing the species composition of stands would surely improve the prediction of total biomass. This species identification was only made possible with the development of spectral resolution increases from multispectral to hyperspectral sensors (see Fig. 2.7). Hyperspectral sensing, also called imaging spectroscopy, samples more regions of the electromagnetic spectrum than multispectral sensors, such that absorption from specific leaf pigments, canopy structure, or leaf water content can be estimated (Curran 1989; Wessman et al. 1989; Yoder and Pettigrew-Cosby 1995;

Kokaly and Clark 1999; Kokaly 2001). The electromagnetic canopy reflectance signature provides enough detailed data to discriminate between signatures of different species. It is therefore possible to apply species specific models to stratify a community into more representative components and calculating biomass by component using for example canopy structure profiles from LiDAR (Chambers et al. 2007). The reader is referred to seminal studies on this topic, including Gong et al. (1997), Martin et al. (1998), Fung et al. (1999), and van Aardt and Wynne (2001), with an extension to operational airborne data (van Aardt and Wynne 2007) and commercial plantations (van Aardt and Norris-Rogers 2008). These approaches also have been borne out in savannah regions by Cho et al. (2010). These studies report classification accuracies >90 % for deciduous species and as high as 85 % for coniferous species, while the commercial species, specifically *Eucalyptus* sp., have proven to be spectral separable with accuracies at approximately 90 %. All of these studies have approached the challenge by subsetting the hyperspectral data to those wavelengths necessary for separation of a specific set of species. In other words, an operational workflow should be preceded by a pilot study: (i) identify the species of interest, (ii) acquire hyperspectral data, (iii) determine which wavelengths are necessary to separate the species on a spectral basis, and (iv) assess the classification accuracy. And herein lies the caveat – hyperspectral data are by design "oversampled" data, i.e., we have more data than we need.

We therefore need to subset the data to the wavelengths required for a specific application in order to develop statistically robust models, or models with a reasonable number of independent or explanatory variables. An example may be warranted: Imagine that a pilot study shows that wavelengths at 452 nm (blue), 622 nm (red), 1,050 nm (near-infrared), 1,452 nm (shortwave-infrared), and 2,248 nm (shortwave-infrared) are essential to separating three coniferous species at an accuracy of 85 %. The logical approach, for an operational implementation, would be to approach a company that specializes in acquiring imagery via airborne detector/s that can be "programmed" to these wavelengths, via the use of wavelength-specific filters. Such companies and sensors do exist, but they typically only operate silicon-based sensors, which are sensitive to the wavelength range of roughly 380–900 nm; the conundrum is obvious: in order to acquire the necessary wavelengths for the species classification application, one would have to build a relatively expensive sensor, unless the subset of wavelengths can be constrained to the silicon range. This latter solution often is viable, but may come at the cost of slight decreases in accuracy. However, this approach is actually not infeasible, given the lower cost and higher prevalence of silicon sensors; many forestry applications, e.g., species classification, nutrient mapping, moisture stress detection, etc., could be constrained to this wavelength range and thus executed on an operational basis. In fact, leaf-level studies have shown extension to airborne cases, thus adding to potential operational implementation.

Results from a study in tropical forests of Costa Rica indicate that there are spectral differences among species that permit classification at leaf to crown scales (Clark et al. 2005), further corroborated by van Aardt and Wynne (2007) in a mixed oak-pine forest in the Virginia piedmont, USA. However there are also temporal, spatial, and spectral variation within populations and even single individuals of forest tree species that will inevitably decrease classification accuracy and need to be assessed on a as-needed basis. A major challenge is to develop classification schemes that can maximize the spectral, spatial, and temporal information content of digital imagery while accommodating inherent variation within species.

2.4 Remote Sensing Methods Estimating Bulk Biomass

2.4.1 RADAR

Several satellite-based RADAR sensors are currently available for biomass studies. There are many papers available on this subject and the reader is advised to study the references provided to gain more insight into the application of RADAR data in biomass studies. Past studies showed that RADAR backscatter does not correlate well with stand parameters like DBH, height or even basal area (Hyyppä et al. 2000). This section thus provides an overview of the technology and application to bulk biomass estimates.

Theory – confirmed by studies – indicate that backscatter signatures at different RADAR frequencies, as well as polarisations of backscatter, result from scattering from different portions of the tree canopy and the ground surface, while slope and aspect also affect backscatter. The portion of the tree with which the RADAR energy interacts is a function of wavelength. Wavelength is described in bands: L-band SAR with 24 cm wavelength (e.g., JERS-1 satellite, HH polarisation), C-band SAR with 5.6 cm wavelength (e.g. ERS-1 and ERS-2 satellites, VV polarisation), and P-band SAR (airborne sensors) are the bands used in remote sensing applications. Longer wavelengths, like L-band and P-band, penetrate deeper into the vegetation canopy, and scattering of the radiation originates from trunks and large branches. At shorter wavelengths, like C-band, scattering occurs in the upper layers of the canopy from leaves and small branches.

SAR interferometry (InSAR) provides information on the topography of the surface and on temporal changes in certain land surface properties. This technique can retrieve both structural information of natural targets, which in some cases can be converted to biophysical parameters, and digital elevation models (DEM), which can be used for geocoding and radiometrically correcting SAR backscatter imagery. Interferometry is based on the principle that two SAR sensors image an area with the same sensor characteristics from different viewing positions. The two sensors are separated by a spatial baseline. From the two signals an interferogram can be computed from which two parameters can be derived: (i) the interferometric coherence as a measure of the correlation between the two signals; and (ii) the interferometric phase, which is related to topographic height (Balzter et al. 2002).

Kasischke et al. (1995) provide detailed discussions on scattering in SAR and the components of woody biomass. They summarise that the variation in stem biomass accounts for more of the variability in RADAR signature in the P-band HH polarisation, while highest correlation in VV polarisation occurred in the case of total biomass in the canopy layer. The authors also concluded that, as RADAR frequencies increases, overall sensitivity to variations in biomass decrease, which is from L-band to P-band to C-band. L-band exhibits similar biomass scattering in HH and VV polarisation to P-band. They conclude that multi-polarisation C-, L-, and P-band SAR data can be used to estimate biomass in pine forests with total aboveground biomass up to 40 kg m^{-2}.

Carreiras et al. (2013) reported that their study used a machine learning algorithm to establish a relationship between in situ forest aboveground biomass (AGB) in Miombo woodlands in Mozambique and L-band Synthetic Aperture Radar (SAR) backscatter intensity (gamma nought, $\gamma°$) data obtained from the Phased Array L-band SAR (PALSAR) sensor, on-board the Advanced Land Observing Satellite (ALOS). The algorithm used, unique bagging stochastic gradient boosting (BagSGB), as it also allows the production of spatially explicit estimates of prediction variability and an indication of the importance of each predictor variable. Estimates of forest AGB with a root mean square error (RMSE) of 5.03 Mg ha^{-1}, based on a tenfold cross validation, were produced with their modelling approach. Also, the coefficient of correlation (r) between the observed and predicted forest AGB value was 0.95, again based on tenfold cross validation. The variable contributing the most to this model was the mean backscatter intensity for the HH polarisation, which was explained by the low tree canopy cover characterising Miombo savannah woodlands, thus invoking scattering mechanisms associated with this polarisation (e.g., trunk-ground scattering).

Saatchi et al. (2007) used RADAR remote sensing data to map biomass distribution of the Amazon basin. The RADAR data were combined with forest inventory plot data and optical remote sensing at 1 km resolution and ranges up to 400 Mg ha^{-1}. Sun et al. (2003) also reported results from the fusion of LiDAR and RADAR data, which hints at more accurate measurement of biomass distribution on a worldwide scale. Studies by Santoro et al. (2002) concluded that stem volume retrieval was possible up to 350 m^3 ha^{-1} in boreal forests, but that forest density still presents a challenge and that this was not possible at the stand level.

2.4.2 LiDAR

Duncanson et al. (2010) reported that space borne LiDAR data is useful for aboveground biomass (AGB) estimation over a wide range of biomass values and forest types, but the application of these data is limited, given their spatially discrete nature. The authors then used an integration of ICESat Geospatial Laser Altimeter System (GLAS) LiDAR and Landsat data to develop methods to estimate AGB in an area of south-central British Columbia, Canada. They compared estimates with a reliable AGB map of the area, derived from high-resolution airborne LiDAR data, to assess the accuracy of satellite-based AGB estimates. GLAS AGB models were shown to reliably estimate AGB ($R^2 = 0.77$) over a range of biomass conditions. A partial least squares AGB model, using Landsat input data to estimate AGB (derived from GLAS), had an R^2 of 0.60 and was found to underestimate AGB by an average of 26 Mg ha^{-1} per pixel when applied to areas outside of the GLAS transect. This study demonstrates that Landsat and GLAS data integration are most useful for forests with less than 120 Mg ha^{-1} of AGB, less than 60 years of age, and less than 60 % canopy cover. These techniques have high associated errors when applied to areas with greater than 200 Mg ha^{-1} of AGB. Airborne studies, however, have shown reasonable accuracies and precisions when it comes to forest biomass or volume estimation, e.g., Lefsky et al. (2002a, b), Popescu et al. (2002, 2004), and van Aardt et al. (2006). In fact, van Aardt et al. (2008) have proven that wall-to-wall enumeration is possible at the taxanomic group level at high accuracies. It is likely obvious that (i) estimation quality improves as stands become more homogeneous, (ii) validation and calibration protocol for remote sensing assessments need to be put in place, and (iii) estimation outcomes are often time, species, and site dependent.

2.4.3 *Multispectral Images*

Wu and Strahler (1994) discusses a method of estimating total standing biomass of coniferous stands, using an inverted canopy reflectance (geometric-optic) model approach with 30 m Landsat TM data. The use of MODIS data is discussed in Cartus et al. (2011), where a MODIS product, vegetation continuous field (VCF), is used in fusion with SAR data. The VCF product provides global estimates of tree canopy cover at 500 m × 500 m pixel size and where tree canopy cover refers to the fraction of skylight obstructed by canopies of trees that are at least 5 m high. Although the VCF product was validated only partially, it was suitable for the fusion since it is only used to identify areas with low and high canopy cover and the accuracy of the canopy cover estimates is not crucial. The study provided an approach for fully automated stem volume estimation with ERS-1/2 tandem coherence although topographic effects still constraints the use of coherence in general. This presents just one of many studies that show the usefulness of multispectral imagery for biomass assessment, with these kinds of sensors often used in tandem with fine-scale assessment to scale estimates to larger areas, e.g., Duncanson et al. (2010) and Roberts et al. (2011). The reader is referred to this latter study for a comprehensive overview of such approaches.

2.5 Case Study of Integrated Approaches and Data Fusion

Increasing applications of combined remote sensing techniques are evidence that integrated approaches are viable options if good accuracies in biomass estimation should be combined with high efficiency. Due to the increasing need for and importance of fusion of multi-modal remote sensing (LiDAR and multispectral imagery) to improve and scale estimates, we present a case study for forest plantations. The selection of an appropriate LiDAR canopy height sample, as representative of the entire study area, was a central focus of this study. The proposed scheme would use LiDAR transects instead of a blanket coverage with the aim of reducing costs and processing time, while maintaining prescribed levels of accuracy and precisions (Hudak et al. 2002). The remainder of the forest area to be inventoried would then be assessed using IKONOS imagery, based on established LiDAR-IKONOS height relationships.

The study was conducted in the Kwazulu-Natal province located in eastern South Africa. The sampled plantation stands are located approximately 50 km south of the town of Pietermaritzburg. The area is known locally as the southern Natal midlands. Rain falls predominantly in the summer months with cold dry winters and warm wet summers.

Mean annual rainfall ranges from 746–1,100 mm (Schulze 1997) and is associated with either frontal systems originating from the south or from thunderstorms generated from convection activity. Temperatures range from 20 °C in summer to below 10 °C in the winter.

Extreme temperature changes are a function of altitude and proximity to the warm Indian Ocean. Soils in the area are characterized by fine sandy clay and humic topsoils, underlain by yellow or red apedal subsoils. The topography of the study area is flat with undulating hills and is classified by Schulze (1997) as being low mountains. Altitude ranges from 362 m aml to over 1,500 m aml with an average altitude of approximately 874 m aml.

2.5.1 Case Study: LiDAR Data

The study area was surveyed in November 2006 using an Optech ALTM 3033 LiDAR sensor mounted on a Cessna Caravan 206 aircraft. The sensor operated in the near infrared portion of the electromagnetic spectrum (1,064 nm) and was configured to capture first and last LiDAR returns. The survey was conducted at an altitude of 550 m aml with a scan angle of 10° and a beam divergence of 0.2 mrad. Scan overlap was fixed to 25 % with a resulting footprint size not exceeding 0.2 m. The resulting point density was greater than 5 points m^{-2} with vertical and horizontal accuracies of 0.15 and 0.28 m, respectively. Pre-processing was undertaken by the vendor who classified returns into ground and non-ground data sets using Terrasolid TerraScan®. The classification process involved an initial automated classification, with the accuracy influenced by the underlying terrain. This was followed by a manual editing process using 0.15 m resolution orthophotos. Ground control was provided by two base stations located on site with errors in the order of 0.031 m, reported for the final digital terrain model. The data were delivered in a local coordinate system known as LO31 (Gauss Conform projection with the central meridian located at 31 °E). The LiDAR data were delivered with an associated 0.15 m colour orthophoto data set.

2.5.2 Case Study: IKONOS Multispectral Imagery

IKONOS was launched in September of 1999 in a 680 km high sun-synchronous orbit with a descending node crossing at approximately 10h15 local solar time. The IKONOS satellite simultaneously collects imagery in four multispectral bands and a single panchromatic band with 11-bit radiometric resolution. Data used in this project were collected by the IKONOS sensor on September 23, 2006. Rational polynomial coefficients (RPCs) were included to aid in the geo-rectification of the imagery. Orthorectification was undertaken using ENVI 4.3. RPCs were used in conjunction with a digital elevation model provided by industry partners. The rectification of the data was undertaken using 20 ground control points collected using a Leica GS 20 DGPS; the resulting root mean square error was less than half a pixel (<2 m). IKONOS radiometric correction began with converting the 11 bit digital numbers to top-of-atmosphere (TOA) radiance (Pagnutti et al. 2003). Following this the data were radiometrically corrected and converted to unitless reflectance using the Fast Line-of-Sight Atmospheric Analysis of Spectral Hypercubes (FLAASH) correction model (Alder-Golden et al. 1999; Matthew et al. 2000). Conversion to reflectance generally results in temporally independent imagery, since the majority of atmospheric influences are corrected for. This theoretically enables extrapolation of results to other seasons and years, as opposed to the temporally restricted use of radiance imagery.

2.5.3 Case Study: Enumeration Data

Field data were collected during spring (October) 2006 in compartments of commercial plantations ranging in age from 4 to 9 years old. A total of 61 plots were sampled in 19 different stands. Plots were established no closer than 50 m to each other to ensure that the sample data set captured as much variability as possible in the chosen stand. The geographic location of the plot centre was recorded using DGPS. A circular plot was then established with an initial radius of 15 m, which was adjusted based on the slope of the compartment. The diameter at breast height (DBH) was recorded for each tree with a DBH larger than 5 cm, in every plot. Tree height was measured using a Vertex hypsometer with at least 15 DBH-height pairs recorded for each plot. The DBH-height pair data were employed to develop regression equations which were used to compute tree height for all trees based on the DBH ($R^2 = 0.98$, $p < 0.001$). Several height metrics were subsequently calculated for each plot. The data collected were used for validation purposes only. Data were collected using a stratified sampling scheme applied to LiDAR height data and informed by an empirical semi-variogram analysis, as described below.

2.5.4 Case Study: Semi-variogram Modelling and Sampling

The selection of an appropriate LiDAR sample size is critical, since the samples should in theory capture as much canopy variability as possible within each compartment. Richards et al. (2000) provide the following criteria when designing a sampling strategy: The scheme should be sensor independent, unbiased estimators and error variance should be computed, areas not of interest should be excluded, the sample should be selected from areas where change is expected to be high, and adjacent plots should not contain redundant information and should be systematically spaced. The area sampled for this study consisted of 1,000 ha of plantation forests made up of *Eucalyptus*, *Pinus*, and Australian *Acacia* species. The species and age groups were defined prior to setting out the sampling strategy, thus the target population had already been defined as *Eucalyptus* compartments between the ages of 4 and 9 years old. Compartments satisfying these initial criteria were selected using a geographical information system based on information from the forest company. LiDAR canopy returns, normalised by a ground digital elevation model, were subsequently selected and added to the analysis data set for the defined compartments. Normalisation of LiDAR returns was necessary in order to account for varying ground height, i.e., the process involved subtraction of the associated DEM values from first return LiDAR point heights. The selection of a suitable sample with which to model tree height was then undertaken using geostatistical methods.

The criteria stipulated by Richards et al. (2000) indicate that the scheme should be sensor independent. In this case the LiDAR sensor was a standard two return system, which constitutes the lower limit in terms of LiDAR technology currently in use. Richards et al. (2000) furthermore state that the variance of the sample estimates should be calculated. We calculated sample and population statistics and evaluated these based on various descriptive statistical measures. However, the sampling scheme first needed to be designed and implemented. The geostatistical methods discussed above, which measured and quantified the spatial dependence of LiDAR canopy height returns, were employed. It was deemed imperative that the sample from the LiDAR height data set should capture as much of the vertical variability as possible present in the compartment of interest. Semi-variograms and their application to LiDAR canopy height returns represent an ideal tool to do just this (Butson and King 1999).

Semi-variograms were derived by first calculating an empirical semi-variogram. Empirical semi-variograms measure the spatial dependence of neighbouring observations for any continuously varying phenomenon and have the advantage of relating key descriptors of the spatial statistics, namely range and sill, of the data (Treitz 2001). Range is especially important in this study as it contributed to determining the width of sample transects. The range value, as calculated by the empirical semi-variogram, quantifies the distance at which height values are no longer statistically related (Curran and Atkinson 1998). This implies that if a sample of points is selected based on the range distance (randomly selected training points), this sample should capture most of the canopy height variability present in the compartment (Woodcock et al. 1988a, b). We recorded the average range from semi-variograms for each of the 61 field plots. Spherical mathematical models were used, which were iteratively optimised using the residual sum of squares (Hiemstra et al. 2009).

The identified optimal LiDAR transect width subsequently was doubled to more accurately reflect an operational use of LiDAR transects and to ensure that the transect width captured the majority of height variation in the study area, with the semi-variogram range serving as an indicator of the minimum width. This resulted in pseudo flight lines of the determined width for the imagery and LiDAR data and a between-transect spacing of 150 m to ensure coverage of the entire study area and all compartments (Richards et al. 2000). The pseudo flight lines were further divided into 10 m^2 blocks, which were sub-sampled to ensure that samples were systematically spaced in order to minimise the potential for redundancy within the sample data set. Training data for modelling purposes eventually were selected from transects located in 19 Eucalyptus compartments. Maximum LiDAR height values (dependent variable) were extracted, since these are recognised as being representative of actual tree height in closed canopy forests (Popescu et al. 2002). Co-located multispectral data were also extracted with the mean spectral reflectance (independent variable) used for modelling maximum LiDAR height.

2.5.5 Case Study: Parametric, Non-parametric, and Non-linear Statistical Modelling

The development of a statistical method with which to relate LiDAR-derived tree height to IKONOS spectral reflectance involved the testing of three statistical paradigms, namely parametric, non-parametric, and non-linear artificial neural networks. Parametric statistical procedures, e.g., regression, make certain assumptions regarding the underlying distribution of the data. It is primarily assumed that the data is normally distributed and that the input variables have similar variances (Cohen et al. 2003). However, as with most environmental data, this is not always the case and certain input variables had to be linearized using statistical transformations (Hudak et al. 2006). Two of the five input variables (IKONOS green and red bands) displayed non-normal distributions and were transformed using standard logarithmic methods. The five input variables were then used as independent variable inputs to a multiple linear regression, where the four IKONOS bands plus the age of the sample compartment were regressed against maximum LiDAR height. Results from this analysis were then interrogated for outliers. Outliers were removed using Cook's distance measure (Cook 1977) and the regression was re-run using the resulting cases. The statistical approach is similar to that employed by Wulder and Seemann (2003); however, we employed only the per-band mean spectral reflectance values as independent variable. The reason for this is that the resultant regression model was applied to the imagery at the pixel level; hence distributional measures would have been unsuitable in this instance (Wulder and Seemann 2003).

The second statistical paradigm employed in this research makes use of non-parametric statistical methods. These approaches are known as "distribution-free" methods and do not rely on the assumption that data are drawn from a given probability distribution. The *k*-nearest neighbour approach is such a non-parametric method that imputes forest inventory variables using reference samples and target mapping units (Reese et al. 2002). Reference samples typically are derived from remote sensing spectral reflectance and co-located forest variables of interest, which in our study was maximum LiDAR height as variable of interest within the reference sample plot. The goal of this approach is consistent with our primary objective, namely to evaluate canopy height estimation at locations not sampled by the LiDAR sensor using spectral reflectance and compartment age as predictor variables. Each target location is assigned a reference value based on the weighted Euclidean distance from its k nearest reference plot(s) according to this approach. The k nearest reference plots are typically defined using weighted Euclidean distance calculated in spectral feature space, while the target variable is estimated by the weighted average of the distances to the k nearest neighbours.

The weighted average distance (in spectral feature space) was calculated in our study using the random-Forest algorithm (Breiman 2001). This algorithm differs from the standard Euclidean distance measure in that it does not make use of a weight matrix, but instead classification and regression trees are used to classify reference and target observations: If a target and reference observation ends up in the same node, they are regarded as being similar. The distance measure is computed as one minus the proportion of trees which contain the same variable, and where a target observation is in the same terminal node as a reference observation. Crookston and Finley (2008) identify two advantages of using random-Forest as opposed to other distance metrics, namely that variables can be a mixture of continuous and categorical types and that the method is non-parametric. Hudak et al. (2008) compared a range of methods using several different error metrics (e.g., root mean square difference) and concluded that the random-Forest method was more robust and flexible than standard distance measures, e.g., Euclidean and Mahalanobis distance (Mahalanobis 1936). Preliminary tests conducted by the authors confirmed this finding, which resulted in the random-Forest approach being chosen as the appropriate distance measure.

2.5.6 Case Study: Non-linear Artificial Neural Network (ANN)

A non-linear artificial neural network (ANN) was the final statistical approach evaluated. Jordan and Bishop (1996) describe neural networks as a graph with

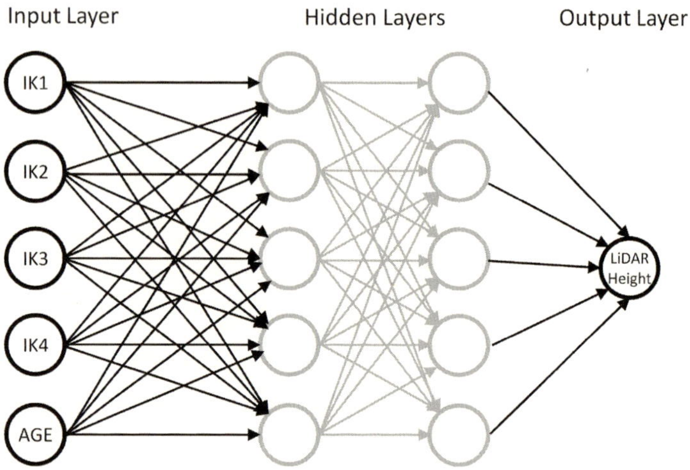

Fig. 2.8 Topological structure of an ANN (Haykin 1994)

patterns, represented by numerical values attached to the nodes of the graph, and transformation between patterns achieved via message passing algorithms. The power of an ANN is rooted in the fact that it is designed to replicate a biological neural system. An advantage of using this methodology is that an ANN has the ability to learn and adapt to the underlying structure of the data set being analysed. The ANN approach is also capable of handling non-normality, non-linearity, and collinearity within a data set of interest (Haykin 1994). Typically, an ANN is trained using a sample or training set that consists of both dependent and independent variables. In this study the training data set was extracted from the 225 sample plots and included maximum LiDAR heights as dependent variable and mean IKONOS reflectance and compartment age as independent variables.

The structure of an ANN is made up of several layers (Fig. 2.8), most notably an input layer (containing the training data), a number of hidden layers, and an output layer. Each layer is in turn made up of a number of neurons, which represent the fundamental processing unit of any ANN. A neuron consists of three parts (Fig. 2.9), namely a set of synapses or connecting links, an adder or transfer function, and an activation function. For this study we first normalised (subtract mean and divide by the standard deviation) spectral reflectance and age before introducing these independent variables to the neurons via the synapses or connection links.

Inputs were standardised (0–1) to facilitate inclusion of the age variable and weights were initially set using a random seed value. Optimisation of the network was undertaken using bootstrapping methods, with 50 % of the training data removed from the training set and used to assess the accuracy of the network weights. Fifty bootstrapped models were calculated for each epoch, with a total of 500 epochs specified. The bootstrapping approach allowed for the calculation of mean effects of each input variable (confidence interval), similar to the standard

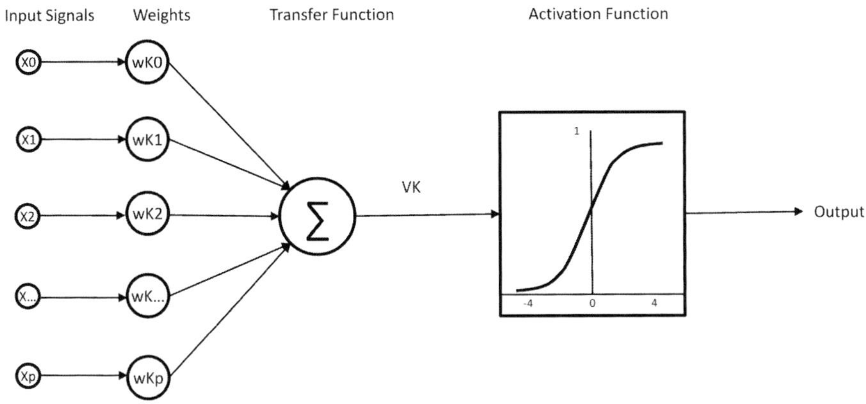

Fig. 2.9 Structure of a neuron (Haykin 1994)

output of a multiple regression analysis. Goodness of fit statistics, similar to those reported in regression models, was used to assess the model. Finally, the neural network was applied directly to the population data to create spatially explicit maps of LiDAR height after model training was completed.

2.5.7 Case Study: Conclusions

The study outlines a methodology to extrapolate LiDAR-measured tree height to areas sampled using only multispectral IKONOS imagery in support of forest management activities. This cost reducing sampling scheme was developed using geostatistical procedures and expert knowledge in order to address the high cost of LiDAR surveys. The scheme sampled 2.25 ha of all LiDAR canopy returns in an area covering approximately 1,000 ha, which equates to 0.25 % of the study area. The premise of the study was that the transect type sampling scheme would reduce the cost of the LiDAR survey and thus make incorporation of the sensor more cost effective within an operational scenario. However, statistical results did not fully support the extrapolation approach proposed here, i.e., using spectral reflectance and age as predictor variables. While the LiDAR accurately measures tree heights, the RMSE values associated with validation data were outside the acceptable error margins employed in operational scenarios (±5 %). Over-prediction of tree heights suggested that the proposed methodology was not perfectly attuned to accurate estimation of tree height for areas external to the LiDAR collection framework. However, given

(i) the fact that goodness-of-fit statistics were promising,
(ii) the documented problems associated with field enumeration techniques,

(iii) the exacerbating factors related the spherical shape of Eucalyptus crowns, it is proposed that future research address the sampling design itself, while also investigating the use of off-nadir (multi-angular) imagery. Such multi-angular imagery has been shown to be more amenable to extraction of vegetation structure than traditional nadir-only imagery. The study concluded that, even though the precision metrics reported here were unsatisfactory for truly operational purposes, specific metrics related to the correlation between LiDAR-derived tree height and extrapolated tree height (multi-spectral imagery) were encouraging. We believe that studies such as this are necessary if we are to eventually provide large scale, operational structural assessments of our forest resources using currently expensive, but accurate LiDAR surveying approaches.

2.6 Conclusions

We believe that future research on biomass inventory should focus on multi-phase inventory with sensor fusion. Research is progressing on methodologies that combine LiDAR and high resolution UAV photography, as well as many other sensor combinations. The synergy of combining sensors like multispectral, LiDAR, and RADAR likely will hold the key to containing cost and errors, as well as covering much larger areas of standing biomass. Treuhaft et al. (2004), for instance, provide a useful summary of RADAR use for forest biomass estimation, as well as fusion of RADAR and optical remote sensing sensor data.

Multi-phase inventories combine the different strengths of different methods and sensors in localisation, as well as quantifying not only the biomass, but also the components of the biomass in terms of leaves, branch wood, and stems above ground. No direct measure of belowground biomass (BGB) is currently possible. Estimates of BGB are possible once proper models exist for an ecotype or species, most typically through an established AGB-BGB conversion factor.

Finally, a table is provided to summarize current forest biomass (inventory) technologies in terms of the level of training required to implement, the cost of execution, and the scale for which it is suitable (Table 2.1). Both the lower limit as well as upper limit of the method is indicated, where small (S) indicate community and stand level, medium (M) indicate farm to catchment level, and large (L) indicates regional to global scale inventory. An almost invariable truth holds: The larger the area required to assess with terrestrial sampling, the higher the cost, while with remote sensing methods, the inverse is true in that the unit cost of assessing a small area is higher than assessment of catchment and larger areas.

Remote sensing is not a golden solution or panacea to all of our forest inventory challenges, whether those be inventory related or focused on the condition (nutrient, moisture) of the resource. But the technologies described above do offer viable solutions to operational assessment needs in the forestry and ecological communities. Some of these technologies are closer to operational implementation

Table 2.1 Matrix comparing inventory methods by level of training needed, cost of execution and size of area which can be covered

			Scale	
Method	Training	Cost	From	To
Sample plot terrestrial survey	Low	Med	S	M
Airborne SAR	High	High	M	L
Terrestrial LiDAR	Med	High	S	
Airborne LiDAR	High	High	S	M
Satellite SAR	High	Low	M	L
Satellite LiDAR	High	Low	M	L
UAV photography	Med	Low	S	
Conventional aerial photography	Med	Med	M	L
Airborne hyperspectral	High	High	S	M
Satellite multispectral hi-res	Med	Low	M	L
Satellite multispectral med-res	Med	Low		L

than other, e.g., airborne LiDAR vs. either space borne or ground-based LiDAR, but in all cases certain commonalities should be observed:

1. Field work will always form an essential component of any inventory, whether for calibration of biomass models developed elsewhere, or for validation of existing models for different sites or seasons.
2. Implementation costs will decrease as new markets develop, vendors' numbers increase, and practitioners adopt novel approaches. However, we should always strive to balance these costs with associated accuracy and precision trade-offs.
3. Finally, fusion of multiple approaches or modalities likely will be key to a successful and comprehensive inventory system – some sensors are best suited to structural assessment (inventory) and others to spectral assessment (species, nutrients, moisture status), while ground-based efforts will always occupy a necessary component in the inventory chain.

There may be other such essential truths, but we trust that this chapter has provided the reader with an overview of what is possible using remote sensing. Research in this field is ongoing, and there is ample evidence that the future of technologically advanced approaches to forest resource assessment is bright.

References

Ackerman S, Ackerman PA, Seifert T (2013) Effects of irregular stand structure on tree growth, crown extension and branchiness of plantation grown *Pinus patula*. South For (in review)

Adler-Golden SM, Matthew MW, Bernstein LS, Levine RY, Berk A, Richtsmeier SC, Acharya PK, Anderson GP, Felde G, Gardner J, Hoke M, Jeong LS, Pukall B, Mello J, Ratkowski A, Burke HH (1999) Atmospheric correction for short-wave spectral imagery based on MODTRAN4. SPIE Proc Imaging Spectrom 3753:61–69

Andersen HE, McGaughey RJ, Reutebuch SE (2005) Estimating forest canopy fuel parameters using lidar data. Remote Sens Environ 94(4):441–449

Attiwil PM, Ovington JD (1968) Determination of forest biomass. For Sci 14:13–15

Baltsavias EP (1999) Airborne laser scanning: basic relations and formulas. ISPRS J Photogramm Remote Sens 54(2–3):199–214

Balzter H, Talmon E, Wagner W, Gaveau D, Plummer S, Yu JJ, Quegan S, Davidson M, Le Toan T, Gluck M, Shvidenko A, Nilsson S, Tansey K, Luckman A, Schmullius C (2002) Accuracy assessment of a large-scale forest cover map of Central Siberia from synthetic aperture radar. Can J Remote Sens 28(6):719–737

Breiman L (2001) Random forests. Mach Learn 45(1):5–32

Bundeswaldinventur (2008) The National Forest Inventory. Bundesministerium für Ernährung, Landwirtschaft und Verbraucherschutz. Berlin. Online: http://www.bundeswaldinventur.de/enid/6a58daffd51c0d46d88bfb4ca088fd05,0/75.html

Butson C, King DJ (1999) Semivariance analysis of forest structure and remote sensing data to determine an optimal sample plot size. In: Proceedings of the 4th international airborne remote sensing conference, Ottawa, 21–24 June 1999. Environmental research institute of Michigan, Ann Arbor, vol II, pp 155–162

Cartus O, Santoro M, Schmullius C, Li Z (2011) Large area forest stem volume mapping in the boreal zone using synergy of ERS-1/2 tandem coherence and MODIS vegetation continuous fields. Remote Sens Environ 115:931–943

Carreiras JMB, Melo JM, Vasconcelos MJ (2013) Estimating the above-ground biomass in Miombo Savanna woodlands (Mozambique, East Africa) using L-band synthetic aperture radar data. Remote Sens 5(4):1524–1548

Chambers JQ, Asner GP, Morton DC, Anderson LO, Saatchi SS, Espırito-Santo FDB, Palace M, Souza C Jr (2007) Regional ecosystem structure and function: ecological insights from remote sensing of tropical forests. Trends Ecol Evol 22(8):389–440

Cho MA, Debba P, Mathieu R, Naidoo L, van Aardt JAN, Asner GP (2010) Improving discrimination of savanna tree species through a multiple endmember spectral angle mapper approach: canopy-level analysis. IEEE Trans Geosci Remote Sens 48(11):4133–4142

Clark ML, Clark DB, Roberts DA (2004) Small-footprint lidar estimation of sub canopy elevation and tree height in a tropical rainforest landscape. Remote Sens Environ 91(1):68–89

Clark ML, Roberts DA, Clark DB (2005) Hyperspectral discrimination of tropical rain forest tree species at leaf to crown scales. Remote Sens Environ 96:375–398

Cohen WB, Maiersperger TK, Gower ST, Turner DP (2003) An improved strategy for regression of biophysical variables and Landsat ETM + data. Remote Sens Environ 84(4):561–571

Cook RD (1977) Detection of influential observation in linear regression. Technometrics 84:561–581

Coops NC, Wulder MA, Culvenor DS, St-onge B (2004) Comparison of forest attributes extracted from fine spatial resolution multispectral and lidar data. Can J Rem Sens 30(6):855–866

Crookston NL, Finley AO (2008) YaImpute: an R package for kNN imputation. J Stat Softw 23(10):1–16

Cunia T (1990) Forest inventory: on the structure of error of estimates. In: LaBau, VJ, Cunia T (eds) State-of-the-art methodology of forest inventory: a symposium proceedings. General technical report PNW-GTR-263. U.S. Department of Agriculture, Forest Service, Pacific Northwest Research Station, Portland. 635 pp

Curran PJ (1989) Remote sensing of foliar chemistry. Remote Sens Environ 30:271–278

Curran PJ, Atkinson PM (1998) Geostatistics and remote sensing. Progr Phys Geogr 22(1):61–78

Duncanson LI, Niemann KO, Wulder MA (2010) Integration of GLAS and Landsat TM data for aboveground biomass estimation. Can J Remote Sens 36(2):129–141

Fung T, Ma FY, Sui WL (1999) Hyperspectral data analysis for subtropical tree species identification. In: Proceedings of ASPRS annual conference, 17–21 May 1999, Portland, 10 pp.

Gaveau DLA, Hill RA (2003) Quantifying canopy height underestimation by laser pulse penetration in small-footprint airborne laser scanning data. Can J Rem Sens 29(5):650–657

Gong P, Pu R, Yu B (1997) Conifer species recognition: an exploratory analysis of in situ hyperspectral data. Remote Sens Environ 62:189–200

Greenberg JA, Dobrowski SZ, Ustin SL (2005) Shadow allometry: estimating tree structural parameters using hyperspatial image analysis. Remote Sens Environ 97:15–25

Haala N, Brenner C (1999) Extraction of buildings and trees in urban environments. ISPRS J Photogramm Remote Sens 54(2–3):130–137

Haykin S (1994) Neural networks: a comprehensive foundation. Prentice Hall, Upper Saddle River, 842 p

Hiemstra PH, Pebesma EJ, Twenhofel CJW, Heuvelink GBM (2009) Real-time automatic interpolation of ambient gamma dose rates from the Dutch Radioactivity Monitoring Network. Comput Geosci 35(8):1711–1721

Hodgson ME, Jensen JR, Schmidt L, Schill S, Davis B (2003) An evaluation of LIDAR- and IFSAR-derived digital elevation models in leaf-on conditions with USGS level 1 and level 2 DEMs. Remote Sens Environ 84:295–308

Hollinger DY (2008) Defining a landscape-scale monitoring tier for the North American Carbon Program. In: Hoover CM (ed) Field measurements for forest carbon monitoring. Springer, New York, pp 3–16

Holmgren J (2004) Prediction of tree height, basal area, and stem volume in forest stands using airborne laser scanning. Scand J For Res 19(6):543–553

Holmgren J, Nilsson M, Olsson H (2003) Estimation of tree height and stem volume using airborne laser scanning. For Sci 49(3):419–428

Hopkinson C, Chasmer L, Lim K, Treitz P, Creed I (2006) Towards a universal canopy height indicator. Can J Rem Sens 32(2):1–14

Hudak AT, Crookston NL, Evans JS, Falkowski MJ, Smith AMS, Gessler PE, Morgan P (2006) Regression modeling and mapping of coniferous forest basal area and tree density from discrete-return lidar and multispectral satellite data. Can J Remote Sens 32(2):126–138

Hudak AT, Crookston NL, Evans JS, Hall DE, Falkowski MJ (2008) Nearest Neighbor imputation of species-level, plot-scale forest structure attribute from lidar data. Remote Sens Environ 112(5):2232–2245

Hudak AT, Lefsky MA, Cohen WB, Berterretche M (2002) Integration of lidar and Landsat ETM + data for estimating and mapping forest canopy height. Remote Sens Environ 82(2–3):397–416

Hyyppä J, Hyyppä H, Inkinen M, Engdahl M, Linko S, Zhu Y-H (2000) Accuracy comparison of various remote sensing data sources in the retrieval of forest stand attributes. For Ecol Manage 128(1–2):109–120

Jordan MI, Bishop CM (1996) Neural networks. ACM Comput Surv 28(1):73–75

Kangas A, Maltamo M (eds) (2006) Forest inventory – methodology and applications, Managing forest ecosystems series. Springer, Dordrecht

Kasischke ES, Christensen NL Jr, Bourgeau-Chavez LL (1995) Correlating radar backscatter with components of biomass in loblolly pine forests. IEEE Trans Geosci Remote Sens 33(3):643–659

Kätsch C, Vogt H (1999) Remote sensing from space—present and future applications in forestry, nature conservation and landscape management. South Afr For J 185(1):14–26

Kokaly RF (2001) Investigating a physical basis for spectroscopic estimates of leaf nitrogen concentration. Remote Sens Environ 75:153–161

Kokaly RF, Clark RN (1999) Spectroscopic determination of leaf biochemistry using band-depth analysis of absorption features and stepwise multiple linear regression. Remote Sens Environ 67:267–287

Lefsky MA, Cohen WB, Parker GG, Harding DJ (2002a) Lidar remote sensing for ecosystem studies. Bioscience 52(1):19–30

Lefsky MA, Cohen WB, Harding DJ, Parker GG, Acker SA, Gower ST (2002b) Lidar remote sensing of above-ground biomass in three biomes. Global Ecol Biogeogr 11:393–399

Magnussen S, Boudewyn P (1998) Derivations of stand heights from airborne laser scanner data with canopy-based quantile estimators. Can J Forest Res 28(7):1016–1031

Mahalanobis PC (1936) On the generalised distance in statistics. Proc Indian Natl Sci Acad Phys Sci 12:49–55

Maltamo M, Tokola T, Lehikoinen M (2003) Estimating stand characteristics by combining single tree pattern recognition of digital video imagery and a theoretical diameter distribution model. For Sci 49(1):98–109

Maltamo M, Mustonen K, Hyyppä J, Pitkänen J, Yu X (2004) The accuracy of estimating individual tree variables with airborne laser scanning in a boreal nature reserve. Can J Forest Res 34(9):1791–1801

Mandallaz D (2008) Sampling techniques for forest inventories, Applied environmental statistics series 4. Chapman & Hall/CRC, Boca Raton

Martin ME, Newman SD, Aber JD, Congalton RG (1998) Determining forest species composition using high spectral resolution remote sensing data. Remote Sens Environ 65:249–254

Matthew MW, Adler-Golden SM, Berk A, Richtsmeier SC, Levine RY, Bernstein LS, Acharya PK, Anderson GP, Felde HW, Hoke MP, Ratkowski A, Burke HH, Kaiser RD, Millerd DP (2000) Status of atmospheric correction using a MODTRAN4-based algorithm. In: Shen SS (ed) SPIE proceeding, algorithms for multispectral, hyperspectral, and ultraspectral imagery VI. Proceedings of SPIE, vol 4049, pp 199–207

Means JE, Acker SA, Fitt BJ, Renslow M, Emerson L, Hendrix CJ (2000) Predicting forest stand characteristics with airborne scanning lidar. Photogramm Eng Remote Sens 66(11):1367–1371

Næsset E (2002) Predicting forest stand characteristics with airborne scanning laser using a practical two-stage procedure and field data. Remote Sens Environ 80:88–99

Næsset E, Gobakken T (2005) Estimating forest growth using canopy metrics derived from airborne laser scanner data. Remote Sens Environ 96(3–4):453–465

Nilsson M (1996) Estimation of tree heights and stand volume using an airborne lidar system. Remote Sens Environ 56(1):1–7

Pagnutti M, Ryan RE, Kelly M, Holekamp K, Zanoni V, Thome K, Schiller S (2003) Radiometric characterization of IKONOS multispectral imagery. Remote Sens Environ 88(1–2):53–68

Popescu SC, Wynne RH, Nelson RF (2002) Estimating plot level tree heights with lidar: local filtering with a canopy-height based variable window size. Comput Electron Agric 37(1–3):71–95

Popescu SC, Wynne RH, Scrivani JA (2004) Fusion of small-footprint lidar and multispectral data to estimate plot-level volume and biomass in deciduous and pine forests in Virginia, U.S.A. For Sci 50(4):551–565

Reese H, Nilsson M, Sandström P, Olsson H (2002) Applications using estimates of forest parameters derived from satellite and forest inventory data. Comput Electron Agric 37(1–3):37–55

Riaño D, Meier E, Allgöwer B, Chuvieco E, Ustin SL (2003) Modeling airborne laser scanning data for the spatial generation of critical forest parameters in fire behavior modeling. Remote Sens Environ 86:177–186

Richards T, Gallego J, Achard F (2000) Sampling for forest cover change assessment at the pantropical scale. Int J Rem Sens 21(6):1473–1490

Roberts JW, van Aardt JAN, Ahmed FB (2011) Image fusion for enhanced forest structure assessment. Int J Remote Sens 32(1):243–266

Saatchi SS et al (2007) Spatial distribution of live aboveground biomass in the Amazon Basin. Global Change Biol 13:816–837

Santoro M, Askne J, Smith G, Fransson JES (2002) Stem volume retrieval in boreal forests from ERS-1/2 interferometry. Remote Sens Environ 81:19–35

Schulze RE (1997) South African atlas of agrohydrology and Climatology. Report TT82/96. Water Research Commission, Pretoria

Seidel D, Albert K, Fehrmann L, Ammer C (2012) The potential of terrestrial laser scanning for the estimation of understory biomass in coppice-with-standard systems Biomass Bioenergy 47:20–25. http://www.sciencedirect.com/science/journal/09619534

Seielstad CA, Queen LP (2003) Using airborne laser altimetry to determine fuel models for estimating fire behavior. J For 101(4):10–15

Smith WB, Miles PD, Vissage JS, Pugh SA (2004) Forest resources of the United States, 2002. General technical report NC-241. USDA Forest Service, Washington, DC
Stauffer HB (1982) A sample size table for forest sampling. For Sci 28(4):777–784
Suarez JC, Ontiveros C, Smith S, Snape S (2005) Use of lidar and aerial photography in the estimation of individual tree heights in forestry. Comput Geosci 31(2):253–262
Sun G, Ranson KJ, Kharuk VI, Kovacs K (2003) Validation of surface height from shuttle radar topography mission using shuttle laser altimeter. Remote Sens Environ 88(4):401–411
Treitz P (2001) Variogram analysis of high spatial resolution remote sensing data: an examination of boreal forest ecosystems. Int J Rem Sens 22(18):3895–3900
Treuhaft RN, Law BE, Asner GP (2004) Forest attributes from radar interferometric structure and its fusion with optical remote sensing. BioScience 54(6):561–571
van Aardt JAN, Norris-Rogers M (2008) Spectral-age interactions in managed, even-aged eucalyptus plantations: application of discriminant analysis and classification and regression trees approaches to hyperspectral data. Int J Remote Sens 29(6):1841–1845
van Aardt JAN, Wynne RH (2001) Spectral separability among six southern tree species. Photogramm Eng Remote Sens 67(12):1367–1375
van Aardt JAN, Wynne RH (2007) Examining pine spectral separability using hyperspectral data from an airborne sensor: an extension of field-based results. Int J Remote Sens 28(1–2):431–436
van Aardt JAN, Wynne RH, Oderwald RG (2006) Forest volume and biomass estimation using small-footprint lidar-distributional parameters on a per-segment basis. For Sci 52(6):636–649
van Aardt JAN, Wynne RH, Scrivani JA (2008) Lidar-based mapping of forest volume and biomass by taxonomic group using structurally homogenous segments. Photogramm Eng Remote Sens 74(8):1033–1044
van Laar A, Akça A (2007) Forest mensuration, Managing forest ecosystems series 13. Springer, Dordrecht
van Laar A, Theron JM (2004) Equations for predicting the biomass of *Acacia Cyclops* and *Acacia saligna* in the western and eastern Cape regions of South Africa: part 1: tree-level models. S Afr For J 201:25–34
van Laar A, van Lill WS (1978) A biomass study in *Pinus radiata*. S Afr For J 107(1):71–76
Waggoner PE (2009) Forest inventories discrepancies and uncertainties. Discussion paper, Washington, DC
Wehr A, Lohr U (1999) Airborne laser scanning – an introduction and overview. ISPRS J Photogramm Remote Sens 54(2–3):68–82
Wessman CA, Aber JD, Peterson DL (1989) An evaluation of imaging spectrometry for estimating forest canopy chemistry. Remote Sens Environ 10(8):1293–1316
Woodcock CE, Strahler AH, Jupp DLB (1988a) The use of variograms in remote sensing: I. Scene models and simulated images. Remote Sens Environ 25(3):323–348
Woodcock CE, Strahler AH, Jupp DLB (1988b) The use of variograms in remote sensing: II. Real digital images. Remote Sens Environ 25(3):349–379
Wu Y, Strahler AH (1994) Remote estimation of crown size, stand density and biomass on the Oregon transect. Ecol Appl 4(2):299–312
Wulder MA, Seemann D (2003) Forest inventory height update through the integration of lidar data with segmented Landsat imagery. Can J Remote Sens 29(5):536–543
Yoder BJ, Pettigrew-Cosby RE (1995) Predicting nitrogen and chlorophyll content and concentrations from reflectance spectra (400–2500 nm) at leaf and canopy scales. Remote Sens Environ 53:199–211
Zimble DA, Evan DL, Carlson DG, Parker RC, Grado SC, Gerrard PD (2003) Characterizing vertical forest structure using small-footprint airborne lidar. Remote Sens Environ 87(2–3):171–182

Chapter 3
Modelling and Simulation of Tree Biomass

Thomas Seifert and Stefan Seifert

3.1 Introduction

A primary objective of sustainable bioenergy production is to quantify the available resource supply because all further planning of the value chain hinges on the available biomass that can be converted. Since biomass is costly to transport, the spatial quantification of the resource is also important. Thus, modern approaches to biomass supply chain management must embrace the resource quantity and location as a key element of the supply chain. Data on resource availability are usually obtained from different sources such as remote sensing and terrestrial inventories, as discussed in Chap. 2, which provide information on the spatial distribution of forests and trees and their dimensions but are, as such, not capable of estimating biomass directly with the necessary accuracy. Thus the main purpose of the application of modelling and simulation techniques in this context is the estimation of the biomass resource from broadly available tree and stand variables. This auxiliary information could be sourced from inventories and remote sensing or could be provided by model projections from growth models to estimate the biomass availability.

Biomass modelling is a typical upscaling process based on statistical modelling. Depending on the modelling and sampling method of choice, different upscaling steps are involved. An upscaling process normally involves two steps: upscaling from the biomass samples to the individual tree and from the tree to the stand (Fig. 3.1).

Each upscaling step is normally characterised by sampling and a regression modelling components. The biomass models are first established on a subset of data of a bigger population with independent variables that are easier to measure than

T. Seifert (✉) • S. Seifert
Department of Forest and Wood Science, Stellenbosch University,
Private Bag X1, 7602 Matieland, South Africa
e-mail: seifert@sun.ac.za

T. Seifert (ed.), *Bioenergy from Wood: Sustainable Production in the Tropics*,
Managing Forest Ecosystems 26, DOI 10.1007/978-94-007-7448-3_3,

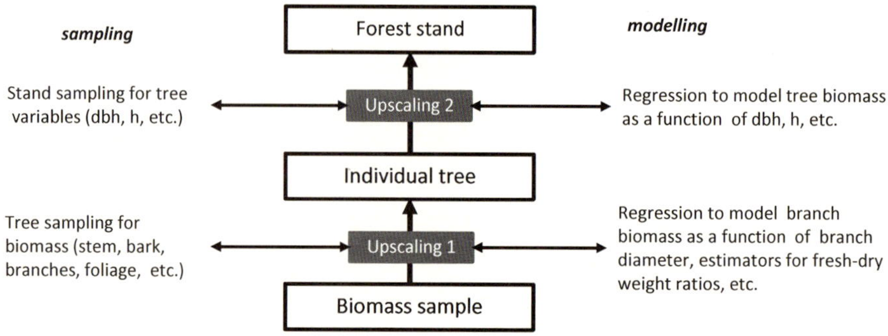

Fig. 3.1 Typical upscaling steps in biomass modelling involve sampling and modelling

the biomass itself. In a next step this auxiliary data from the bigger population of interest is entered into the model to estimate the biomass for the entire population. This applies equally to upscaling from samples to the individual tree and from trees to the stand.

Each upscaling step is subject to a specific error. To assess the uncertainty of the biomass estimation the error of the biomass has to be estimated as well. Since upscaling can be complex as a result of the combined involvement of various regression models and sampling procedures that all contribute to a specific error budget, the calculation of the error-propagation forms an essential part of biomass estimation.

This chapter follows the structure of the upscaling concept, describing the sampling and modelling for upscaling from samples to trees in Sect. 3.2 and details the modelling process for upscaling from a tree to a stand in Sect. 3.3. This is followed by Sect. 3.4 on error-propagation where the error budgets of both upscaling steps are integrated. Finally, some implications of using biomass models in growth simulators are discussed in Sect. 3.5.

3.2 Upscaling from Biomass Samples to Tree Biomass

3.2.1 Selection of the General Method for Destructive Tree Sampling

A first step to create a biomass model typically involves destructive sampling of trees. There is a multitude of biomass sampling methods which would warrant a book on their own. In the following section a pragmatic approach is followed since biomass sampling does not form the major focus of this chapter. The different approaches in assessing the biomass of trees can be roughly classified in bulk sampling approaches where more than one tree is sampled and individual tree sampling (Fig. 3.2). The first method is usually based on in-field chipping and is a

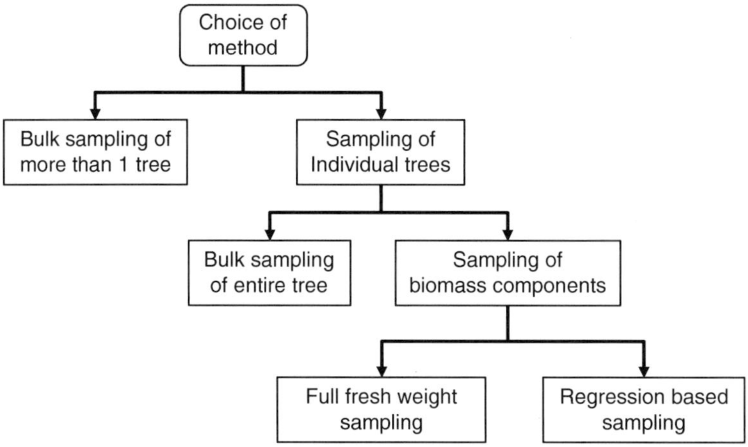

Fig. 3.2 Archetypes of methods for biomass sampling

more industrial than scientific practice. Bulk sampling is frequently applied in short rotation plantations when e.g. the biomass of coppiced trees is to be measured and entire rows or stands are harvested and chipped for fresh weight calculation (e.g. Hytönen et al. 1987). Another typical application of bulk sampling is harvesting studies of invasive woody vegetation with a high proportion of multi-stemmed trees and bushes, where only a biomass value per area is required (Kitenge 2011). However, in most cases dry weight is usually not determined from all chips but from a randomised sub-sample, in order to minimise transport and handling losses and to increase the efficiency of the method. The advantage of the technique is that bulk sampling of many trees is highly efficient and mimics real harvesting conditions rather closely. But the results obtained by the bulk method are sometimes difficult to compare with other studies because the chipping usually involves as loss which is specific to the machine equipment used (see Chap. 6). A more scientific sampling approach is based on individual trees. In this instance a bulk sampling of the entire tree or the aboveground part can be done with all the positive and negative aspects mentioned above. To avoid chipping losses a full tree fresh weight can be determined for example by a harvester mounted scale (Pettersson and Njordfjell 2007). However, in this instance representative sub-sampling for dry weight determination might prove to be a challenge.

In many biomass studies an estimation of the different biomass components such as bark, wood, leaves and branches is a main objective because different proportions of biomass components will influence ash contents and calorific values of the tree (see Chap. 8). The quantification of biomass components is also a prerequisite to assess the impact on nutrient balances as a result of biomass export from the stand (see Chap. 10).

The practical choice of the individual tree sampling method is mainly determined by the number of components that should be differentiated between, but also by the

Table 3.1 Suitability of biomass sampling methods according to the objective of the study

	Bulk method for more than one tree	Bulk method for individual trees	Full fresh weight sampling	Regression based sampling
Productivity studies	+	+	o	−
Short rotation coppice or multi-stemmed alien invasive vegetation	+	+	o	−
Carbon sequestration studies	−	−	+	o
True production, partitioning, nutrient export, biofuel quality	−	−	o	+

The icons mark a low (−), average (o) and high (+) suitability of a method for a purpose

tree size and the available time and work force. A full measurement of the fresh weight of the stem is always preferable since it eliminates an upscaling step (part of the upscaling step 1 in Fig. 3.1) and thus a possible source of error. The method has been successfully applied in several biomass studies for example in short rotation Eucalypt plantations (du Toit 2008; Dovey 2009). While harvester mounted hanging scales can support the measurement of stem fresh weight, the separation of twigs and foliage on site is only feasible for smaller trees under the normal time and money constraints. A regression based sampling approach, which is introduced in detail in the following section, is thus necessary for bigger trees to maximise efficiency. In the end the choice of the method always depends on the scope and specific objectives of the biomass study (Table 3.1). In addition to the listed method archetypes a multitude of mixed approaches can also be applied.

3.2.2 Example of a Regression Supported Sampling Method

In the following example of a feasible approach to biomass sampling for above-ground components difficulties are outlined and possible solutions shown. The sampling is based on the selection of a number of representative trees from a population according to the principles of statistical sampling as described in Chap. 2. After this step further decisions about the sampling method that have to be made will include:

- the relevant biomass components;
- the sampling design of the stem biomass;
- the sampling design of the crown biomass; and
- the model to scale up from samples to the individual trees.

A plethora of different ways can be identified in the published literature to define the biomass components of a tree. A straightforward approach to calculate the aboveground biomass of trees, that can be used successfully in a good proportion of applications, is to divide the tree hierarchically. It is first divided in stem and crown, the stem is then further subdivided into bark and wood biomass and the crown into foliage and branches. Depending on the objective it might be useful to further differentiate fruits/cones, heart and sapwood, merchantable branches, dead branches or to add twigs as finer branch material to the components. High differentiation levels are usually warranted if nutrient budgets for sustainability assessments of biomass harvesting have to be established (see Chap. 10 and Seifert et al. 2006; Block et al. 2008; Dovey 2009; Ackerman et al. 2013b). In this example of a mid-rotation *Pinus radiata* tree (age 14 years) only the bark and wood of the stem, branches (bark and wood aggregated) and needles were used as components.

3.2.3 Sampling and Upscaling of Stem Biomass

Stem biomass is reconstructed from samples based on the fact that oven-dry biomass is a function of stem volume and basic density (Eq. 3.1)

$$BM = V \cdot R \tag{3.1}$$

Here *BM* is the oven-dry biomass (kg), *V* is the stem volume (m^3), and *R* is the basic density defined as dry weight divided by green volume (kg/m^3).

Volume calculation, although a long time established part of forest mensuration, is not always trivial in biomass studies because unlike in stem volume determination the non-merchantable branch volume matters in biomass studies. The predicament is that there is no standardised definition for a twig as opposed to a branch. Nor is it always clear in broad-leaved species that tend to grow multi-stemmed or with forked stems what is to be labelled as stem and what as branch. As indicated in Fig. 3.3 a solution is, for example, to select the largest branch as a stem and define all other parts as branches. It must, however, be noted that this practice is not compatible with most taper and volume models that are based on volume calculated from the sum of all stems of a merchantable size and thus are based on a virtual, single stemmed tree. With redefining all other forks of a stem as big branches the merchantable stem volume and the derived stem biomass will be underestimated compared to estimations based on available volume functions and basic density. Thus a method compatible to stem volume determination, where the basal area of all merchantable stems and branches is added up to a defined cut-off diameter might be more useful.

A large variety of sampling methods were tested for biomass sampling such as systematic sampling in absolute or relative heights or more recently randomised branch sampling techniques (Jessen 1955; Valentine and Hilton 1977; Valentine

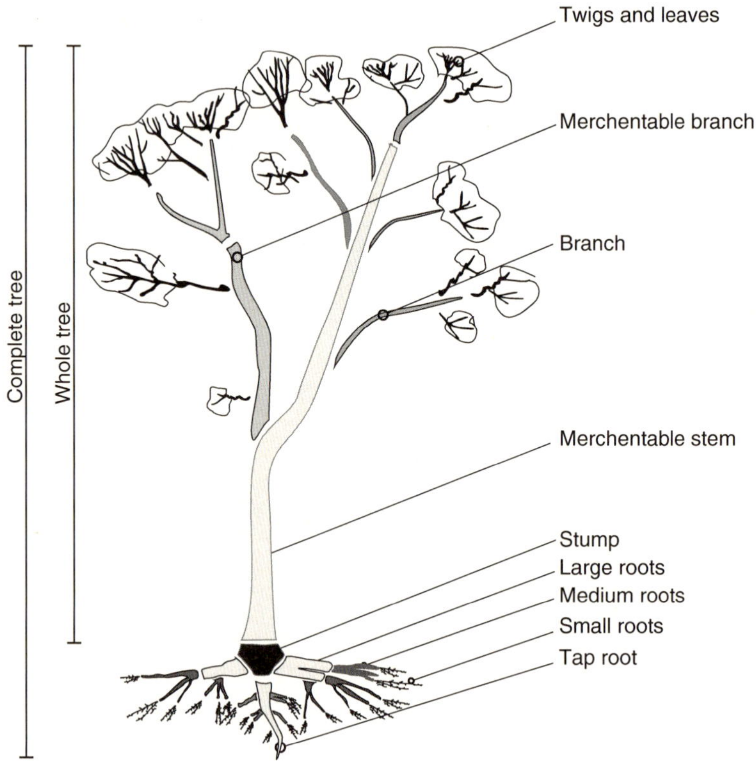

Fig. 3.3 Biomass components of a broad-leaved tree (modified from Young 1964)

et al. 1984; Gregoire et al. 1995; Gaffrey and Saborowski 1999; Saborowski and Gaffrey 1999; de Gier 2003), which have been proven to improve accuracy for a set sampling effort. Experience shows that the choice of the sampling method is not in the first place driven by aspects of the sampling theory, but rather by the level of sophistication and experience of the sampling team. In developing countries computers for field work are often unavailable. Field teams for biomass sampling are frequently recruited from an inexperienced body of persons rather than from scientific personnel or experienced field technicians. Therefore highly sophisticated sampling designs that rely on computer software in the field are often not feasible. Thus, often the principle "the simpler the better" pays off. A gain in efficiency as a result of the use of e.g. randomised sampling techniques can be easily negated by the incorrect application of the sampling method by inexperienced field crews, while straightforward systematic sampling methods (e.g. samples taken every 3 m) have a better chance to be implemented correctly. However, a decrease in sampling efficiency is to be taken into account.

The volume calculation of stem sections between measuring points is typically based on Smalians's formula (van Laar and Akça 2007) or alternatively on the geometric formula of a frustum of a cone (Eq. 3.2).

$$V = \frac{\pi h}{3}\left(R^2 + Rr + r^2\right) \tag{3.2}$$

Here V is the section volume (m^3), R is the bigger end radius of the stem section and r the smaller top end radius. In case no cut-off diameter was defined, the volume of the stem tip can be determined as a cone by setting the top end radius r to zero.

The initially introduced Eq. 3.1 implies that stem volume and basic density calculation are equally important for the successful determination of stem biomass. Trees are known to vary considerably in wood density within the stem in radial and longitudinal direction and also between trees and sites. Thus, pre-information on longitudinal density gradients in the species of interest should be used wherever possible to adjust the sampling design of discs beforehand accordingly. To base biomass upscaling merely on mean literature values of basic density is a very crude approach that might, due to density variations within and between trees lead to seriously biased estimates.

The upscaling from sample discs typically contains a measurement component where basic density is determined at disc level and a modelling component based on the estimation of fresh weight/dry weight ratios or a regression approach to obtain information for the entire stem. In general, biomass should be provided as dry mass. Fresh mass is site and species specific (Marden et al. 1975; Kokkola 1993) and also subject to a substantial intra-annual and inter-tree tree variation (Kokkola 1993). In addition, the harvesting technique (debarking) and weather conditions after felling can modify the fresh weight (Adams 1971; Kokkola 1993) rendering it a suboptimal variable for tree biomass characterisation. A comparative quantification of biomass is only possible when an equilibrium in moisture content is reached that can be determined in accordance with scientific lab standards. The rich body of publications on biomass prove that there is a considerable variation in the definition of dry mass between studies that varies from about 40 to 105 °C. In nutrient studies with nitrogen content in phytomass as a response variable the drying temperature is usually limited to a maximum of 60–65 °C to avoid nitrogen loss as a result of volatility of some chemical components. Cones are often dried below 40 °C, for germination to remain possible. Oven-dry weight of wood refers per convention in wood science to drying to a state of constant weight at about 103 ± 2 °C (see e.g. DIN EN13183-1). Using drying temperatures of 70 °C and below, for example to maintain the volatile nitrogen parts for further nutrient analysis, leads to higher biomass values of a magnitude of 2–3 % in wood (Forrest 1969; Barney et al. 1978). Seifert and Müller-Starck (2009) reported similar weight changes for Norway spruce cones. The cone weight was 84 % (dried at 38 °C), 80 % (at 60 °C) and 78 % (at 105 °C) in proportion to the fresh weight. Evidence shows that it is paramount that differences in drying temperature are taken into account when pooling data and also when comparing established functions. Furthermore the drying regime

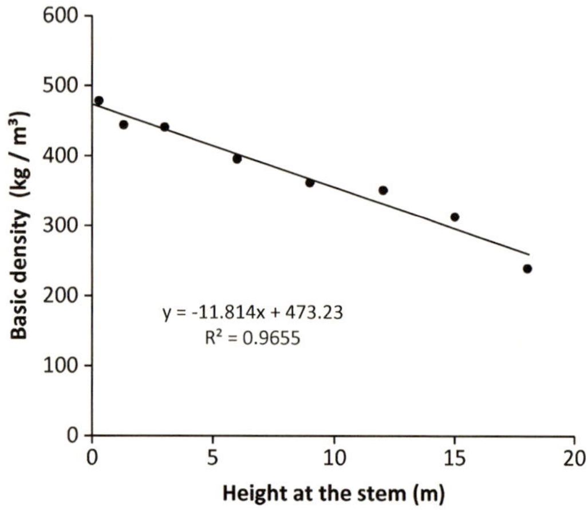

Fig. 3.4 Regression function of basic density over stem length based on eight sample discs from a young *Pinus radiata* tree. Higher degree polynomials or spline fits can be used as well if sufficient sample numbers are available

should be provided in publications for a reliable comparison of biomass studies. However, the establishment of further drying series at different temperatures seems to be warranted to obtain transfer functions for the different tree components, which would facilitate a conversion of biomass at different drying temperatures and would thus facilitate a comparison of results obtained in different biomass studies.

The most common principle in density determination of sample discs is the Archimedean principle of water displacement (American Society for Testing and Materials 1999). A basic physical principle is used that relates the buoyant force of a body directly to its volume. The method is simple and implies establishing of weight of a sample in air and in water with a scale. The density of water is 1 g/cm^3 and that of air is negligible, which results in Eq. 3.3 for the final density (Gerthsen 1997, p. 97).

$$\frac{W_{water}}{W_{air}} = \frac{\left(\rho_{sample} - \rho_{water}\right) \cdot V}{\left(\rho_{sample} - \rho_{air}\right) \cdot V} = 1 - \left(\frac{\rho_{water}}{\rho_{sample}}\right) \tag{3.3}$$

Here W is weight (g), ρ is density (g/cm^3) and V is volume (cm^3).

The weight in water is determined by full immersion of the wood sample in a water basin, which is placed on top of a scale. The scale reading in gram after immersion equals the volume of the sample in cm^3. It is important that the sample is fully saturated with water (to constant weight), so that no additional water is taken up into the sample during the displacement measurement.

Based on individual disc density a longitudinal and partially a radial density variation can be taken into account, if a regression is applied that models basic density as a function of height in the stem. This procedure also averages out measurement errors to a certain degree (Fig. 3.4).

Based on the assumption that tree stems are rotationally symmetric along their longitudinal axis, basic density of a stem can be obtained by mathematical integration of the density regression function multiplied with the cross-sectional area along the tree height from the lower end to the upper end of the stem section (Eq. 3.4).

$$\int_{hl}^{hu} A(x)\rho(x)\mathrm{d}x \tag{3.4}$$

Here *hu* and *hl* represent the height of the top and bottom end of the section, $A(x)$ is the cross-sectional area of the stem at height x between *hu* and *hl* and $\rho(x)$ is the basic density at the same height determined by a regression function.

A mathematically simpler, but still sufficiently accurate solution can be achieved by determining the height of the centroid (centre of gravity) of each section if enough sections are taken. The centroid of the frustum of a cone is calculated according to Weisstein (2013), based on Eshbach (1975), Harris and Stocker (1998) and Kern and Bland (1948) as indicated in Eq. 3.5.

$$z = \frac{l\left(R^2 + 2Rr + 3r^2\right)}{4\left(R^2 + Rr + r^2\right)} \tag{3.5}$$

where z is the height of the center of gravity, R is the radius of the bottom end of the stem section (m), r is the radius at the top end (m), and l is the length of the stem section.

This height of the centroid z is then recalculated into the absolute height at the stem of that tree and then used to obtain a basic density value from the regression of density over height, as a representative value for the section. The obtained basic density is finally multiplied with the section volume. Using the regression example illustrated in Fig. 3.4 and a log of 3 m length and $R = 0.40$ m, $r = 0.36$ m, one would come to a height of the centroid (z) within the section of 1.44 m. If we assume that this log was the second 3 m section of the stem we have to add the stump height (0.3 m) and the first log (3.0 m) to z, resulting in an absolute height of the centroid that is 4.74 m. For this stem section an average basic density of 417 kg m^{-3} would be determined according to the tree-specific regression obtained in Fig. 3.4. Using this calculation template section by section will result in a reliable approximation of stem biomass.

3.2.4 *Sampling and Upscaling of Crown Biomass*

As stated earlier sampling of foliage and branch biomass is usually done as a regression sampling process, because foliage and branches should be separated

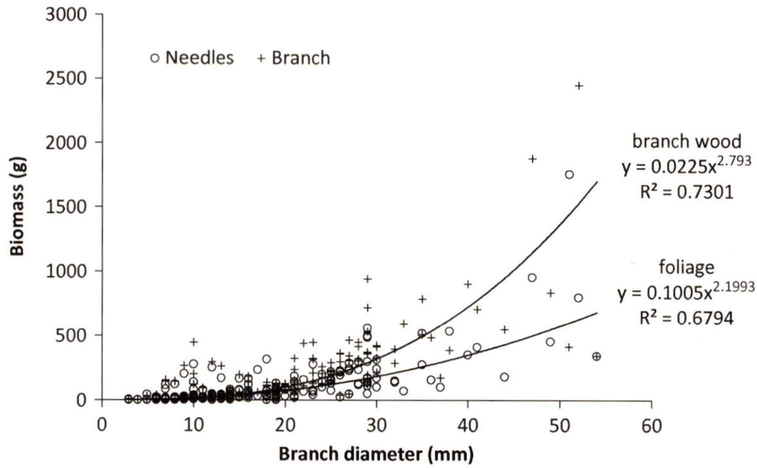

Fig. 3.5 Branch level regression to model branchwood and foliage biomass from branch diameter (N = 245 branches of *P. radiata* from 11 trees; Seifert unpublished)

for biomass, nutrient/ash content, and eco-physiological calculations. However, manually separating a tree's foliage from the twigs is a task that can only be done for small trees with reasonable effort. Therefore a regression sampling is employed to estimate branch and foliage biomass from branch diameter. Consequently, all branches of the tree are measured in diameter for upscaling later on. A sub-sample is then cut and sampled in detail in the lab or field lab. In this context it is relevant whether the branch diameters are measured perpendicularly to the main stem axis or with the axis, since the measurement perpendicular to the stem axis usually yields smaller values. The difference in branch diameter for 14-year-old *Pinus radiata* was about 2.0 % (estimated from on 245 branches of 11 trees; Seifert, unpublished). It is therefore important that the measurements of the sample branches and the total branches are taken consistently in the same direction. Depth of branches in the crown, defined as the distance of the branch insertion from the tip, provides an additional variable that can improve the estimation of the proportions of branchwood and foliage. Thus, the position along the stem could be a second variable to be determined in the field. After separating the foliage from the branches and drying it, a regression model is established to estimate foliage and branch biomass from branch diameter or branch circumference. A possible method to limit branch sampling is to establish the branch biomass functions based on a data set where all sample branch data of one plot are pooled. An example is presented in Fig. 3.5. This way a more robust function can be established if several trees per plot are sampled. The clustered data structure with the inherent inter- and intra-tree variability can be taken into account with mixed models (Pearce et al. 2010).

Based on this methods a dry biomass for the stem, the branches and the foliage can be obtained.

3.2.5 *Reporting Standards for the Upscaling to Individual Trees*

For the upscaling step from the samples to the individual tree the following variables should be reported as a minimum set of information: Drying temperature and regime used on the biomass, sampling and reconstruction method, definition of the biomass compositions and if regression models were used, the modelling technique, R^2 and standard deviation of the residuals, and preferably the model parameters including parameter statistics.

3.3 Upscaling from the Tree to the Stand Level

3.3.1 *Challenges*

The construction of statistical biomass models encompasses several challenges. Applicable models have to be accurate, which means precise and unbiased and should provide confidence limits to assess their applicability. Other typical issues are clustered data structures, heteroskedasticity of data and the desired additivity of biomass components. If sufficient data is available nonparametric *k*-NN approaches have also shown to provide good results in estimating tree biomass (Fehrmann et al. 2008) but a constant challenge remains the scarcity of biomass data. This frequently leads to models based on a small number of trees from a few stands only. Chave et al. (2004) showed the exponential decline of the error in the estimation of total aboveground biomass for tropical tree species. Their data provides evidence that a minimum of about 50 trees would be required to achieve an error of 10 % in the determination of aboveground biomass; 5 % were achieved with a minimum of about 150 sampled trees. A logical conclusion seems to be to pool all available data on a tree species, to increase sites and tree numbers and achieve a more generic model (e.g. Wirth et al. 2003). This mostly valid option is unfortunately often impaired by the manifold of different ways used to measure biomass (cut-off diameters, drying prescriptions, definition of components etc.), all of which complicates the compilation of coherent data sets.

3.3.2 *Heteroskedasticity*

An apparent challenge in the modelling of biomass is heteroskedasticity. This effect of growing variance with growing dimension of trees is characteristic of tree data. The problem is that heteroskedasticity (in-homogeneity of variances) violates a basic assumption of ordinary least squares regression (van Laar and Akça 2007).

To solve this problem, transformations are often applied to the data, which have the positive side effect that the well-established framework of linear regression can be used instead of nonlinear statistics. However, this transformation process comes with the flaw that the estimation is biased after back-transformation and has to be corrected. Different methods have been suggested for bias correction (Finney 1941; Baskerville 1972; Yandle and Wiant 1981; van Laar and Akça 2007). Alternative estimation methods to ordinary least squares regression have been proposed to deal with the problem of heteroskedasticity. Linear weighted least squares and iterative non-linear least squares regression (for nonlinear relationships) methods showed good results (Verwust 1991; Parresol 2001). With growing computing capacity weighted nonlinear least squares estimation might be the most elegant choice to avoid transformation bias and heteroskedasticity (Schabenberger and Pierce 2002).

3.3.3 Additivity

Additivity of biomass equations means that the different equations for biomass fractions (foliage, branches, stem, bark, etc.) should add up to the same biomass as if an equation for the total aboveground biomass is used for biomass estimation to have a consistent set of models (Kozak 1970; Cunia and Briggs 1984, 1985). This consistency between model estimates for biomass fractions and total biomass is not trivial since regressions are not perfect and even a slight bias for a model of a biomass component will affect the sum and define a deviation from an estimation of the total model. In addition, contemporaneous correlations between equations must be taken into account. If this is not done the efficiency of the model is reduced and biased parameter estimates could be the result as shown by Parresol and Thomas (1996).

For this reason different methods have been applied to ensure additivity (Parresol 1999). The most widely applied approach is a multivariate regression procedure based on the simultaneous estimation of all equations with joint-generalized least squares (Cunia and Briggs 1984, 1985), also called seemingly unrelated regression (SUR) in the more recent literature (Parresol 2001; Saint-André et al. 2005; Brandeis et al. 2006). Segmented regression and penalised spline approaches have also been tested to force additivity in biomass models (Goicoa et al. 2011).

3.3.4 Modelling Example

In this section an example of tree biomass modelling is presented, introducing several key techniques. All models were parameterised using the freely available statistical software package R (R Development Core Team 2012).

The starting point for this modelling section is a set of data obtained from 99 *Eucalyptus grandis* trees from the Karkloof experiment at different ages, ranging

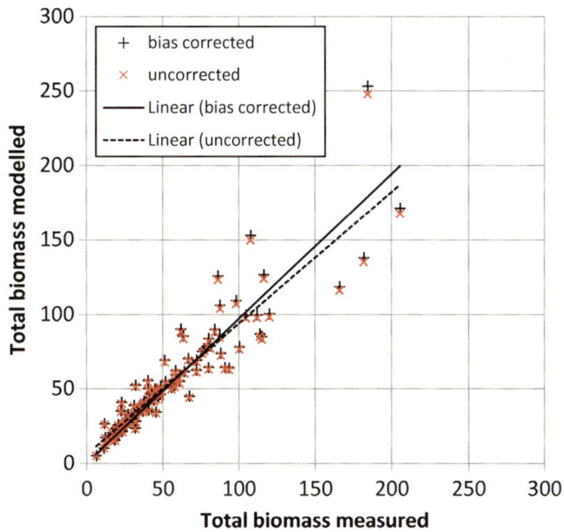

Fig. 3.6 The effect of back-transformation bias correction on the estimation of total biomass. After transformation a nearly unbiased model was achieved (*full line*), while the uncorrected model showed a clear bias (*dotted line*)

from 0 to 11 years. The data is described in du Toit (2008). The biomass has been scaled up to tree level and four biomass fractions that should be modelled (stemwood, stem bark, branches including bark, and foliage). The objective is to create models for the different biomass fractions as well as for the total biomass under the constraint of additivity.

A method for forcing additivity was introduced, based on a separated estimation of total aboveground biomass of the tree and the compositional estimation of proportions for the biomass fractions. The total biomass was estimated with a traditional ln-transformed linear model to account for heteroskedasticity (Eq. 3.6). As independent variables DBH, height and the compound variable D^2H were tested. The best model was selected based on the Akaike Information Criterion, which has theoretic foundations in information theory (Burnham and Anderson 2004). The AIC penalises the inclusion of additional variables and thus helps to keep the models parsimonious. The best model fit in this case was achieved by Eq. 3.6.

$$\ln\left(ABM_{total}\right) = a + b\ln\left(D^2H\right) + c\ln(H) \tag{3.6}$$

Here ABM_{total} is the total above ground biomass (kg) of the tree, D is the DBH (cm) and H is the tree height (m). For back-transformation the variance of the residuals σ^2 was calculated, multiplied by 0.5 and added to Eq. 3.6 for back-transformation bias correction (Eq. 3.7) as proposed by Baskerville (1972).

$$ABM_{total} = e^{\left(a+b\ln\left(D^2H\right)+c\ln(H)+\sigma^2 0.5\right)} \tag{3.7}$$

An almost unbiased model was achieved (Fig. 3.6).

The next step was to model the proportional distribution of the biomass fractions in a simultaneous approach based on the Aitchison-Simplex and an isometric log-ratio transformed (ILR) model (Aitchison 1982; van den Boogaart and Tolosana-Delgado 2008). The model foundations for this approach have been laid down by Aitchison (1982, 1986) and the method was successfully applied for the estimation of compositions in Geosciences (van den Boogaart and Tolosana-Delgado 2008). The data are modelled as a closed composition with a relative geometry. This means that the individual components are forced to add up to 1 and their relative proportions are of interest rather than their absolute values. Similar to the logarithmic transformation the ILR-transformation takes care of the effect of heteroskedasticity. Expressed in simple words the compositional relative proportions are transformed into a euclidian orthogonal coordinate system. The dimensionality (D) is hereby reduced by one, because the compositions have to add up to 1. This means, that the last component in the euclidian space is not directly predicted but automatically derived in the back-transformation step. Classical multivariate analysis is applied in the euclidian space and then the results are back-transformed into the original coordinates to provide meaningful composition parts.

The ILR-transformation is based on a CLR-tansformation, and a multiplication with a triangular Helmert matrix (van den Boogaart and Tolosana-Delgado 2008). The matrix multiplication does the dimensional reduction from D to D-1. The triangular (D, D-1)-Helmert matrix can be derived from a normalised Helmert contrast matrix as shown in van den Boogaart et al. (2013). Further details on the theoretical framework are provided by Aitchison et al. (2002) and van den Boogaart and Tolosana-Delgado (2008). One convenient side-effect of this method is that the solution in the transformed space can be found by applying well known linear statistics. The "compositions" package of R was used (van den Boogaart and Tolosana-Delgado 2007; R Development Core Team 2012). The code for modelling compositions in R is available from the web page of this book.[1]

After ILR-transformation of the data combinations of the following independent variables were tested: DBH, H, D^2H, ABM_{total} (see Eq. 3.8).

The best model in the transformed space was achieved by a linear combination of DBH, height and total aboveground biomass (Eq. 3.8)

$$ilr\,(composition_{ABM}) = bBHD + cH + dABM_{total} \tag{3.8}$$

The regression parameter output is provided in Table 3.2.

Table 3.2 Regression output for the fit of Eq. 3.8 for the Eucalypt data set in R

	DF	Pillai	Approx. F	num Df	den Df	Pr(>F)
(Intercept)	1	0.98300	1,792.48	3	93	<2.2e−16***
DBH	1	0.13955	5.03	3	93	0.002835**
H	1	0.65681	59.33	3	93	<2.2e−16***
Total	1	0.33864	15.87	3	93	2.059e−08***
Residuals 95						

(<0.01 significance level), *(<0.001 significance level)

[1] Code for additive estimation of biomass components for the freely available statistical language R (R Core Team 2012) and a set of example data can be downloaded from http://www.springer.com/life+sciences/forestry/book/978-94-007-7447-6

These are the parameters in the ILR-transformed space. Backtransformation is conveniently available in the statistical software package compositions. The model can be evaluated, based on the two different model parts separately and on the total model, where the compositions are multiplied with the total biomass estimation.

3.4 Model Evaluation and Model Error

Statistical models are usually evaluated with reference to their accuracy, which is defined as the deviation of an estimated quantity from a true value. Accordingly, accuracy is a compound function consisting of the precision (repeatability of estimates or variation around a true value) and of the bias (directed deviation from a true value). An accepted measure of accuracy is the mean squared error (MSE) as used in Eqs. 3.9, 3.10, 3.11 and 3.12 (Hellman and Fowler 1999).

$$MSE\left(\widehat{x}\right) = var\left(\widehat{x}\right) + bias(\widehat{x})^2 \tag{3.9}$$

Here $MSE\left(\widehat{x}\right)$ is the mean squared error of a model (accuracy), calculated according to Eq. 3.10.

$$MSE\left(\widehat{x}\right) = \mathrm{var}\left(\widehat{x}\right) + \left[\frac{\sum_{i=1}^{n}\widehat{x_i}}{n} - \mu\right]^2 \tag{3.10}$$

and,

$\widehat{x_i}$ = Model estimate of biomass
μ = True mean value
n = sample size
and Eqs. 3.11 and 3.12 apply

$$\mathrm{var}\left(\widehat{x}\right) = \frac{\sum_{i=1}^{n}\widehat{x_i^2} - \frac{\left(\sum_{i=1}^{n}\widehat{x_i}\right)^2}{n}}{n-1} \tag{3.11}$$

$$bias\left(\widehat{x}\right) = \frac{\sum_{i=1}^{n}\widehat{x_i}}{n} - \mu \tag{3.12}$$

In addition, to model evaluation the error propagation of the full upscaling process should be addressed to provide error budgets for the biomass estimation. The fact that biomass modelling typically involves a combination of different sampling and modelling steps complicates error budgeting. Detailed background information on error budgeting in biomass estimation can be found in Cunia (1987, 1990), Wharton and Cunia (1987), Yang and Cunia (1989), and van Laar and Akça (2007). Cunia (1990, p. 169) points to three general sources of error: "*There is first the sampling error: the same sampling procedure applied repeatedly to the same forest population leads generally to selection of different sample units and, thus, to different estimates. And then there is the measurement error when the same sample units (trees or plots) measured by different people lead to different recorded values and, thus, to different estimates. Finally, the third error component is that of the statistical model used in deriving estimates; same inventory data analyzed and interpreted by different statisticians may lead to different estimates*". Cunia (1990) adds the application error as a fourth source of error. This is based on the fact that biomass models are usually parameterised from data of a different population than the population where they are applied for estimation. However, the sampling and measurement error are usually assessed in combination (Cunia 1990). An assumption, which is often made in error budgeting of biomass models is that the models are unbiased, which is strictly speaking not always true but it reduces the focus to the variance as the only source of error (Cunia 1990). The following section will be based on this assumption as well. It is a frequently made mistake to exclude the upscaling from the sample to the tree from the error budget, which can only be done if the trees were fully harvested and their dry weight calculated without harvesting losses and sampling or ratio-modelling steps involved, which is rarely the case in most bioenergy studies. In all other cases error budgets for both upscaling steps ('sample to tree' and 'tree to stand') have to be determined.

Error propagation equations can be found for example in Ku (1966) or Bevington and Robinson (1992). Also van Laar and Akça (2007, p. 264) stipulate functions for variance estimation for different forms of linear equations and point out the most frequently used ones in biomass upscaling to be the additive and the multiplicative combination (Eqs. 3.13 and 3.14).

Equation	Variance	
$\overline{z} = \overline{x} + \overline{y}$	$s_{\overline{z}}^2 = s_{\overline{x}}^2 + s_{\overline{y}}^2 + 2s_{\overline{xy}}$	(3.13)
$\overline{z} = \overline{x} \cdot \overline{y}$	$s_{\overline{z}}^2 = \overline{y}^2 s_{\overline{x}}^2 + \overline{x}^2 s_{\overline{y}}^2 + 2 \cdot \overline{x} \cdot \overline{y} \cdot s_{\overline{xy}}$	(3.14)

Here $s_{\overline{xy}}$ is the covariance of $\overline{x}$ and $\overline{y}$, and $s_{\overline{x}}^2$ and $s_{\overline{y}}^2$ are the variances.

Equation 3.13 is used in biomass estimation if, for example, crown and stem biomass are added, while Eq. 3.14 is applied in allometric biomass models or for multiplying plot biomass estimates with plot areas (van Laar and Akça 2007).

Equation 3.14 can also be rewritten as relative error (Chave et al. 2004), as indicated in Eq. 3.15. The last covariance term may be omitted if $\overline{x}$ and $\overline{y}$ are independent (van Laar and Akça 2007, p. 264).

$$\frac{s_{\overline{z}}^2}{\overline{z}^2} = \frac{s_{\overline{x}}^2}{\overline{x}^2}\left(\frac{\delta \ln(f)}{\delta \ln(\overline{x})}\right)^2 + \frac{s_{\overline{y}}^2}{\overline{y}^2}\left(\frac{\delta \ln(f)}{\delta \ln(\overline{y})}\right)^2 + 2\frac{s_{\overline{xy}}\delta \ln(f)\delta \ln(f)}{\overline{x}\overline{y}\delta \ln(\overline{y})\,\delta \ln(\overline{x})} \tag{3.15}$$

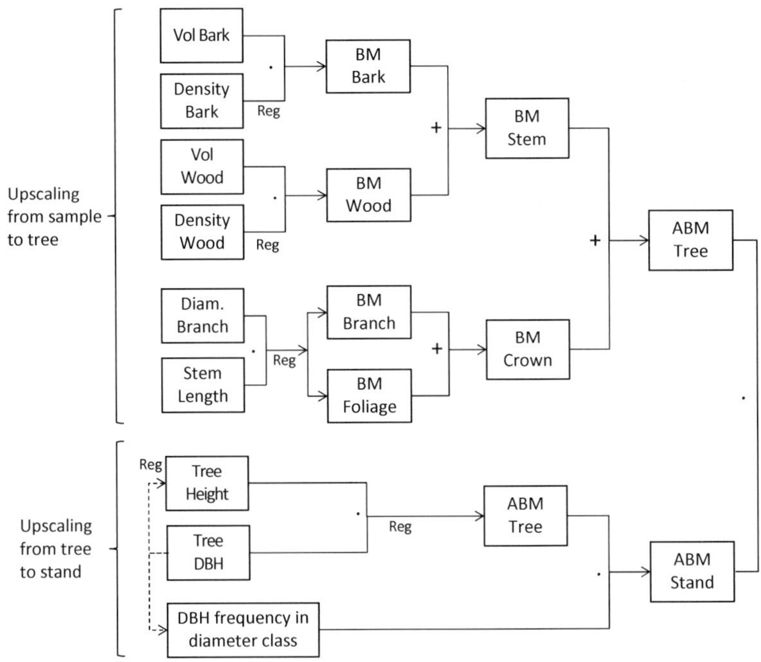

Fig. 3.7 Error propagation for an upscaling example. Indicated are multiplicative (·) and additive (+) error combinations and if a regression is the base for error calculation (Reg)

The term $\delta \ln(f)/\delta \ln(\overline{x})$ is the partial derivate of $\ln(f)$ with respect to $\ln(\overline{x})$ and is added to increase the accuracy with aid of a Taylor series (Chave et al. 2004). The error ${s_{AGB}}^2$ for a typical allometric model that predicts aboveground tree biomass from diameter at breast height and tree height $f(D,H) = aD^{\alpha}H^{\beta}$ is calculated as indicated in Eq. 3.16; here noted without the partial derivatives to simplify of notation (*vide* Chave et al. 2004).

$$s_{AGB}^2 = \langle AGB \rangle \cdot \sqrt{\alpha^2 \frac{s_D^2}{D^2} + \beta^2 \frac{s_H^2}{H^2} + 2\alpha\beta \frac{s_{DH}^2}{DH}} \tag{3.16}$$

Finally, the errors of the different upscaling steps have to be combined. This results in a series of different terms that are combined according to variances determined by Formulae 3.12 and 3.13. It is essential that all the different upscaling steps involved are taken into account. The combination of errors for this example involves the upscaling from sample to the tree and the upscaling from tree to stand. The latter is basically also the standard procedure for error quantification in volume based forest inventories and is expressed by Eq. 3.16. The major error sources including their combination rules are illustrated in Fig. 3.7.

The final combination of the upscaling steps is attributed to the fact that biomass estimation should be viewed as a two-stage sampling process (Cunia 1990). Further upscaling steps might be added, such as from stands to strata according to the same combination rules under the inclusion of forest area information as can be received by remote sensing (Chap. 2). It should once again be emphasised here that it is critical to include the upscaling from the samples to the tree in this context. Ignoring this upscaling step, as frequently found in literature, is only warranted if the complete drymass of the tree was measured as indicated before. A good example to underpin this statement is the regression of the foliage dry mass from branch diameter as an essential part of the upscaling. The degree of determination of only $R^2 = 0.68$ (Fig. 3.5) shows a considerable error potential and is only one of several error sources in the first upscaling step. Thus ignoring the first upscaling step would lead to a crude underestimation of the error.

3.5 Simulation of Biomass in Growth Models

Simulation of biomass in individual tree growth models adds a level of complexity to the modelling steps described above. A seamless integration of biomass models in the simulation environment is required and one of the major objectives is to project biomass consistently with other outputs like volume. This means, for example, that simulated stem biomass should be consistent with the product of simulated stem volume and wood density. The application of allometric equations to estimate biomass from diameter and height usually leads to inconsistencies with taper and volume models as far as stem biomass is concerned. The reason is that biomass models are naturally parameterised on much smaller data sets because of the huge effort it takes to gather biomass data. In addition, the data sets are frequently not even a sub-sample of the trees sampled for to establish the taper and volume functions.

Three different avenues could be followed to achieve a consistent estimation of biomass: (1) The use of expansion factors, (2) the application of stem biomass models, or (3) the combination of taper and density models.

A straightforward method to integrate biomass in growth simulators is based on expansion factors. Biomass expansion factors (BEF) serve as multipliers to convert timber volume to biomass and are widely applied in tropical and subtropical regions (e.g. Brown et al. 1989; Chhabra et al. 2002; Dovey 2009). Since the stem is the main contributor to tree biomass this approach usually produces plausible results. However, expansion factors linked to the stem volume only are constrained to the use within the same silvicultural treatment of the parameterisation data. They are not particularly suited to adapt to changes in the relationship between stem volume and aboveground tree biomass. Examples of such a situation are trees in the understorey, that have overproportionally more stem biomass as a result of suppressed crowns or also trees growing with single-sided competition that might branch earlier and have substantially more branch biomass (Ackerman et al. 2013a). This is the reason why the use of constant BEFs without further adaptations has been criticised (Soares and Tomé 2004).

A second, widely used modelling approach is based on stem biomass-models, which are constructed according the established technique of taper functions. An appropriate modelling method was introduced by Parresol and Thomas (1991) and further refined by Jordan et al. (2006). A definite advantage of such an approach is that it is easy to keep it consistent with existing taper functions and volume models. Another advantage is that stem biomass can be predicted for any cut-off diameter. But one disadvantage is that the wood density is fixed in the model and cannot be adapted easily to new growing conditions of the tree which might limit the generality of the model. It must also be noted that the crown biomass is not covered by the model and has to be modelled additionally, e.g. with allometric models.

A third alternative is the modelling approach based on linked taper and density models (Seifert et al. 2006). Stem biomass is simulated spatially based on taper-equations for volume determination and a separate wood density model, which predicts basic wood density as a function of tree ring growth. While the disadvantage is that the taper and density models might originate from different data sets this model can adapt to different silvicultural treatments and extreme growth conditions can be simulated reliably as well. However, this approach does also not cover crown biomass. A possible solution, also applicable to stem biomass models as well, is to develop a model for the estimation of biomass proportions, e.g. the ILR-transformed model described earlier, and simulate foliage and branch biomass according to the obtained proportions simply by multiplying the obtained stem biomass with the proportion. All other biomass fractions can then be determined because an absolute value for the stem biomass is fixed and the proportions are known. This way it is possible to establish a fully consistent modelling system for the simulation of volume, biomass and biomass composition with all the advantages of flexibility with respect to silvicultural treatment (including wood density changes) and flexible cut-off diameters in the tree harvesting process.

3.6 Conclusions

An important observation in biomass modelling is the manifold of different approaches. Despite numerous efforts no real or *de facto* standards for sampling and modelling have been established so far, which complicates a comparison of studies and might confuse people wanting to inform themselves about biomass modelling. This point is particularly valid for error budgeting. Different methods and descriptors of error are prevalent. It is no wonder that Cunia (1990) warned from accepting existing methods of error budgeting without prior reflexion. However, a diversity of methods also gives evidence for a dynamic field of science that is still in its evolution. Many aspects, such as the error budgeting and the seamless integration of models in growth and yield simulators still have scope for future improvement.

This chapter is intended to act as a structured guideline for interested researchers and practitioners through the quagmire of biomass modelling and will hopefully be a valuable support despite its many places where explanatory text could not be extended by examples but had rather to be based on references.

References

Ackerman S, Ackerman PA, Seifert T (2013a) Effects of irregular stand structure on tree growth, crown extension and branchiness of plantation grown *Pinus patula*. South For (4)

Ackerman P, Ham C, Dovey S, du Toit B, de Wet J, Kunneke A, Seifert T, Meincken M, von Doderer C (2013b) The use of forest residues for bioenergy in southern Africa, ICFR bulletin 03/2013. Forest Engineering Southern Africa (FESA) and The Institute for Commercial Forestry Research (ICFR), Pietermaritzburg, p 173

Adams EL (1971) Effect of moisture on red oak sawlog weight. Research Note NE-133. USDA Forest Service, Northeastern Forest Experimental Station, Upper Darby

Aitchison J (1982) The statistical analysis of compositional data (with discussion). J R Stat Soc Ser B 44(2):139–177

Aitchison J (1986) The statistical analysis of compositional data, monographs on statistics and applied probability. Chapman and Hall, London

Aitchison J, Barcelo-Vidal C, Egozcue JJ, Pawlowsky-Glahn V (2002) A concise guide for the algebraic–geometric structure of the simplex, the sample space for compositional data analysis. In: Bayer U, Burger H, Skala W (eds) Proceedings of IAMG'02—the 8th annual conference of the International Association for Mathematical Geology, Berlin, pp 387–392

American Society for Testing and Materials (1999) D 2395-93 (Reapproved 1997): standard test methods for specific gravity of wood and wood-based materials, 127. In: 1999 Annual book of ASTM standards, section 4, vol 04.10: wood. ASTM, West Conshohocken

Barney RJ, Vancleve K, Schlentner R (1978) Biomass distribution and crown characteristics in two Alaskan *Picea mariana* ecosystems. Can J For Res 8(1):36–41

Baskerville GL (1972) Use of logarithmic regression in the estimation of plant biomass. Can J For Res 2(1):49–53

Bevington PR, Robinson DK (1992) Data reduction and error analysis for the physical sciences. WCB/McGraw-Hill, Boston, p 328

Block J, Schuck J, Seifert T (2008) Einfluss unterschiedlicher Nutzungsintensitäten auf den Nährstoffhaushalt von Waldökosystemen auf Buntsandstein im Pfälzerwald. Forst und Holz 63(7/8):66–70

Brandeis TJ, Delaney M, Parresol BR, Royer L (2006) Development of equations for predicting Puerto Rican subtropical dry forest biomass and volume. For Ecol Manage 233:133–142

Brown S, Gillespie AJR, Lugo AE (1989) Biomass estimation methods for tropical forests with applications to forest inventory data. For Sci 35:881–902

Burnham KP, Anderson DR (2004) Multimodel inference: understanding AIC and BIC in model selection. Sociol Methods Res 33:261–304

Chave J, Condit R, Aguilar S, Hernandez A, Lao S, Perez R (2004) Error propagation and scaling for tropical forest biomass estimates. Philos Trans R Soc Lond B 359:409–420

Chhabra A, Palria S, Dadhwal VK (2002) Growing stock-based forest biomass estimate for India. Biomass Bioenergy 22(3):187–194

Cunia T (1987) On the error of biomass regressions and the corresponding inventory estimates. In: Proceedings of the 9th annual southern forest biomass workshop, Mississippi State University, Biloxi

Cunia T (1990) Forest inventory: on the structure of error of estimates. In: LaBau VJ, Cunia T (eds) State-of-the-art methodology of forest inventory: a symposium proceedings, General Technical Report PNW-GTR-263. U.S. Department of Agriculture, Forest Service, Pacific Northwest Research Station, Portland, 635pp
Cunia T, Briggs RD (1984) Forcing additively of biomass tables: some empirical results. Can J Forest Res 14(3):376–384
Cunia T, Briggs RD (1985) Forcing additively of biomass tables: use of the generalized least squares method. Can J Forest Res 15(1):23–28
De Gier A (2003) A new approach to woody biomass assessment in woodlands and shrublands. In: Roy PS (ed) Geoinformatics for tropical ecosystems. Bishen Singh Mahendra Pal Singh, Dehra Dun, pp 161–198
Dovey SB (2009) Estimating biomass and macronutrient content of some commercially important plantation species in South Africa. South For 71:245–251
du Toit B (2008) Effects of site management on growth, biomass partitioning and light use efficiency in a young stand of *Eucalyptus grandis* in South Africa. For Ecol Manage 255(7):2324–2336
Eshbach OW (1975) Handbook of engineering fundamentals. Wiley, New York
Fehrmann L, Lehtonen A, Kleinn C, Tomppo E (2008) Comparison of simple linear and mixed effect regression models and a *k*-nearest neighbour (*k*-NN) approach for estimation of single-tree biomass. Can J For Res 38(1):1–9
Finney DJ (1941) On the distribution of a variable whose logarithms is normally distributed. J R Stat Soc Ser B 7:155–166 (cited from van Laar and Akça 2007)
Forrest WG (1969) Variations in the accumulation, distribution and movement of mineral nutrients in radiata pine plantations. Ph.D. thesis, Australian National University, Canberra, 276pp
Gaffrey D, Saborowski J (1999) RBS, ein mehrstufiges Inventurverfahren zur Schätzung von Baummerkmalen. I. Schätzung von Nadel- und Asttrockenmassen bei 66-jährigen Douglasien. AFJZ 170(10/11):177–183
Gerthsen C (1997) Physik. 19th edn revised by Vogel G. Springer, Berlin, pp 1262
Goicoa T, Militino AF, Ugarte MD (2011) Modelling aboveground tree biomass while achieving the additivity property. Environ Ecol Stat 18:367–384
Gregoire TG, Valentine HT, Furnival GM (1995) Sampling methods to estimate foliage and other characteristics of individual trees. Ecology 76(4):1181–1194
Harris JW, Stocker H (1998) Frustum of a right circular cone. §4.7.2. In Handbook of mathematics and computational science. Springer, New York, p 105
Hellman JJ, Fowler GW (1999) Bias, precision, and accuracy of four measures of species richness. Ecol Appl 9(3):824–834
Hytönen J, Lumme I, Törmälä T (1987) Comparison of methods for estimating willow biomass. Biomass 14:39–49
Jessen RJ (1955) Determining the fruit count on a tree by randomized branch sampling. Biometrics 11:99–109
Jordan L, Souter R, Parresol B, Daniels RF (2006) Application of the algebraic difference approach for developing self-referencing specific gravity and biomass equations. For Sci 52(1):81–92
Kern WF, Bland JR (1948) Frustum of right circular cone. §29. In Solid mensuration with proofs, 2nd edn. Wiley, New York, pp 71–75
Kitenge EM (2011) Harvesting of invasive woody vegetation (*Eucalyptus lehmanii*, *Leptospermum laevigatum*, *Acacia cyclops*) as energy feedstock in the Cape Agulhas Plain of South Africa. M.Sc. thesis, Department of Forest and Wood Science, Stellenbosch University, Stellenbosch. Online: http://hdl.handle.net/10019.1/17873
Kokkola J (1993) Drying of pulpwood in Northern Finland. Silva Fennica 27(4):283–293
Kozak A (1970) Methods for ensuring additivity of biomass components by regression analysis. For Chron 46(5):402–404
Ku H (1966) Notes on the use of propagation of error formulas. J Res Natl Bur Stand C Eng Instrum 70C(4):263–273

Marden RM, Lothner DC, Kallio E (1975) Wood and bark percentage and moisture contents of Minnesota pulpwood species. Research Paper NC-114. USDA Forest Service, North Central Forest Experimental Station, St. Paul

Parresol BR (1999) Assessing tree and stand biomass: a review with examples and critical comparisons. For Sci 45(4):573–593

Parresol BR (2001) Additivity of nonlinear biomass equations. Can J For Res 31(5):865

Parresol BR, Thomas CE (1991) Econometric modeling of sweetgum stem biomass using the IML and SYSLIN procedures. In: Proceedings of the sixteenth annual SAS users group international conference, SAS Institute Inc., Cary, pp 694–699

Parresol BR, Thomas CE (1996) A simultaneous density-integral system for estimating stem profile and biomass: slash pine and willow oak. Can J For Res 26:773–781

Pearce HG, Anderson WR, Fogarty LG, Todoroki CL, Anderson SAJ (2010) Linear mixed-effects models for estimating biomass and fuel loads in shrublands. Can J For Res 40(10):2015–2026

Pettersson M, Njordfjell T (2007) Fuel quality changes during seasonal storage of compacted logging residues and young trees. Biomass Bioenergy 31:782–792

R Development Core Team (2012) R: a language and environment for statistical computing. R Foundation for Statistical Computing, Vienna. http://www.R-project.org. Accessed 1 Mar 2012

Saborowski J, Gaffrey D (1999) RBS, ein mehrstufiges Inventurverfahren zur Schätzung von Baummerkmalen. II. Modifizierte RBS-Verfahren. AFJZ 170(12):223–227

Saint-André L, M'Boub AT, Mabiala A, Mouvondy W, Jourdan C, Roupsard O, Deleporte P, Hamela O, Yann Nouvellon Y (2005) Age-related equations for above- and below-ground biomass of a Eucalyptus hybrid in Congo. For Ecol Manage 205:199–214

Schabenberger O, Pierce FJ (2002) Contemporary statistical models for the plant and soil sciences. CRC Press, Boca Raton

Seifert T, Müller-Starck G (2009) Impacts of fructification on biomass production and correlated genetic effects in Norway spruce (*Picea abies* L. [Karst.]). Eur J For Res 128(2):155–169

Seifert T, Schuck J, Block J, Pretzsch H (2006) Simulation von Biomasse- und Nährstoffgehalt von Waldbäumen mit dem Waldwachstumssimulator SILVA. In: Proceedings of the annual meeting of the section yield science in the Deutscher Verband Forstlicher Forschungsanstalten, Staufen, pp 208–223

Soares P, Tomé M (2004) Analysis of the effectiveness of biomass expansion factors to estimate stand biomass. In: Proceedings of the international conference on modeling forest production, Vienna, 19–22 Apr 2004

Valentine HT, Hilton SJ (1977) Sampling oak foliage by the randomized-branch method. Can J For Res 7:295–298

Valentine HT, Tritton LM, Furnival GM (1984) Subsampling trees for biomass, volume, or mineral content. For Sci 30:673–681

van den Boogaart KG, Tolosana-Delgado R (2008) "Compositions": a unified R package to analyze Compositional Data. Comput Geosci 34(4):320–338

van den Boogaart KG, Tolosana-Delgado R, Bren M (2013) R package "Compositional Data Analysis" version 1.30-1. 2013-04-12 16:23:59. Online: http://cran.r-project.org/

Van Laar A, Akça A (2007) Forest mensuration, Managing forest ecosystems series 13. Springer, Dordrecht

Verwust T (1991) Logarithmic transformations in biomass estimation procedures: violation of the linearity assumption in regression analysis. Biomass Bioenergy 1(3):175–180

Weisstein EW (2013) Conical frustum. From MathWorld – a wolfram web resource. http://mathworld.wolfram.com/ConicalFrustum.html. Accessed 07 Apr 2013

Wharton HE, Cunia T (1987) Estimating tree biomass regressions and their error. Publication Nr. NEGTR-117. USDA Forest Service, Northeastern Forest Experiment Station, Broomall

Wirth C, Schumacher J, Schulze ED (2003) Generic biomass functions for Norway spruce in Central Europe—a meta-analysis approach toward prediction and uncertainty estimation. Tree Physiol 24:121–139

Yandle DO, Wiant VJ (1981) Estimation of plant biomass based on the allometric equation. Can J For Res 11:833–834

Yang X, Cunia T (1989) Construction and application of biomass regressions to size classes of trees. In: Proceedings of the "State of the Art Methodology of Forest Inventory", Syracuse, July 30–August 5, 1989

Young HE (1964) The complete tree concept – a challenge and an opportunity. Reprint of Proc Soc Ann Forum Denver, 3pp (cited from Hallika 1989)

Chapter 4
Managing Southern African Woodlands for Biomass Production: The Potential Challenges and Opportunities

Paxie W. Chirwa, Stephen Syampungani, and Coert J. Geldenhuys

4.1 Introduction

The Southern African vegetation is generally referred to as the Zambezian Phytoregion. The region covers over ten countries in Central and Southern Africa lying between latitudes 3° and 26° south with a total area of 377 million ha (White 1983). The region falls within the tropical summer-rainfall zone with a single rainy season (November–April) and two dry seasons, a cool season from May to August and a hot season from September to November (Geldenhuys and Golding 2008). Annual rainfall ranges from 500 to 1,500 mm which decreases from north to south (Chidumayo 1997). The dominant soils of the region are rhodic and Haplic Nitosols and Chromic Xerosols with Calci-Chromic Cambisols and Pellic Vertisols in some places (Chidumayo 1997). The flora includes more than 8,500 species of which 54 % are endemic (Geldenhuys and Golding 2008). Based on variation in rainfall and soils, various distinct vegetation types are observed in Southern Africa.

P.W. Chirwa (✉)
Forest Science Postgraduate Programme, University of Pretoria, 5-15 Plant Sciences Complex, Pretoria 0028, South Africa
e-mail: paxie.chirwa@up.ac.za

S. Syampungani
School of Natural Resources, Copperbelt University, P.O. Box 21692, Kitwe, Zambia
e-mail: syampungani@cbu.ac.zm

C.J. Geldenhuys
Department of Forest and Wood Science, University of Stellenbosch, P/Bag X1, 7602 Matieland, South Africa
e-mail: cgelden@mweb.co.za

T. Seifert (ed.), *Bioenergy from Wood: Sustainable Production in the Tropics*, Managing Forest Ecosystems 26, DOI 10.1007/978-94-007-7448-3_4,

4.2 Vegetation Types of Southern Africa

The major vegetation types of Southern Africa include Miombo woodland, undifferentiated woodland, Mopane woodland and Semi-arid Shrubland.

4.2.1 Miombo Woodlands

Miombo woodland is a significant biome covering about 10 % of the African land masses (ca 2.5–4 million km^2 depending on definition, White 1983; Millington et al. 1994). It is the prevalent vegetation throughout the Zambezian region where soil is freely drained but rooting environment is restricted (see Campbell 1996; Fig. 4.1). Floristically and physiognomically it is very different from other woodland types, and is nearly always dominated by *Brachystegia* species, either alone or with *Julbernadia* species or *Isoberlinia* species (Geldenhuys and Golding 2008). Most miombo woodlands are semi-deciduous, but some are completely deciduous while others are almost evergreen (White 1983; Geldenhuys and Golding 2008). Most miombo dominant species are light demanding and showing some degree of fire resistance, but cannot survive repeated fires (Syampungani 2008). A distinction is made between Wetter Miombo (rainfall > 1,000 mm, canopy height > 15 m, floristically rich) and Drier Miombo (rainfall < 1,000 mm, canopy height < 15 m, floristically poor) (Geldenhuys and Golding 2008). Wet miombo woodland occurs over much of eastern Angola, northern Zambia, south-western Tanzania and central Malawi in areas receiving more than 1,000 mm rainfall per year (Frost 1996). The canopy height is usually 15 m, reflecting generally deeper and moister soils which create favorable conditions for growth (White 1983; Frost 1996). *Brachystegia floribunda*, *B. glaberrima*, *B. taxifolia*, *B. wangermeeana* and *Marquesia macroura* are widespread in this vegetation type (White 1983). It is also associated with widespread vegetation namely; dry evergreen forest and thicket, swamp forest, evergreen riparian forest and wet dambos (Fanshawe 1971).

Most miombo species are semi-heliophilous, show some degree of fire-resistance, but however, the dominants cannot survive repeated fierce fires (Lawton 1978). The principal canopy associates are *Afzelia quanzensis*, *Anisophyllea pomifera*, *Erythrophleum africanum*, *Faurea saligna*, *Marquesia macroura*, *Parinari curatellifolia*, *Pericopsis angolensis* and *Pterocarpus angolensis* (Chidumayo 1997). Several species of *Uapaca* and *Monotes* occur scattered in miombo as shrubs less than 10 m tall (White 1983). They are frequently dominant on shallow soils and in secondary miombo, and are abundant in the scrub woodland that represents the ecotone between miombo and the edaphic grassland of waterlogged depressions (White 1983).

The dry miombo woodland occurs in southern Malawi, Mozambique and Zimbabwe (White 1983; Frost 1996). *Brachystegia spiciformis*, *Brachystegia boehmii*, and *Julbernardia globiflora* are often the only dominants present (White 1983). The dry miombo has three subtypes namely: (i) *Brachystegia boehmii-*

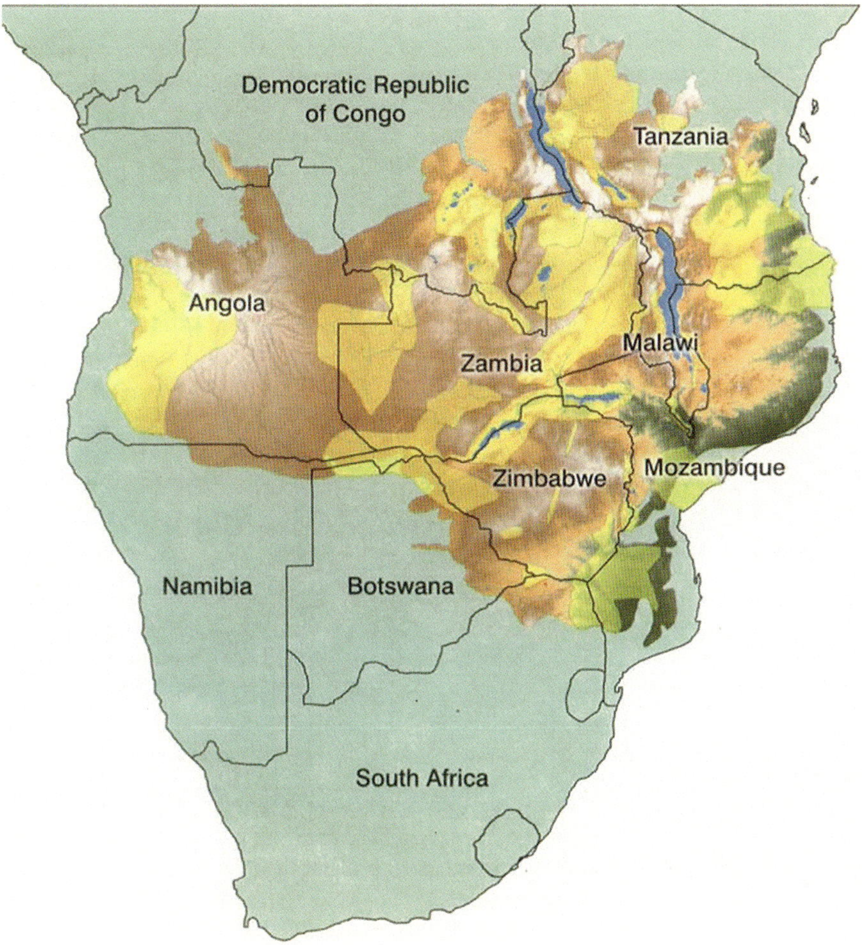

Fig. 4.1 Miomboecoregion (Source: Timberlake and Chidumayo (2011))

Brachystegia spiciformis-Brachystegia utilis (ii) *Brachystegia manga-Julbernadia* woodland and (iii) *Brachystegia spiciformis-Julbernadia* woodland (Chidumayo 1997). *B. boehmii-B. spiciformis-B. utilis* is associated with *Diplorhynchus condylocarpon*, *Lannea* spp. *Ochna* spp., and *Pseudolachnostylis maprouneifolia* as understorey species while the *B. manga-Julbernardia* woodland has *Diospyros* spp., *D. condylocarpon* and *Ochna spp.*, as the common understorey species.

4.2.2 Undifferentiated Woodlands

The most extensive undifferentiated woodlands are the teak and acacia woodlands. The so called Zambezi teak woodland is dominated by the *Baikiaea plurijuga* that occurs in the Kalahari sands especially associated with head waters of the Upper Zambezi and Okavango rivers (Timberlake et al. 2010). Other species associated with Baikiaea include *Pterocarpus angolensis*, *Guibourtia coleosperma* and *Schinziophyton rautanenii* (mungongo). Within the Zambezi teak forest, are five woodland sub-types distinct in species composition; (i) *Guibourtia* woodland, (ii) *Burkea-Erythrophleum* woodland (iii) *Burkea-Diplorhynchus* scrub (iv) *Diplorhynchus* scrub and (v) *Parinari* suffrutex savanna. The dominance of species varies from one sub-type to another. However, *Amblygonocapus andongesis*, *Baikiaea plurijuga*, *Brachystegia floribunda*, *B. longifolia*, *B.spiciformis*, *Burkea africana*, *Combretum spp.*, *Cryptosepalum exfoliatum ssp. pseudotaxus*, *Dialium engleranum*, *Erythrophleum africanum*, *Guibourtia coleosperma*, *Isoberlinia angolensis* and *Parinari curatellifolia* seem to be associated with dominance from one sub-type to another. In the drier southeastern parts of the warm dry forest region, open mixed acacia woodlands, dominated by Acacia species (*A. nigrescens*, *A. nilotica*, *and A. gerrardii*) and Combretum are found. Other species include *Burkea africana*, *Terminalia sericea*, *Kirkia acuminata*, *Pseudolachnostylis maprouneifolia*, *Sclerocarya birrea and Zizyphus mucronata.* It is floristically richer than either miombo or mopane woodland and are more easily defined by the absence of miombo or mopane dominants (Geldenhuys and Golding 2008).

4.2.3 Mopane Woodland and Semi-arid Shrubland

Mopane woodlands, with *Colophospermum mopane* as the dominant tree species, are mostly confined to the lower lying areas on heavier textured soils in the wide, flat valley bottoms of lower Okavango, Cunene, Zambezi, Shire, Limpopo and Luanga in southern Angola, northern Namibia, northern Botswana, Zimbabwe, northern South Africa, southern Zambia, southern Malawi and south central Mozambique (Timberlake et al. 2010). It occurs at an elevation of 200–1,200 m although normally it ranges between 300 and 900 m. The rainfall in these areas ranges from 400 to 700 mm per year, although the species itself may be found in drier areas in north-western Namibia (Timberlake et al. 2010). Sometimes the woodland forms pure stands of *Colophospermum mopane.* Mopane woodland can be 10–20 m tall, with stands up to 25 m, but in some areas scrub mopane of 2–3 m tall covers extensive areas, in mosaic with taller stands (Geldenhuys and Golding 2008). It is capable of growing under a range of climatic and edaphic conditions but is restricted in distribution by fires and competition from other species (Geldenhuys and Golding 2008). Mopane has a shallow root system with a dense concentration of fine roots in the top soil (Fanshawe 1971). Some scattered *Munga* woodland elements occur in places represented chiefly by *Acacia nigrescens*, *Adansonia digitata*, *Combretum imberbe*, *Kirkia acuminata* and *Lannea stuhlmannii*. The canopy dominants are *Colophospermum mopane*, *Acacia luedertzii*, *Acacia nigrescens*, *Adansonia digitata*, *Afzelia quanzensis*, *Albizia harveyi*, *Albizia amara*, *Brachystegia boehmii*,

Combretum imberbe, *Hyphaene ventricosa*, *Kirkia acuminata*, *Lannea stulhmannii*, *Philenoptera violacea*, *Pterocarpus angolensis*, *Sclerocarya birrea* and *Strychnos potatorum*.

4.2.4 Combretaceae Woodlands and Semi-arid Shrubland

Combretum and *Acacia* woodlands are found in patches in some parts of Southern Africa. In South Africa, Combreteaceae woodlands are dominated by *Terminalia sericea* which is usually found in association with other species. *Terminalia sericea* typically occurs on deep (>1 m) sandy soils, and may be nearly monodominant following disturbance (Scholes 2004). Shallower or rocky, infertile soils are dominated by *Combretum apiculatum* in semi-arid situations, giving way to *Combretum collinum* in slightly wetter areas (Scholes 2004).

Semi-arid shrubland vegetation formation covers over 900,000 km^2 and comprises either microphyllous wooded grassland or shrubland in which there is a more-or-less continuous grass layer (Timberlake et al. 2010). Due to low rainfall, drought, low temperatures, exposure to wind or salinity and toxicity or extreme oligotrophy of the soil, operating singly or in various combinations, the trees are sparsely distributed and are around 5–8 m high (White 1983; Timberlake et al. 2010). The characteristic species include various *Acacia* species (*A. erioloba*, *A. luederitzii*, *A. fleckii*, *A. hebeclada*, *A. mellifera*, *A. tortilis*), *Boscia albitrunca*, *Dichrostachys cinerea* and *Terminalia sericea* (Scholes 2004; Timberlake et al. 2010). Although the characteristic species are mostly Acacias, only one or two of these species may dominate the vegetation formation (Scholes 2004) depending on the availability of moisture. For example, *Acacia mellifera* and *A. erioloba* are widespread on the arid fringe; *A. karroo* in the southeast; *A. tortilis* and *A. nilotica* in the northeast, and *A. robusta* and *A. sieberiana* in the moister areas (Scholes 2004).

4.3 Woodland Utilization and Associated Impacts

More than 80 % of the rural population in sub-Saharan Africa is poor and traditionally relies on forests for their livelihoods (Schreckenberg et al. 2006); including slash and burn agriculture, charcoal and timber production. Fuel wood is one of the primary sources of energy for domestic use and processing (curing tobacco, drying fish, etc.) throughout the SADC region (Shackleton and Clarke 2007). It accounts for the highest percentage of the national energy budget in many Southern African countries namely; 85 % in Mozambique (Brigham et al. 1996); 76 % in Zambia (Chidumayo 1997); 95 % in Malawi (PROBEC 2009) and 52 % in Zimbabwe (Griffin 1999).

Charcoal production and the use of land for agriculture have resulted in deforestation and forest cover loss (Brown 2001), and massive loss of fauna, flora and

Table 4.1 Deforestation rates in miombo woodland predominated countries

Country	Total forest cover 2005 (000 ha)	Annual change rate 1990–2000 (000 ha/year)	%/year	2000–2005 (000 ha/year)	%/year
Angola	59,104	−125	−0.2	−125	−0.2
Malawi	3,402	−33	−0.9	−33	−0.9
Mozambique	19,262	−50	−0.3	−50	−0.3
Tanzania	35,257	−412	−1.0	−412	−1.1
Zambia	42,452	−445	−0.9	−445	−1.0
Zimbabwe	17,540	−313	−1.5	−313	−1.7

Source: FAO (2005)

high productive ecosystems (Syampungani 2008; Chirwa et al. 2008a; Syampungani et al. 2009). Large tracts of forestland are being converted to either agricultural fields or abandoned charcoal production sites. For example, the annual rate of deforestation per year between 1990 and 2000 ranged from 33,000 ha in Malawi to 445,000 ha in Zambia (FAO 2005). Mayaux et al. (2004) estimated forest area under agriculture to be 15.2 % in the Southern African region. Because of the nature of shifting cultivation that is widely practiced in the region, land under agriculture represents a number of cover types, including cropland, abandoned fields and fallow at various stages of recovery present; forming a typical mosaic landscape (Timberlake et al. 2010). Such mosaic landscapes have forest formations that are represented by tall, almost closed-canopy stands and many areas of cleared woodland for shifting cultivation and charcoal production (Syampungani 2008). This, to some extent, has affected the spatial integrity of most of the woodlands and forests in the region. When converting woodlands to agriculture not all trees are necessarily cleared and, as a result, cultivated land is often dotted with trees such as *Adansonia digitata*, *Vitellaria paradoxa*, *Sclerocarya* spp., *Borassus* spp., and *Faidherbiaalbida* (Timberlake et al. 2010).

Population pressure coupled with the drive to monocultures increased deforestation in the region thereby contributing to loss of biodiversity (Geist 1999). The low rates of economic transformation to individual levels have led to widespread poverty in the region. This in turn has increased the demand for woodland resources, and therefore deforestation (Table 4.1). This is because the economies of the sub-Saharan Africa are still dependent on the premises of charcoal production, shifting cultivation and to some extent timber harvesting (see Syampungani 2008).

4.4 Productivity Potential in the Different Biomes with Case Study Data from the Sub-region

Many parameters may be used to express productive potential of biomes. Productivity may be expressed in terms of stocking, mass per unit area and basal area. However, not many studies have been conducted for estimating total biomass and therefore productivity of Southern African vegetation. Partly, this could be attributed to the fact that the ecoregion has too many species with different rooting habits that need to be studied. Most of the species in the ecoregion are generally deep rooted with long lateral roots; the average tap root depth of canopy and understorey trees is 2.4 and 1.5 m, respectively, and lateral roots which are diffused at different depths may extend for up to 15 m from tree canopy (Savory 1963). However, Mistry (2000) observed that most tap roots in most miombo dominants extend up to a depth of about 5 m. Excavating and recovering roots for biomass estimation in the Southern African vegetation is rather a very difficult and laborious task, and also costly. Additionally, any study that attempts to employ excavative methods for estimating biomass would prove destructive and costly in such ecosystems. As such, there is some kind of limitation in the use of allometric equations to convert external measurements, such as trunk diameter, and sometimes height to estimate total biomass and therefore woodland productivity. Notwithstanding, studies conducted in Southern African vegetation show variability in productivity potential. The differences in productivity across Southern African woodlands/vegetation are as a result of inherent climatic and edaphic variation across the region; and to an extent the resultant land use. Most of the woodlands across Southern Africa are capable of recovering following disturbance cessation (Geldenhuys 2005; Chirwa et al. 2008a; Syampungani 2008). The re-growth may be from either coppices of stumps or root origin or stunted seedlings present in the herbal layer at the time of clearing (Chidumayo and Frost 1996). Savanna woodland species generally have both vertically and horizontally extensive root systems which facilitate recuperation after cutting (Mistry 2000). These extensive root systems tend to produce root suckers and coppices once the above-ground parts are removed. Syampungani (2008) observed that a number of species tend to coppice namely; *Brachystegia* spp., *Isoberlinia angolensis*, *Julbernadia paniculata*, *Pseudolachnostylis maprouneifolia*, etc. in stands that were previously under charcoal production and slash and burn agriculture.

4.4.1 Stocking and Basal Area

The available literature on Southern African woodland indicates that stem density vary significantly from one woodland type to another and also whether the woodland is a regrowth or not. Density ranged from 837 to 1,131 stems ha^{-1} in old growth Kalahari woodlands of the undifferentiated woodland while in Miombo old growth, the stocking was reported to be 2,434–2,773 stems ha^{-1} on average (Syampungani 2008). Higher variations in stocking (stems/ha) have also been observed in mesic, semiarid and arid localities in central lowveld of the South African savannas; Shackleton and Scholes (2011) recorded higher values in mesic (18,530 stems ha^{-1}) compared to semi-arid (3,996 stems ha^{-1}) and arid (3,978 stems ha^{-1}) localities. A comparison of regrowth stand densities between Kalahari woodlands and miombo woodlands indicated a significant variation from 1,131 to 6,685 stems ha^{-1} in

Table 4.2 Biomass related parameters of Southern Africa woodlands

Range of variables		Vegetation type	Authors
Density (stems ha^{-1})	1,121–6,926	Re-growth (miombo)	Syampungani et al. (2010), Strang (1974)
	2,434–2,773	Uneven aged mature woodland (miombo)	Syampungani (2008)
	3,978–18,530	Uneven aged mature woodland (South African Central lowveld)	Shackleton and Scholes (2011)
	837–1,131	Uneven aged mature woodland (Kalahari)	Timberlake et al. (2010)
	7,264–9,700	Regrowth (Kalahari)	Timberlake et al. (2010)
Basal area (m^2)	7–22	Uneven aged mature woodland (Miombo)	Lowore et al. (1994), Freson et al. (1974)
	7.12–12.44	Uneven aged mature woodland (South African Central lowveld)	Shackleton and Scholes (2011)
Mean biomass (Mg ha^{-1})	1.5–90	Uneven aged mature woodland (Miombo & Mopane woodland)	Chidumayo (1990, 1991), Tietema (1993)
	22–44.47	Uneven aged mature woodland	Guys (1981), Martin (1974)
	18.41–37.9	Uneven aged mature woodland (South African Central lowveld)	Shackleton and Scholes (2011)
Growth rate (Mean annual ring width, mm)	4.4–5.6	Regrowth (Miombo)	Syampungani et al. (2010)
	2.3–4.8	Uneven aged mature woodland (Miombo & South African savannas)	Shackleton (2002), von Maltitz and Rathogwa (1999), Chidumayo (1988a, b)

Miombo woodland to 7,264 to 9,700 stems ha^{-1} in Kalahari woodlands (Table 4.2). A variation in stocking per recovery stage/disturbance factor was reported by Strang (1974). Initially, a steady increase in stocking from 925 to 5,810 stems ha^{-1} was observed between 1.5 and 18 years since cutting in the Rhodesian (now Zimbabwean) Highveld which was protected against fires (Strang 1974). However, lower stocking levels were observed in the same locality which was constantly experiencing fires (Strang 1974).

The basal area ranged from 7 to 22 m^2 ha^{-1} in old uneven aged stands of various woodland types (Table 4.2) with the lowest being recorded on lithosols in Southern Malawi at about 650 mm mean annual precipitation and the highest being recorded in wet miombo woodland deep soils of the Democratic Republic of Congo at 1,270 mm rainfall (Lowore et al. 1994; Freson et al. 1974). Lower basal area (e.g. 9.81 m^2 ha^{-1}) was mostly associated with young regrowth stands of up to 20 years old since cutting (Chidumayo 1987). However, higher values of stand basal areas of between 30 and 50 m^2 ha^{-1} have been recorded in wet miombo and dry miombo of Zambia and Zimbabwe respectively, in small sized plots (Chidmayo 1985; Grundy 1995).

4.4.2 Tree Growth Rate and Wood Biomass

Tree growth is influenced by many factors, including genetics, climate and soils, as well as levels of disturbance of fires, diseases, slash-&-burn agriculture and charcoal production. Among the most studied and utilized species in Southern Woodlands are *Brachystegia spiciformis*, *Pterocarpus angolensis*, *Julbernadia paniculata*, *Isoberlinia angolensis*, *Brachystegia floribunda* and *Sclerocarya birrea* (Shackleton 2002; Grundy 2006; Syampungani et al. 2010; Helm 2011). The majority of tree species studied have diameter increments ranging from 0.03 to 2.6 cm per annum (Helm 2011; Syampungani et al. 2010; Timberlake et al. 2010). *Brachystegia spiciformis* in western Zambia grew by 0.24–0.33 cm diameter per annum while in Zimbabwe, the species was reported to grow by 0.03–0.27 cm per annum (Grundy 2006; Trouet et al. 2006). A study in Zambia (Syampungani et al. 2010) revealed that there is significant difference in annual ring width between species (Table 4.3); but the increment in annual ring width of similar species did not differ significantly between disturbance factors.

Woody biomass is said to increase with increase in mean annual rainfall across the Southern African region (Frost 1996). Aboveground biomass in old-growth, uneven aged stands has been reported to be 55 Mg ha^{-1} in dry miombo woodland of Zambia and Zimbabwe on average (Chidumayo 1991; Guys 1981) while the average biomass in wet miombo woodland has been observed to be about 90 Mg ha^{-1}(Table 4.2). Additionally, lower values have been observed in young re-growth stands of miombo woodlands (1.5 Mg ha^{-1}) in a 3–6 year old coppice stands. Similar variations have also been observed in Mopane woodland namely; 1.1 Mg ha^{-1} in south eastern Zimbabwe to 79 Mg ha^{-1} in northern Botswana (Tietema 1993). Shackleton and Scholes (2011) also observed an increase in basal area and biomass from an arid locality to mesic one. They attributed this to differences in stocking between the mesic locality, and the arid and semi-arid localities. This also suggests that moisture availability has an influence on biomass accumulation across ecosystems.

Root biomass studies have been limited in Southern African woodlands. However, what has been documented clearly is that the Zambezian woodland species have horizontally and vertically extensive root systems. Maximum recorded lateral distance ranges from 15 to 27 m in dominant miombo species namely *Julbernadia paniculata* and *Brachystegia spiciformis* (Strang 1965; Savory 1963). In dry

Table 4.3 Growth rates of key Miombo species under different disturbances

Species	Mean annual ring width, mm							
	Stand category and age							
	Slash and burn regrowth stands/age (years)				Charcoal regrowth stands/age (years)			
	7–8	10	15+	Mean	7–8	10	15+	Mean
Brachystegia floribunda	4.8 ± 0.3	5.8 ± 0.2	4.7 ± 0.2	5.1 ± 0.6	3.8 ± 0.3	4.9 ± 0.3	4.6 ± 0.1	4.4 ± 0.6
Isoberlinia angolensis	5.7 ± 0.4	5.8 ± 0.1	4.6 ± 0.6	5.4 ± 0.7	5.6 ± 0.3	6.6 ± 0.4	4.6 ± 0.2	5.6 ± 0.9
Julbernadia paniculata	5.0 ± 0.2	5.0 ± 0.2	4.2 ± 0.2	4.7 ± 0.5	3.6 ± 0.2	4.8 ± 0.2	4.7 ± 0.2	4.4 ± 0.7

Source: Syampungani et al. (2010)

miombo woodland of Central Zambia, an average of 32.7 Mg ha^{-1} as root biomass was observed by Chidumayo (1993). In the Transvaal region of South Africa, Roux et al. (1994) recorded total root biomass of 29.79 t Mg ha^{-1} he in dense Mopane woodland.

4.5 Management Practices for Improved Productivity in Extensively Managed Woodlands in Southern African

The management systems are designed to increase the woody biomass of different woodland systems. The response of the woodlands in terms of biomass accumulation varies from place to place as it is a function of a number of factors which include rainfall, soil type, temperature and management practices. Because of the low returns from the dry forest and woodlands in terms of physical products, some of the management practices are deemed uneconomic. The management practices are designed to meet specific tangible products (see Chidumayo et al. 1996; Chirwa et al. 2008a). Wood production, for example, in miombo woodland is also affected by the way miombo trees respond to harvesting. Responses depend on the phenological state, degree of resistance to fire, ability to resprout, seeding patterns, seed germination characteristics and seedling development (Chidumayo et al. 1996). Miombo woodland usually responds to wood harvesting by coppice regeneration, but the rate of regeneration is affected by human activities (Chirwa et al. 2008b). The best harvesting techniques and management practices in the post-harvest era are those that promote regeneration.

4.5.1 Silvicultural Management

The emphasis of silvicultural systems is on wood products, traditionally timber, but recently extended to incorporate other wood products such as firewood and poles of various sizes (Lowore and Abbot 1995; Chidumayo et al. 1996). However, very little research has been done on harvesting rates and designing management systems for non-wood products. The argument has been that products which are seasonally available such as fruits do not require harvesting limits, and that provided no damage is done to the trees during harvesting, the impact of fruit removal is minimal (see Shackleton and Clarke 2007). However, harvesting of bark for various products including medicine, rope fibre and for making beehives can be highly destructive and result in increased tree mortality (Chidumayo et al. 1996). A number of methods for reducing the negative impacts of bark harvesting have been proposed and tested, including obtaining bark from woody material that has already been cut for other purposes and improved harvesting methods that prevent ring barking and reduce fungal infestation; substitutions such as the use of leaves to obtain medicinal products rather than bark, and the provision of timber beehives (see Geldenhuys et al. 2006). Three basic silvicultural systems have been employed in harvesting extensively managed woodlands, especially miombo woodlands, namely; coppice with standards, selection system and complete coppice or clear cutting. Employing either of these systems requires that some management mechanisms are put in place to ensure high productivity. For example, adhering to optimum diameter classes within which particular species have high coppicing effectiveness would provide for enhanced coppicing ability for many woodland species. Handavu et al. (2011) observed that *Brachystegia longifolia*, *B. spiciformis* and *Isoberlinia angolensis* tend to have high coppicing ability in the diameter range of 15–36 cm DBH. Additionally, increased stump heights during woodland clearing have been observed to enhance the survival of stumps and coppicing. Grundy (1990) observed a reduction in coppices in lower stumps (<5 cm) compared to higher stumps (>1.3 m) in *Brachystegia spiciformis*. Shackleton (2001) made a similar observation in the study of indigenous savanna tree species (*Terminalia sericea*) for fuelwood production. According to Shackleton (2001), this may be attributed to lower potential impacts of browsers and fires. As such, the consideration of cutting heights will provide for marked effects on the resultant coppice number and regrowth rate, and hence harvest turnover time. However, too high stumps may result in instability of coppices as they develop.

Other factors that may be considered in enhancing productivity include plant age and surface area. Furthermore, thinning of regrowth stands to reduce competition for nutrients between often many coppices may result in increased survival and vigour of coppices. Lastly, management especially of young stands should also include protection against fires and drought.

4.5.1.1 Coppice with Standards Systems

The system involves leaving behind few trees of different species over a harvested area. The regrowth has the same pole and timber qualities as that from complete clear cutting near the ground. The high value species are left until maturity whilst the other species are clear cut and the regeneration is managed to produce a range of small dimension wood products such as firewood and poles (Tuite and Gardiner 1990; Lowore and Abbot 1995; Shackleton and Clarke 2007). The trees that are left may be important for either timber or fruit or fodder. Additionally, trees that do not coppice may also be left behind to ensure regeneration perpetuation of such species. For example, in Zambia most of the trees that were left over a cut area were timber species. However, the system has been proposed for use by small-scale farmers in Malawi as a means of maximizing the production of firewood and poles whilst retaining high value species that produce non-wood products and other services, such as fruit trees with spiritual significance (Lowore and Abbot 1995). The system has an advantage of retaining a portion of tree cover and protecting the site from erosion and sun scorch. This silvicultural system requires that fires are excluded in the early stages so that the coppices are not affected by fire.

4.5.1.2 Selection System

The impacts of selective and clear cutting on miombo woodland regeneration and indeed in many extensively managed types of woodland depends on the characteristics of the trees. Most inventories in miombo woodland have tended to exclude plants in the pre-sapling phase which probably form the largest reservoir for the future tree crop. Chidumayo et al. (1996) reported that about 95 % of the plants at the sites they sampled were suppressed saplings of which canopy species constituted 44 and 84 % of total species and plants, respectively. Shoot growth among suppressed miombo saplings is slow; but the plants tend to accumulate a relatively large belowground biomass which constitutes the perennating organ that regenerates new shoots following repeated shoot die-back during the dry season (Chidumayo 1993). It has therefore been suggested that shading by canopy trees contributes to slow shoot growth of suppressed saplings in miombo woodland (Lees 1962; Werren et al. 1995); although competition for nutrients and water stress during the long dry season, fires and herbivory probably also contribute to this retardation. Notwithstanding, removal of canopy trees accelerates the rate of recruitment from suppressed saplings to sapling and tree phases (Chidumayo 1993; Werren et al. 1995; Chidumayo et al. 1996). While there may be additional tree recruitment from suppressed saplings after tree cutting, resulting in tree density in young regrowth being usually very high, this is a temporary phenomenon because density eventually returns to the pre-felling level at maturity (Chidumayo et al. 1996). Studies (Trapnell 1959; Chidumayo 1997) carried out at different ages of the regrowth stands demonstrated that recruitment is a temporal phenomenon in the Miombo woodland. Chidumayo (1997) demonstrated that although high tree densities in the early stages of woodland recovery like the ones observed by Trapnell (1959) in 11 year-old regrowth may be observed, the density tend to reduce drastically up to 95 % as the regrowth stand attains maturity. Thinning of stems in regrowth woodland is therefore a desirable silvicultural practice that replaces the slow natural thinning process during woodland maturation from regrowth.

4.5.1.3 Complete Coppice or Clear Cutting System

This silvicultural system is appropriate when harvesting firewood for the charcoal industries (Shackleton and Clarke 2007). Since a large area is cleared the system results in the highest rate of regrowth of the three systems. This is because the stumps of most miombo trees have the ability to produce coppice shoots once they are cut (Boaler and Sciwale 1966; Strang 1974). It tends to produce rapid regeneration that can be managed according to the product requirement. Additionally, the system provides for maximum light for light woodland demanding species. However, the system will yield very low levels of shade tolerant species in the early stages of woodland recovery. In Malawi, research has shown that the best woody regrowth after felling is obtained after clear cutting (Werren et al. 1995). This is therefore the best way of encouraging regeneration from stumps and suppressed saplings. However, this management technique is unsuitable for large areas, steep slopes, or in riverine vegetation because of the increased erosion which occurs in the first years after cutting. Chidumayo et al. (1996) have advocated the use of shelterbelts in sensitive areas that are especially demarcated across the slope. Once leaf-area development in coupes reaches the pre-felling level, which usually takes 10 years or more, the shelterbelts may be cleared while the regrowth strips act as shelterbelts.

4.5.1.4 Cutting Cycles

Rotational harvesting is recommended for both the complete coppice or clear cutting and coppice with standard systems. This is where a particular woodland is divided into enough blocks to allow for continuous cutting cycle to be established in such a way that by the time the last block is cut, the first is ready for re-harvesting. The number of felling coupes, then, is equal to the number of years it takes for the tree to regenerate and reach harvesting size also known as cutting cycle. Cutting will depend on the type of product to be harvested. Some guidelines from certain countries have been developed. For example, based on results from a relatively moist site in Malawi, cutting cycles of between 3 and 5 years are recommended for the production of firewood and small poles, 10–15 years for medium sized poles and roofing struts, 25 or more years for large poles and timber and 40 or more years for saw log timber (Abbot and Lowore 1999). However, it is difficult to apply the concept of *cutting cycles* in selectively cut areas since trees in old-growth extensively managed woodlands vary in age/size (Chidumayo et al. 1996; Geldenhuys 2010). In selective cutting, each tree is selected on the basis of its size and suitability for the intended product. Thus selection criteria will vary by species and tree within species. Other factors can also influence selectivity: such as an increase in the net price of a wood product, improvement in the harvesting or processing technology. Increased poverty among harvesters also tends to reduce selectivity, affect cutting cycles and may cause over-exploitation of a species (Hosier 1993). Cutting cycles for regrowth woodland in clear-cut sites depends on product type and requirements and the wood product to be harvested, within the constraints of requirements for ecological sustainability. Chidumayo et al. (1996) contend that in the absence of data, on natural regeneration and productivity that is highly variable in space and time, it is difficult to prescribe precise cutting cycles.

4.5.2 Fire Management

Much of the current knowledge of fire and its effects on Zambezian Phytoregion structure and functioning has come from anecdotal evidence supplemented by information derived from limited number of experiments. The interest of focus among the early researchers has been more on later stages in changing of vegetation communities with emphasis on ecological groups (see Lawton 1978). However, several studies attempted to determine the effects of fire frequency and burning period on the structure and function of the miombo woodlands (see Trapnell 1959; Kikula 1986; Chidumayo 1988a, b; Zolho 2005). In Zambia, Trapnell (1959) and Chidumayo (1988a, b) studied the response of miombo species to varying burning regimes. Similar studies were conducted in Tanzania (Kikula 1986) and Mozambique (Zolho 2005). All recent studies (Kikula 1986; Chidumayo 1988a, b; Zolho 2005) confirmed the findings of Trapnell (1959) about sprouting behavior of miombo species under fire influence. The effects of fire on forest composition have also been observed in other vegetation types such as the forest in Okavango in Namibia (Geldenhuys 1977) and Kruger National Park in South Africa (Higgins et al. 2007). These studies demonstrated that species dominance and coppice effectiveness can be influenced by fire frequency and intensity. This is because fire may affect the perennating organs and root food reserves. Fire attack on perennating organs and food reserves usually results in die-back of shoots as a result of depletion of root food reserve of parent plants due to systematic and continuous effect of fire (Kennard et al. 2002). From a management perspective, fire management in extensively managed woodland should take into account the age of the woodland, the phenology of the dominant and/or desirable species, the type of land use and the management objectives of the area. Burning may not be necessary where livestock grazing or litter harvesting removes most of the fuel biomass. If burning is carried out in woodland areas where wood is a desired product, it should be done at the end of the rainy season, when the moisture levels in both grass and tree layer are relatively high (Trapnell 1959; Chidumayo et al. 1996).

4.6 Institutional Frameworks and Policy Directions for Managing Extensively Managed Woodlands in Southern Africa

Management of forest resources and other natural resources in the miombo ecoregion is governed by several acts and policies (Kayambazinthu et al. 2003). These regulatory systems provide access to forest resources through regulations, which involve the issuance of permits (Oduol et al. 2008). Reinforcement of conservation by-laws by traditional leaders has been known in some communities (Chirwa et al. 2008a; Syampungani et al. 2009). For example, in northern Zambia, by-laws regulating bush burning and the opening and closing dates of caterpillar collection are enforced by the chiefs or the paramount chief (Holden 1991; Chidumayo and Mbata 2002). In Tanzania, some chiefs declared some forests as traditional reserves (Kowero et al. 2003). Although the government acts and local regulatory systems could in principle help to protect natural resources, the drawback is that their implementation has been ineffective due to the independent nature of their operations and subtle competition between institutions (Kowero 2003). Additionally, the management style emanating from the enforcement of the legal instrument has been restrictive in nature (Kowero 2003). For example, policies governing the management of protected areas have so far stressed the non-consumptive utilization of protected resources (Munthali and Mughogho 1992). This has hampered the opportunity for the rural communities to manage and live in harmony with their natural resources. Some government acts and policies also tend to take away some functions and responsibilities of the traditional leadership, central in an African setting (Virtanen 1999); yet traditional institutions provide for sustainable forest management. If policies have to be effective, they need to be inclusive by taking all stakeholders on board and provide for an opportunity for stakeholders to meet their needs as they contribute towards sustainable forest management. Policies that provide an opportunity for residents to have access to forest products tend to encourage or enhance participatory forest resource management.

Participatory forest management using communities in proximity to the resource, seems to be the most plausible way of ensuring sustainable management, provided issues of benefits and access are clearly defined from the onset (see Arnold 2001; Ham et al. 2008). Most policy and legal framework point to the need for devolution of powers (Dewee et al. 2011). Devolution where the rights of access, use, control and ownership are completely devolved to the local communities is perceived to be more effective than where there is artificial devolution of power (see Dewee et al. 2011). Furthermore, decentralized forest management can be enhanced if the subsistence and commercial use of the resources can be integrated through sustainable resource management systems for long-term socio-economic benefits of all stakeholders. Some local examples exists where participatory approaches have been successfully implemented including the Communal Areas Management Programme for Indigenous Resources (CAMPFIRE) in Zimbabwe and the Administrative Management Design (ADMADE) for Game Management Areas (GMA), in Zambia (see Van Rijsoort 2000).

On bioenergy, policies conducive for promotion of small- scale, more sustainable technologies such as alternative (clean) energy development and agroforestry are lacking (see Ambali et al. 2011). Programs and policies which take into account the complex economic forces that influence energy production and consumption are urgently needed. It is therefore important that forestry and energy policies are complementary in order to foster the achievement of maximum benefits which wood can offer. Additionally, the promotion of more efficient energy use and the development of more modern production systems such as the use of wood for electricity production may contribute towards sustainable forest management.

4.7 Conclusions

The Zambezian phytoregion is a significant biome accounting for a large proportion of the African land masses. The Woodlands/forests of this region are widely exploited for charcoal production, slash and burn agriculture and timber. It is a source of livelihood for the majority of the rural poor of Southern Africa. Consequently, this has had implications on the vegetation cover in the region.

However, various studies have demonstrated that most of the woodland species of the region are capable of recovering once the disturbances arising from utilization for charcoal production, slash and burn agriculture ceases. This is because these species have both vertically and horizontally extensive root systems which facilitate recuperation after cutting. Additionally, a large number of seedlings and or saplings tend to be present at the time of clearing. The recovery rate and productive potential of the woodlands vary across the region. Limited data that indicates the productivity of the Southern African woodlands is influenced by several factors among which are (i) inherent climatic and edaphic factors; (ii) the resultant land use; (iii) the influences of fire; and (iv) standard assessment methods for both above and below ground biomass. The variation in productivity from one woodland system to another implies that management systems need to be developed specific to each woodland system if woody biomass is to be increased. This would also require that the influence of several factors on the productivity of the woodland systems is well understood. Furthermore efforts to develop capacity for the Forest Service or other government bodies responsible for forest management must be undertaken.

There is also need to formulate policies and legal framework that are inclusive. Currently, the local regulatory and the government regulatory systems do not work closely as the government system is considered to be superior over the local system (traditional system). There is need to understand how best the factors that make the two regulatory system operate independently. Formulation of policies conducive for promotion of small scale sustainable technologies such as alternative (clean) energy production and Agroforestry need to be encouraged.

References

Abbot PG, Lowore JD (1999) Characteristics and management of some indigenous firewood species in Malawi. For Ecol Manage 119:111–121

Ambali A, Chirwa PW, Chamdimba O, van Zyl WH (2011) A review of sustainable development of bioenergy in Africa: an outlook for the future bioenergy industry. Sci Res Essays 6(8):1697–1708

Arnold JEM (2001) Forests and people: twenty-five years of community forestry. Food and Agriculture Organization of the United Nations, Rome

Boaler SB, Sciwale KC (1966) Ecology of a Miombo site, Lupa North Forest Reserve, Tanzania. III. Effects on the vegetation of local cultivation practices. J Ecol 54:577–587

Brigham T, Chihongo A, Chidumayo E (1996) Trade in woodland products from Miombo region. In: Campbell B (ed) The Miombo in transition: woodlands and welfare in Africa. CIFOR, Bogor

Brown L (2001) Eco-economy: building an economy for the earth. Earthscan and Earth Policy Institute, London/Washington, DC

Campbell B (ed) (1996) The Miombo in transition: woodlands and welfare in Africa. CIFOR, Bogor

Chidmayo EN (1985) Structural differentiation of contiguous savanna woodland types in Zambia. Geo Eco Trop 9:51–66

Chidumayo EN (1987) Woodland structure, destruction and conservation in Copperbelt area of Zambia. Biol Conserv 40:89–100

Chidumayo EN (1988a) A re-assessment of effects of fire on miombo regeneration in the Zambian Copperbelt. J Trop Ecol 4:361–372

Chidumayo EN (1988b) Regeneration of *Brachystegia* woodland canopy following felling for tsetse fly control in Zambia. Trop Ecol 29:24–32

Chidumayo E (1990) Above-ground woody biomass structure and productivity in a Zambezian woodland. For Ecol Manage 36(1):33–46

Chidumayo EN (1991) Woody biomass structure and utilization for charcoal production in a Zambian miombo woodland. Bioresour Technol 37:43–52

Chidumayo EN (1993) Responses of miombo to harvesting: ecology and management. Stockholm Environment Institute, Stockholm

Chidumayo EN (1997) Miombo ecology and management: an introduction. IT Publications/Stockholm Environment Institute, Stockholm

Chidumayo E, Frost P (1996) Population biology of Miombo trees. In: Campbell B (ed) The Miombo in transition: woodland and welfare in Africa. CIFOR, Bogor

Chidumayo EN, Mbata KJ (2002) Shifting cultivation, edible caterpillars and livelihoods in the Kopa area of northern Zambia. For Trees Livelihoods 12:175–193

Chidumayo EN, Gambiza J, Grundy I (1996) Managing Miombo woodlands. In: Campbell B (ed) The Miombo in transition: woodland and welfare in Africa. CIFOR, Bogor

Chirwa PW, Akinnifesi FK, Sileshi G, Syampungani S, Kalaba FK, Ajayi OC (2008a) Opportunity for conserving and utilizing agrobiodiversity through agroforestry in Southern Africa. Biodiversity 9:45–48

Chirwa PW, Syampungani S, Geldenhuys CJ (2008b) The ecology and management of the Miombo woodlands for sustainable livelihoods in Southern Africa: the case for non-timber forest products. South For 70(3):237–245

Dewee PA, Campbell BM, Katerere Y, Sitoe A, Cunningham BA, Angelsen A, Wunder S (2011) Managing the miombo woodlands of Southern Africa: policies, incentives and options for the rural poor. J Nat Resour Policy Res 2:57–73

Fanshawe DB (1971) The vegetation of Zambia. Government Printers, Lusaka

Food and Agriculture Organization (2005) Forestry outlook study for Africa: subregional report: Southern Africa. African Development Bank/European Commission/FAO, Rome

Freson R, Goffinet G, Malaisse F (1974) Ecological effects of the regressive succession in muhulu-miombo-savanna in Upper Shaba, Zaire. In: Proceedings of the first international congress of ecology. Structure, functioning and management of ecosystems, The Hague, September 1974. PUDOC, Wageningen, pp 365–371

Frost PGH (1996) The ecology of miombo woodlands. In: Campbell B (ed) The Miombo in transition: woodland and welfare in Africa. CIFOR, Bogor

Geist JH (1999) Global assessment of deforestation related to tobacco farming. Tob Control 8:18–28

Geldenhuys CJ (1977) The effect of different regimes of annual burning on two woodland communities in Kavango. S Afr For J 102:32–42

Geldenhuys CJ (2005) Basic guidelines for silvicultural and management practices in Mozambique. Report FW-04/05, FORESTWOOD cc, Pretoria

Geldenhuys CJ (2010) Managing forest complexity through application of disturbance–recovery knowledge in development of silvicultural systems and ecological rehabilitation in natural forest systems in Africa. J For Res 15:3–13

Geldenhuys CJ, Golding JS (2008) Resource use activities, conservation and management of natural resources of African savannas. In: Faleiro FG, Lopes A, Neto D (eds) Savannas: Desafios e estastegiaspara o equilibrio entre sociedade, agronegocio e recursos naturais. Embrapa Cerrados, Planaltina

Geldenhuys CJ, Syampungani S, Meke G, Vermeulen WJ (2006) Response of different species on bark harvesting for traditional medicine in South Africa. In: Bester JJ, Seydack AHW, Vorster T, van der Merwe IJ, Dzivhani S (eds) Multiple use management of natural forests and woodlands: policy refinements and scientific progress. Department of Water Affairs and Forestry, Pretoria

Griffin J (1999) Study on the development of transboundary natural resources management area in Southern Africa: main report. Biodiversity Support Program, Washington, DC
Grundy IM (1990). The potential for management of the indigenous woodland in communal farming areas of Zimbabwe with reference to the regeneration of *Brachystegia spiciformis* and *Julbernardia globiflora*. Unpublished M.Sc. thesis, University of Zimbabwe, Harare
Grundy I (1995) Regeneration and management of *Brachystegiaspiciformis* (Benth) and *Julbernadia globiflora* (Troupin) in miombo woodland, Zimbabwe. D.Phil. thesis, University of Oxford, Oxford
Grundy I (2006) Age determination of miombo species *Brachystegia spiciformis* (Leguminoseae-Caesalpinoideae) in Zimbabwe, using growth rings. South Afr For J 206:5–12
Guys PR (1981) Change in the biomass and productivity of woodlands in the Sengwa Wildlife Research Area, Zimbabwe. J Appl Ecol 18:507–519
Ham C, Chirwa PW, Theron F (2008) The forester as a change agent: from trees between the people to people between the trees. In: Theron F (ed) The development change agent: a micro-level approach to development. JL van Schaik Publishers, Pretoria
Handavu F, Syampungani S, Chisanga E (2011) The influence of stump diameter and height on coppicing ability of selected key Miombo woodland tree species of Zambia: a guide for harvesting for charcoal production. J Ecol Nat Environ 3(14):461–468
Helm CV (2011). Investigating the life history strategy of an African Savanna tree, *Sclerocarya birrea* subsp. *caffra* (marula). Ph.D. thesis, University of Witwatersrand, Johannesburg
Higgins SI, Bond WJ, February EC, Bronn A, Euston-Brown DIW, Enslin B, Govender N, Rademan L, O'Regan S, Potgieter ALF, Scheiter S, Sowry R, Trollope L, Trollope WSW (2007) Effects of four decades of fire manipulation on woody vegetation structure in savanna. Ecology 88(5):1119–1125
Holden S (1991) Edible caterpillars – a potential agroforestry resource? Food Insects Newsl 4(3):3–4
Hosier RH (1993) Charcoal production and environmental degradation. Energy Policy 21:491–509
Kayambazinthu D, Matose F, Kajembe G, Nemarundwe N (2003) Institutional arrangements governing natural resource management of the miombo woodland. In: Kowero G, Campbell BM, Sumaila UR (eds) Policies and governance structures in woodlands of Southern Africa. CIFOR, Bogor
Kennard DK, Gould K, Putz FE, Frederickesen TS, Morale F (2002) Effect of disturbance intensity on regeneration mechanisms in tropical dry forest. For Ecol Manage 162:197–208
Kikula IS (1986) The influence of fire on the composition of miombo woodland of Southwest Tanzania. Oikos 46:317–324
Kowero G (2003) The challenges to natural forest management in sub-Saharan Africa rural development: experiences from the Miombo woodlands of southern Africa. In: Kowero G, Campbell BM, Sumaila UR (eds) Policies and governance structures in woodlands of Southern Africa. CIFOR, Bogor
Kowero G, Kaoneka RSA, Nhantumbo I, Gondo P, Jumbe CBL (2003) Forest policies in Malawi, Mozambique, Tanzania and Zimbabwe. In: Kowero G, Campbell BM, Sumaila UR (eds) Policies and governance structures in woodlands of Southern Africa. CIFOR, Bogor
Lawton RM (1978) A study of dynamic ecology of Zambian vegetation. J Ecol 66:175–198
Lees HMN (1962) Working plan for the forests supplying the Copperbelt Western Province. Government Printer, Lusaka
Lowore J, Abbot PG (1995) Initial regeneration of miombo woodland under three silvicultural systems. Forestry Research Institute of Malawi, Zomba
Lowore JD, Abbot PG, Werren M (1994) Stackwood volume estimations for miombo woodlands in Malawi. Commonw For Rev 73:193–197
Martin RB (1974) Structure, biomass and utilization of vegetation in mopane and miombo woodlands of Sengwa Wildlife Research Area. Certificate in Field Ecology thesis, University of Rhodesia, Salisbury
Mayaux P, Bartholome E, Fritz S, Belward A (2004) A new land-cover map of Africa for the year 2000. J Biogeogr 36:1181–1193

Millington AC, Chritchley RW, Douglas TD, Ryan P (1994) Prioritization of indigenous fruit tree species based on farmers evaluation criteria: some preliminary results from central region, Malawi. In: editors? Proceedings of the regional conference on indigenous fruit trees of the Miombo ecozone of Southern Africa, ICRAF, Nairobi

Mistry J (2000) World Savannas: Ecology and Human Use. Pearson Education Limited, Edinburgh

Munthali SM, Mughogho DEC (1992) Economic incentives for conservation: bee-keeping and Saturinidae caterpillar utilization by rural communities. Biodivers Conserv 1:143–154

Oduol PA, Ajayi OC, Matakala P, Akinnifesi FK (2008) The role of institutional arrangements and policy on the conservation, utilization and commercialization of indigenous fruits in Southern Africa. In: Akinnifesi FK, Leakey RRB, Ajayi OC, Sileshi G, Tchoundjeu Z, Matakala P, Kwesiga FR (eds) Indigenous fruit trees in the tropics: domestication, utilization and commercialization. ICRAF/CABI, Wallingford

PROBEC (2009) Programme for basic energy and conservation in Southern Africa: basic energy facts: Malawi profile. http://www.probec.org/displaysection.php?czacc=&zSelectedSectionID=sec1242301660. Accessed 31 July 2011

Roux A, Smit GN, Swart JS (1994) Root biomass of a dense stand of *Colophospermum mopane*. Bull Grassland Soc South Afr 5(1):50

Savory BM (1963) Site quality and tree root morphology in Northern Rhodesia. Rhod J Agric Res 1:55–64

Scholes RJ (2004) Woodlands of South Africa. In: Lawes MJ, Eeley HAC, Shackleton CM, Geach BGS (eds) Indigenous forests and woodlands in South Africa: policy people and practices. University of KwaZulu-Natal Press, Pietermaritzburg

Schreckenberg K, Awono A, Degrande A, Mbosso C, Ndoye O, Tchoundjeu Z (2006) Domesticating indigenous fruit trees as a contribution to poverty reduction. For Trees Livelihoods 16:35–51

Shackleton CM (2001) Managing regrowth of an indigenous savanna tree species (*Terminalia sericea*) for fuelwood: the influence of stump dimensions and post-harvest coppice pruning. Biomass Bioenergy 20:262–270

Shackleton CM (2002) Growth patterns of *Pterocarpus angolensis* in savannas of the South African low veld. For Ecol Manage 166:85–97

Shackleton CM, Clarke JM (2007) Research and management of Miombo woodlands for products in support of local livelihoods. Available at: http://www.cifor.cgiar.org/miombo/doc. Accessed on 14 Feb 2008

Shackleton CM, Scholes RJ (2011) Above ground woody community attributes, biomass and carbon stocks along a rainfall gradient in the savannas of Central lowveld, South Africa. S Afr J Bot 77:184–192

Strang RM (1965) Bush encroachment and secondary succession in the southern Rhodesian Highveld with special reference to soil moisture relations. Ph.D. thesis, University of London, London

Strang RM (1974) Some man-made changes in successional trends on the Rhodesian highveld. J Appl Ecol 111:249–263

Syampungani S (2008) Vegetation change analysis and ecological recovery of the Copperbelt miombo woodlands of Zambia. Ph.D. thesis, University of Stellenbosch, Stellenbosch

Syampungani S, Chirwa PW, Akinnifesi FK, Ajayi OC, Sileshi G (2009) The miombo woodlands at the cross roads: potential threats, sustainable livelihoods, policy gaps and challenges. Nat Resour Forum 33:150–159

Syampungani S, Geldenhuys C, Chirwa PW (2010) Age and growth rate determination using growth rings of selected miombo woodland species in charcoal and, slash and burn regrowth stands in Zambia. J Ecol Environ 2(8):167–174

Tietema T (1993) Biomass determination of fuelwood trees and bushes of Botswana, Southern Africa. For Ecol Manage 60:257–269

Timberlake J, Chidumayo E, Sawadogo L (2010) Distribution and characteristics of African dry forests and woodlands. In: Chidumayo E, Gumbo DJ (eds) The dry forests and woodlands of Africa: managing for products and services. Earthscan, London, UK

Timberlake J, Chidumayo E (2011) Miombo ecoregion vision report. Occasional publications in biodiversity no. 20. Biodiversity Foundation for Africa, Bulawayo

Trapnell CG (1959) Ecological results of woodland burning experiments in Northern Rhodesia. J Ecol 47:129–168

Trouet V, Coppin P, Beeckman H (2006) Annual ring patterns in Brachystegia trees of miombo woodland reveal climatic influence. Biotropica 38:375–382

Tuite P, Gardiner JJ (1990) A survey of farm regrowth in south-western Tanzania. Department of Forestry, University College, Dublin

Van Rijsoort J (2000) Non-timber forest products (NTFPs): their role in sustainable forest management in the tropics. National Reference Centre for Nature Management, International Agricultural Centre, Wageningen

Virtanen P (1999) Community in context: chiefs and councils in Mozambique. Paper presented at the seminar and workshop on Governance, Property Rights and Rules for Woodland and Wildlife Management in Southern Africa, Harare

von Maltitz GP, Rathogwa NR (1999) Dynamics of *Pterocarpus angolensis* (Kiaat) in South Africa. CSIR Division of Forest Science and Technology, unpublished report

Werren M, Lowore J, Abbot P, Siddle B, Hardcastle P (1995) Management of miombo by local communities. University of Aberdeen and Forestry Research Institute of Malawi, Zomba

White F (1983) The vegetation of Africa. A descriptive memoir to accompany the UNESCO/AETFAT/UNSO vegetation map of Africa. Maintain, UNESCO, Paris

Zolho R (2005) Effect of fire frequency on the regeneration of Miombo woodland in Nhambita, Mozambique. M.Sc. thesis, University of Edinburgh, Edinburgh

Chapter 5
Biomass Production in Intensively Managed Forests

Ben du Toit

5.1 Introduction

Intensively managed plantations of fast-growing trees, planted on a short rotations at high stand densities, is arguably one of the most productive and energy efficient ways to produce biomass. In this chapter we discuss silvicultural options to establish and manage highly productive bio-energy plantations, based on case studies from short-rotation plantation forests in warm-climate regions. Our focus is on the growing of biomass as a main product. We also explore the energy and green house gas balances from such intensively managed systems.

A topic that has been extensively researched in short-rotation pulpwood plantations in tropical and warm-climate countries, is that of intensive management to boost stand productivity (Schönau 1984; Nambiar 2008; Stape et al. 2008, 2010; Gonçalves et al. 2007; Fox et al. 2007b; du Toit et al. 2010). This management style has been dubbed "Intensive, Site-specific Silviculture" in Southern Africa (Schönau 1984; du Toit et al. 2010), and has been responsible for large improvements in productivity. It has also been categorised under the more general field of "Precision Forestry" by some authors because the management philosophy hinges on choosing and implementing a suite of management operations that are specifically suited to the ecological capability of a specific site type, or to alleviate constraints to productivity on a specific site type (Pallett 2005; du Toit et al. 2010). This production system, with some adaptations, is arguably the most suitable starting point to design silvicultural regimes for intensively managed, highly productive biomass plantations. In its current form, it is usually a man-made (afforested) monoculture tree crop managed under a clear felling system. However, it has some important

B. du Toit (✉)
Department of Forest and Wood Science, Stellenbosch University,
Private Bag X1, Matieland 7602, South Africa
e-mail: ben@sun.ac.za

T. Seifert (ed.), *Bioenergy from Wood: Sustainable Production in the Tropics*,
Managing Forest Ecosystems 26, DOI 10.1007/978-94-007-7448-3_5,

differences to conventional agricultural monocultures, namely minimum soil cultivation where feasible, relatively low dependence on artificial chemical inputs for sustained productivity, prolonged periods of minimal cultural interventions, a net carbon footprint that is strongly positive, and the incorporation of significant biodiversity/conservation areas for the maintenance of ecosystem services (usually in the form of ecological networks or corridors) within the broader system in the landscape (cf. Chap. 10). In Sects. 5.2, 5.3, 5.4 and 5.5, we will discuss the most important elements of "intensive, site-specific silviculture" from the existing body of knowledge which essentially consist of a number of strategic choices (e.g. choice of genetic material and planting density/rotation length combination) as well as cultural practices (e.g. site preparation, vegetation management and fertilization). We will also focus on the most promising adaptations of conventional systems may be necessary to optimise this kind of silvicultural system for bio-energy plantations.

5.2 Matching Highly Productive Tree Taxa with Specific Site Types and Bio-energy Production Systems

There can be little doubt that the selection and genetic improvement of fast-growing tree taxa (in this context referring to species, provenances, families within provenances, hybrids or clones) have strongly boosted productivity on intensively managed plantation forests in the tropics (Zobel and Talbert 1984; Verryn 2002, 2008; Pallett and Sale 2002; Kanzler et al. 2003; Wu et al. 2007; Boreham and Pallett 2007). Particularly impressive tree breeding successes in short-rotation pulpwood plantation forestry also include improvements in properties that enhance processing (e.g. wood properties, stem form and ease of debarking – Malan and Verryn 1996; Dvorak et al. 2008) or properties that allow for better survival and productivity of a specific taxon under adverse circumstances (e.g. disease resistance, cold/drought/frost tolerance, improved water use efficiency and herbicide resistance – Hodge and Dvorak 2007; Herbert 2012; du Toit 2012; Mitchell et al. 2013). It is therefore imperative for any intensive bio-energy production system to invest in a focussed tree improvement programme that can conserve a broad genetic base of fast-growing tree families, constantly breed for resistance to newly emerging pests and diseases, and constantly improve quality of the biomass and its suitability for the particular production process. The word-wide trend in highly productive short-rotation plantation forests is to move increasingly towards planting a variety of genetically improved, vegetatively propagated hybrid clones (rather than raising seedlings from half-sib or full sib families within species), which are deployed in a mosaic of small blocks to minimise risk. Some of the more important reasons for this trend revolve around the following facts: (a) genetic gains are large and guaranteed, (b) hybrid vigour can be obtained, (c) disease resistance can be obtained through hybridisation, (d) large improvements in stand uniformity, with which comes ease of mechanisation and an increase in the partitioning to above-ground biomass (Stape et al. 2010).

Another advantage of planting clones rather than pure species of fast-growing exotics has to do with the invasive potential of some taxa when planted in a non-native environment. There is an increasing body of evidence showing that several hybrid plants are sterile or do not produce large quantities of viable seedlings (Eldridge et al. 1994; Lopez et al. 2000; Owens and Miller 2009; Chen 2010). It follows that highly bred hybrid tree clones that are less fertile or sterile can potentially be selected for planting biomass crops, so that they do not pose an invasive threat. This aspect needs further testing and experimentation, but holds promise for the establishment of "greener" bio-energy crops.

Equally important to the genetic tree improvement process that may improve stand productivity is intensive experimentation with site-taxon matching. Matching the planted taxa to site conditions is obviously important in biophysically complex landscape where climatic and edaphic conditions differ markedly in space and time. There are several examples all over the world where species/provenances or families that were not well adapted to specific climate conditions have been devastated by a single risk factor, e.g. the infection of pine species that have little resistance to *Diplodia pinea* when planted on sites that experience hailstorms (Swart et al. 1988), and the stem breakage caused in *Acacia mearnsii* and *Eucalyptus grandis* stands in the KwaZulu-Natal Midlands of South Africa following episodic, heavy snowfalls (Gardner and Swain 1996). However, new pests or diseases can easily be introduced to regions of exotic plantations, e.g. *Phytophthora pinifolia* in Chile or *Sirex noctilio* in South Africa (Tribe and Cillié 2004; Durán et al. 2008; Hurley et al. 2012). Furthermore, there are documented evidence of pests and diseases of indigenous trees that infect or attack distantly-related exotics, e.g. *Crysoporthe austroafricana* stem canker that recently spread to infect exotic stands of *Eucalyptus grandis* stands in sub-tropical parts of South Africa (Wingfield et al. 2008). It follows that regional enterprises with fairly uniform climates (and therefore chiefly rely on just one or perhaps a few species) are at risk. In such cases, it pays to invest in research on several taxa that could potentially be suited to a site: (a) to allow for the planting a mosaic of different families/clones and so to minimise risk, (b) to have alternative taxa that can be deployed rapidly and effectively in case of the introduction of a significant pest or disease to the region, and (c) to improve site-taxon matching for lesser known taxa.

Several authors have been successful in matching species or provenances to a broader region using climatic similarities, often with a computer-aided approach, e.g. bioclimatic parameters. However, finer scale matching of taxa to specific site types require local knowledge of tree response to risk factors, as well as stand productivity and quality of biomass that can be obtained under the range of prevailing site conditions. This involves planting and testing all promising taxa across a wide range of site types in the region where it could be grown. It also presupposes that a sophisticated yet simple site classification and site evaluation system exists (e.g. Louw and Scholes 2002; Smith et al. 2005; Louw et al. 2011; Louw and Smith 2012) – sophisticated enough to take both risk factors and the drivers of resource availability to stands into account, yet simple enough to implement in practice. Information gathered from studies such as the aforementioned examples is the only reliable foundation upon which detailed site-taxon matching can be based. Accurate site-taxon matching becomes even more important when (a) the genetic base of the planting material becomes increasingly narrow, e.g. in the sequence provenance → family → clone, and (b) when the silvicultural regime tends toward short rotations of unthinned crops. A stand consisting of a several families of half-sibs that are only moderately suited to a specific site will still contain some individuals that are well suited to the prevailing conditions, and those individuals could be the final crop trees remaining after thinning in a medium to long rotation system. This is often the case in plantation forests grown for sawtimber on 20–35 year rotations in Southern Hemisphere countries. However, if a single clone (that is not particularly well suited to the site conditions) is planted in a short, unthinned rotation, the productivity of that stand is guaranteed to be sub-optimal, and the productivity loss will be compounded over successive rotations if a coppice system is used. Furthermore, in the event of climatic extremes, an entire crop could be affected by pests or diseases, brought about by stress. It is therefore essential to have a risk profile as well as a "response surface" for the productivity potential of each important or potentially important taxon across a the broadest possible range of site types, to aid the silviculturalists in their decision making.

5.3 Selecting the Optimum Combination of Stand Density, Harvesting System and Rotation Length

Strategic planning for the most appropriate silvicultural and harvesting system for bio-energy crops should be done simultaneously for maximum economic benefit. The reasons are: (1) The profitability of plantation systems are often strongly influenced by harvesting and transport costs, the latter commonly constituting the biggest share of all expenses in the value chain from plant to mill, and (2) Different harvesting systems are designed to work optimally within specific ranges of individual tree volumes (Ackerman and Pulkki 2004), for example, (a) mechanised conventional timber harvesting with individual tree volumes from approximately 0.1 to 0.9 m^3, (b) clearfelling with chainsaws from 0.01 to 0.1 m^3, and (c) modified agricultural harvester <0.01 m^3. We will explain this relationship with data from *Eucalyptus grandis* crops grown in South Africa, where the aforementioned volume ranges would translate into diameters at breast height (dbh) classes of approximately 16–32 cm; 8–15 cm and <8 cm, respectively. From an economic perspective, it is thus imperative to design the silvicultural system in such a way that it could deliver mature crops falling within a specific range, and to match this with the capabilities of the chosen harvesting system. To a large degree, this can be achieved by manipulating the relationship between stand density and rotation length in short-rotation crops. However, (Coetzee 1999) has shown that this relationship is strongly dependent on the site index (or similar measure of site production potential). An example of mean annual increment development in South African *E. grandis* crops, grown on various stand densities across three different site indices are shown in Fig. 5.1, based on the data produced by Coetzee et al. 1996; Coetzee and Naicker 1998; Coetzee 1999, with key data points summarised in Table 5.1. The site index in this study is defined as the mean height of trees per compartment that fall into the 80th percentile with respect to dbh, at a reference age of 5 years (hereafter SI_5).

From Fig. 5.1 and Table 5.1, it is clear that the peak MAI on a site with $SI_5 = 26$ can be achieved (a) as early as 3.6 years with 2,000 stems ha (possibly even at 3 years if more than 2,000 sph had been tested), however, (b) it will take up to 5.0 years if only 800 stems were established per hectare. The quadratic mean dbh of scenario (a) in the aforementioned text would be 13.3 cm; while that of scenario (b) would be 19.6 cm. On a low productivity site ($SI_5 = 15.5$) the MAI will culminate at 7.0 years with 2,000 stems (scenario c) and will only culminate beyond 12 years with 800 stems per hectare (scenario d). These data sets clearly show that MAI and individual tree size are strongly related to the interactive effects of rotation length, stand density and site index. It follows that site-specific management regimes should be developed for rotation length by stand density combinations. Scenario (b) lends itself to harvesting with a mechanised system, whereas scenario's (a) and (c) are more suited to a chainsaw system. The data of Coetzee (1999) did not test very dense stocking levels, but it appears that systems with between 3,000 and 4,000 sph could yield slightly higher peak MAI's, with the volume carried on small stems which lend themselves to harvesting with a modified agricultural harvester. Sochacki et al. 2007, working on a low productivity site in Australia, showed that stand densities of up to 4,000 sph yielded the largest volume production at age 3 years in that study. Stand densities of 3,000–4,000 sph could thus be considered if harvesting with modified agricultural harvesting equipment is envisaged.

If the silviculturalist opts for very high stand densities (more than 2,000 sph) with the aim to utilise a modified agricultural harvesting system, there will be additional factors that have to be considered when deciding on the optimum stand density by rotation length combinations across a range of site indices. These considerations will include the following:

- Increased cost of establishment because more trees have to be planted
- Increased tree stress due to intraspecific competition with high stand densities (in Sochacki et al. (2007) study, tree mortality was an important factor affecting final biomass production on some treatments).
- Less flexibility around the felling age (especially on higher site indices), because productivity may decline sharply if the rotation over-matures (see Fig. 5.1).
- Early canopy closure, leading to lower weed management costs.
- Lower levels of inter-specific competition (*i.e.* between competing vegetation and trees), which will improve tree uniformity and an increasing fraction of NPP being partitioned to above-ground tissues (Little et al. 2003; Stape et al. 2010).
- Changes in wood characteristics such as density and fibre properties. Shorter rotations will have an increased proportion of juvenile wood in the final volume of biomass harvested.
- Increases in nutrient depletion from the site due to intensive biomass harvesting.

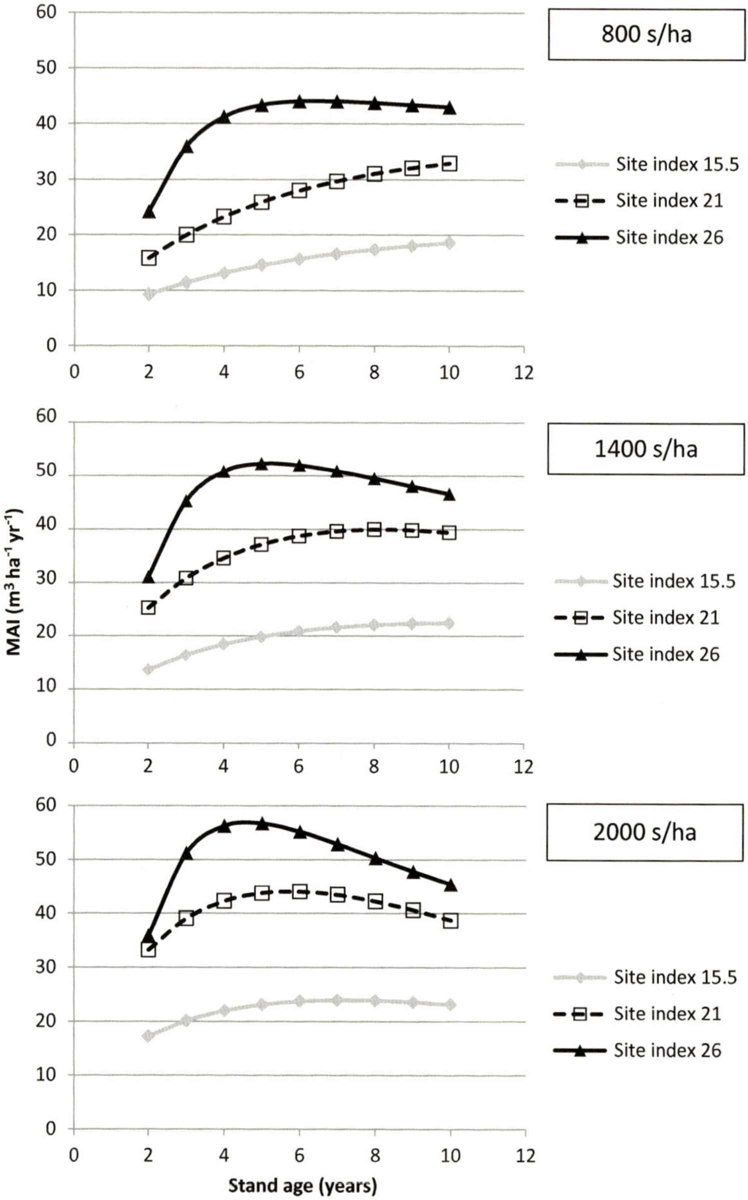

Fig. 5.1 Development of mean annual increment (MAI) in *Eucalyptus grandis* stands over time on sites with varying productivities (indicated by site indices (SI_5)) and stand densities ranging from 800 to 2,000 stems/ha (Adapted from Coetzee et al. 1996; Coetzee and Naicker 1998; Coetzee 1999)

Table 5.1 The culmination age of MAI, the actual MAI at the culmination point and the quadratic mean dbh (Qdbh) of the trees at that specific age and stand density, for the three site indices (base age 5), based on data in Fig. 5.1

	800 S/ha			2,000 S/ha		
Site index	Peak age	Peak MAI	Qdbh	Peak age	Peak MAI	Qdbh
26	5.0	44.0	19.6	3.6	57.3	13.3
21	>10	n.d.	>20	5.7	44.2	13.5
15.5	>12	n.d.	>22	7.0	24.0	12.9

5.4 Optimising Growth Conditions at Time of Establishment Through Harvest Residue (Slash) Management and Soil Tillage

Site preparation techniques (soil tillage and/or slash management) as well as harvesting impacts management aim to improve conditions for early tree and root growth by improving some or all of the following properties: soil aeration in waterlogged soils (Zwolinski et al. 2002), water infiltration, microclimatic conditions near the transplant (Carlson et al. 2004), soil nutrient mineralisation rates (du Toit and Dovey 2005; du Toit et al. 2008), or by ameliorating root growth impediments such as hard-setting soils (Gonçalves et al. 2008), semi-impenetrable or compacted layers (Smith et al. 2000, 2001) or competing vegetation (Little et al. 2001). Experimental evidence show that, where site-appropriate slash management or site preparation techniques have been applied, it had a significant impact on transplant survival and eventually on stand productivity at time of harvesting (du Toit and Dovey 2005; Gonçalves et al. 2007; du Toit et al. 2010). Conversely, the application of intensive tillage operations to situations where it did not alleviate growth-limiting situations have been shown to result in poor long-term growth responses, which, considering the high input costs, may be uneconomical (Smith et al. 2001; Lacey et al. 2001; Carlson et al. 2006; Lincoln et al. 2007). The key is thus to recognise opportunities where site preparation activities can successfully be applied and to rather implement minimum cultivation and tillage and slash conservation measures on site where the risks are high or where responses are likely to be small (Smith et al. 2000, 2001; Gonçalves et al. 2008). Some of the most important findings emanating from intensive experimentation on these issues in South Africa, Brazil, South-eastern USA and Australia are highlighted below.

When afforesting for the first time into dense native vegetation, such as grassland, it is advisable to implement intensive surface cultivation techniques, provided that the slope is not too steep and that so-called duplex soils are avoided (i.e. light textured topsoils such as sands/loamy sands with an abrupt transition to a heavy texture such as clays or silty clays). The surface cultivation eliminates the competing vegetation and stimulates an increase in nutrient mineralization after tillage, which will boost early tree growth. Basal area improvements at maturity of 11–52 % have been recorded in South Africa for a range of eucalypt stands with this treatment (Smith et al. 2000; du Toit et al. 2010).

In *re*-establishment situations, significant responses to surface cultivation are far less likely due to the beneficial effect of old root channels from previous crops, especially if soil structure or consistency does not limit root growth (Smith et al. 2001; Nambiar and Sands 1990), and the (generally) lower levels of competing vegetation that can form a dense root mat. In all these cases, minimum cultivation is recommended (Gonçalves et al. 2008). In exceptional cases, where soils have a hard-setting consistency (cohesive soils) or have suffered compaction, surface cultivation techniques such as shallow ripping could improve tree growth significantly. This approach is especially attractive likely on short rotations where drought risk in the mature phase of the crop is less likely (Gonçalves et al. 2008).

Early growth responses to deep subsoiling have been recorded on many soil types, only to diminish over time and becoming insignificant or uneconomical during drought periods or by rotation end (Smith et al. 2001; Gonçalves et al. 2008). Deep subsoiling is only economically justifiable in highly specialised situations, e.g. where inaccessible layers of soil or highly weathered saprolite can be made accessible to tree roots following subsoiling. Additional examples of inappropriate soil tillage operations are deep subsoiling operations in soils that have no macrostructure or hard-setting attributes, or excessive tillage and cultivation of duplex soils that are highly erodible. Soil quality can also be degraded by nutrient depletion (e.g. high-intensity slash burning on nutrient poor soils) or by excessive and frequent tillage of soils that will speed up mineralisation and subsequent leaching losses of soil carbon and nitrogen in young stands (Smith et al. 2001; du Toit et al. 2001, 2010) This issue is discussed more fully in Chap. 10.

Land surface modifications such as ridging and trenching can be highly beneficial where permanent or prolonged waterlogging (e.g. on flat slopes) limits root aeration and nutrient availability (Zwolinski et al. 2002; Kyle et al. 2005), especially in short rotation crops. However, this practice is certainly not suitable for moderate to well drained soils (especially those in dry climates), as it will render stands more vulnerable to drought stress.

In the preceding paragraphs, we discussed the effects of slash management operations in the inter-rotation period on nutrient supply to newly established tree stands. In very intensive biomass production systems, where most of the above ground biomass is harvested in ultra-short rotations, minimal harvest residue will remain on site. Furthermore, ultra short rotations will mean that between roughly 20 and 50 % of the stand's lifespan is spent in a pre-canopy closure state, where litterfall is nil or minimal. It follows that the forest floor will most probably be greatly reduced in size (compared to longer rotations) because litterfall inputs are lower and mineralisation rates in the semi-open canopy are usually faster due to increased temperature and water availability. An international trial series in tropical climates simulating such an intensive utilization scenario has recently been completed where harvest residue plus forest floor material was removed in certain treatments. Results showed that removal of residue plus forest floor will almost certainly result in depressed growth of subsequent rotation(s) of trees on nutrient poor soils (Deleporte et al. 2008; Gonçalves et al. 2007; Mendham et al. 2008), but interestingly, also on sites that are nutrient rich by forestry standards (du Toit et al. 2008; Mendham et al. 2008). The reason for this seems to be not just impact of removing a certain percentage of the total nutrient pool in the system, but rather the removal of a substantial percentage of the *readily mineralizable nutrient pool* (du Toit and Dovey 2005; du Toit et al. 2008). This finding has serious implications for the long-term nutritional sustainability of biomass harvesting systems that collect all (or most of) the above-ground biomass, and will be discussed further in Chap. 10. The removal of predominantly woody material from the harvesting residue (i.e. leaving the fine twigs and foliage plus the forest floor in situ) has shown much smaller impacts and can potentially be managed sustainably with much smaller external nutrient inputs (Dovey 2012).

5.5 Intensive Cultural Management to Maximise Growth Resource Utilization

To understand the fundamental processes and mechanisms driving stand productivity, we need to introduce the so-called production ecology equation (after Landsberg and Waring 1997), which states that

$$\mathrm{NPP} = \mathrm{iPAR} * \alpha_c * \mathrm{R},$$

where

NPP = net primary production
iPAR = intercepted photosynthetically absorbed radiation
α_c = canopy quantum efficiency (mol of C sequestered per mol radiation absorbed)
R = Respiration (a fairly constant value in young tree crops).

Figure 5.2 shows the theoretical development of leaf area index over time in short rotation eucalypt pulpwood crops with relatively high stand densities (after data from du Toit and Dovey 2005; du Toit et al. 2008; White et al. 2009). The three scenario's in Fig. 5.2 are: (i) low-level silvicultural inputs, (ii) intensive silvicultural inputs that temporarily improve resource availability, and (iii) Intensive site or silvicultural treatments that ensured a prolonged improvement in resource availability. The latter two responses have also been labelled as Type A (Type II) and Type B (Type I) responses (Snowdon and Waring 1984; Snowdon 2002; Rubilar et al. 2008). The areas under the curves represent the cumulative leaf area that is deployed over the rotation, which is responsible for radiation interception and hence, photosynthesis.

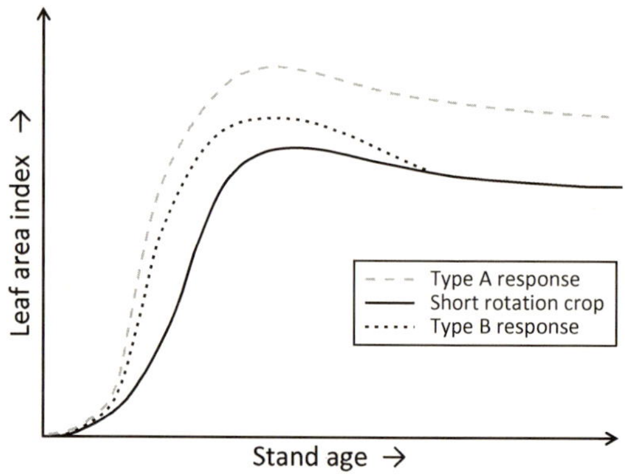

Fig. 5.2 Theoretical development of Leaf area index in short-rotation eucalypt crops (After du Toit and Dovey; du Toit et al. 2008; White et al. 2009). Scenario (a) is a control, scenario (b) experienced a temporary boost in resource availability and scenario (c) was subjected to a long-term improvement of growth resource availability

5.5.1 Type B Responses (a.k.a. Type I Responses)

The leaf area deployed governs the interception of solar radiation, which is pivotal in the production ecology equation presented above. The sooner that the leaf area index (LAI) can reach a peak value, the sooner optimal growth can take place because radiation interception is linearly related to biomass production (Linder 1985; Turnbull et al. 1988; Dovey 2005). In short rotations, such as bio-energy crops, the time period from planting until deployment of peak LAI may make up a substantial portion of the full rotation length, and it is thus of critical importance minimise this period. The rapid deployment of a peak LAI is aided by high planting densities, but as described in Sect. 5.3, expensive harvesting operations also place a limit on the stand density that can be used, because of harvesting piece size constraints.

Rapid deployment of peak LAI can be achieved by intensive silvicultural management which will boost the availability of soil water and nutrients to young transplants. Management of competing vegetation and fertilization at time of establishment are two critically important operations in this regard (Little and Van Staden 2003; Wagner et al. 2006; Little et al. 2007; du Toit et al. 2010). Fertilization should be site- and crop specific to ensure best economic returns.

Hardwood stands usually have a very high demand for nutrients in the period up to and including canopy closure, e.g. Laclau et al. (2003). Research on fertilization of short-rotation *Acacia* and *Eucalyptus* tree crops in warm climates initially focussed on relatively small applications (of mainly N and P) that would boost stand productivity (Williams 1928; Beard 1952; Schönau 1983, 1984; Herbert and Schönau 1989). Fertilization at (or soon after) establishment is commonly done by commercial tree growers because of the relatively low input costs and large gains on highly weathered, P deficient sites, or alternatively, gains due to Type B responses eluded to earlier in this chapter (Barros et al. 1992, 2004; Herbert 1996; du Toit 2002; Gonçalves et al. 2008; Bennett et al. 1997; du Toit et al. 2010; Maree et al. 2012). Application rates for this type of fertilizer application usually include P at

10–40 g per tree (Gonçalves et al. 2004; du Toit et al. 2010). In specific cases, responses to additional N (0–30 g N per tree) K (0–15 g per tree) and small quantities of B has been observed (Gonçalves et al. 2004; du Toit et al. 2010). This need for N applications depends on the soil conditions (Noble and Herbert 1991) and site preparation/slash management options (du Toit and Dovey 2005; Smith and du Toit 2005; du Toit et al. 2008). Gonçalves et al. (2004, 2008), as well as Gava (1997) make the point that N and K applications are becoming more common in eucalypt plantations that have undergone several crop cycles, apparently due to increase nutrient losses in harvesting the possible depletion of readily mineralisable N.

In pine plantations, P fertilization during the establishment phase (commonly at levels between 20 and 60 kg/ha) may lead to large growth responses, but this is mainly limited to highly weathered, P deficient soils (Donald 1987; Xu et al. 1995a, b; Fox et al. 2007a; Kotze and du Toit 2012). Furthermore, pines in subtropical and warm temperate climates have generally shown the biggest responses to nutrient additions during mid or late rotation periods (12–20 years of age), when nutrient demand is much larger than supply (Donald 1987; Payn et al. 1988; Turner et al. 1996; Carlson 2000; Fox et al. 2007a; Kotze and du Toit 2012). Levels of 200–400 kg of N and 50–100 kg of P commonly give good results in mid-rotation pines. Fertilization after 12 years of age may be too late in very short rotations grown for biomass, especially if they are planted at higher stand densities. However, it appears that younger pine stands may have sufficient capacity to take up moderate nutrient applications, judging from the responses to P, K and Mg applications have been documented at time of canopy closure (age of first pruning in most stands under conventional management regimes) where acute deficiencies existed (Kotze and du Toit 2012). This finding may be of importance to short rotation pine stands that are under pressure from nutrient depletion: Economic responses can be obtained by such early fertilization efforts.

The application of hydrogels can also improve water availability during a critical period following planting (Viero and Little 2006). In addition to minimising competition, vegetation management also ensures a more homogenous crop, and this will result in the greater partitioning of NPP to above-ground biomass production, which is an added benefit.

5.5.2 *Type A Responses (a.k.a. Type II Responses)*

Silvicultural treatments that ensure the prolonged improvement of growth resources to tree stands usually have the greatest improvement on stand productivity, and these have been dubbed Type A effects. This happens because not only the LAI development process is accelerated, but also because of changes to the canopy quantum efficiency and carbohydrate partitioning to above-ground parts (Stape 2002; Giardina et al. 2003; du Toit 2008). Examples are: Fertilization of a nutrient deficient stand where water availability is not limiting (Giardina et al. 2003); Irrigation of a stand where water is limiting growth (Stape 2002); P fertilization that is efficiently re-cycled and remains in the system for subsequent rotations (Snowdon 2002; Crous et al. 2007, 2008) or site preparation options improving the rooting volume accessible to trees (Zwolinski et al. 2002). The mechanism for responsible for the (usually large) Type A responses appears to revolve around an increase in canopy quantum efficiency (often accompanied by improvements in partitioning of carbohydrates to above-ground tissues), rather than a primary reliance on an accelerated LAI development, as is common in Type B responses (du Toit 2008). Where it is thus possible to implement operations that would elicit a Type A response (e.g. operations that can fundamentally change resource availability) consideration should be given to the fact that it may greatly improve stand productivity and resource use efficiency, and that these improvements are likely to recur over several future rotations. This may offset the costs of fertilizer or even the (often higher) costs of operations such as trenching, land surface modifications or subsoiling. However, as described in Sect. 5.4, conditions where intensive land preparation options are effective are limited to specific site and soil types.

5.6 Interactions Among Intensive Silvicultural Operations

The interactive effects of tree improvement, site-taxon matching, stocking, and cultural practices has been tested across five site types of different SI's on the eastern seaboard of South Africa (Boreham and Pallett 2007; du Toit et al. 2010). As a departure point, a control treatment was selected that represented many short-rotation pulpwood production stands in warm climates: A species that was well adapted to each specific site (but not the very best match possible, i.e. the second best species on that particular site), which had not undergone genetic improvement, planted at a stocking of 1,111 stems/ha with no fertilization at planting and an intermediate level of weed control. The average productivity in this experimental series could be increased by 46 % above the control treatment at 5 years of age, by using genetically improved planting stock, the best site-taxon matching, a stocking of 1,667 stems/ha and intensive weed control plus fertilization at establishment. The most important finding of this study was that, although there were no significant interactions between major factors across all sites, the response to individual factors (genetics, site-species matching, stand density, fertilization and weed control) were additive. It follows *each element* of the intensive silviculture system contributes substantially to the overall gain in productivity, and that all elements should be implemented at the higher (or more intensive) level to realise the gain that had been achieved. Bio-energy production systems should be planned to integrate all the above elements of intensive silviculture into the production system.

5.7 Intensified Silviculture, Fertilization and the Carbon Footprint

One of the main reasons for growing bio-energy plantations is a reduction of the carbon emissions whilst obtaining energy benefits. Most of the intensive silvicultural treatments needed to ensure high levels of productivity in plantations (e.g. site preparation, weeding and fertilization – see Sect. 5.5) require some energy inputs. For example, fertilizer application uses fossil fuel based energy (and is responsible for some carbon dioxide emissions) during manufacture and/or mining, transport and application. In this section we will contrast the carbon costs of fertilization and other silvicultural treatments with the potential carbon gains from increased growth in short-rotation plantations.

Energy and carbon budgets for specific scenario's in bio-energy crop systems are usually calculated through a life cycle analysis (LCA) approach, which takes account of energy inputs and outputs, as well as carbon gains and losses associated with every step of the production process. The major steps to construct carbon and energy budgets have been proposed by Schlamadinger et al. (1997) and subsequently implemented by *inter alia* Matthews (2001) Heller et al. (2003) and include:

- Definition of the system boundaries to do all calculations. A reference system is normally chosen, against which different alternative management scenario's are contrasted.
- Estimating the total energy benefits from a specific scenario (including waste products).
- Estimation of total energy inputs in the production system (including energy investments in infrastructure and energy losses along the fuel chain).
- Estimating the carbon sequestration that takes place.
- Estimating of total carbon emissions from each specific scenario.
- Estimating the net emission of other greenhouse gasses, e.g. N_2O, which can be presented as CO_2 equivalents for the purposes of the budget.

Most authors stress the importance of detailed carbon and energy budgets for every step in the LCA because changes in management regime of biomass production systems can lead to large variations in the carbon and energy budget. Mead and Pimentel (2006) make a good case to show that individual silvicultural operations should be evaluated to decide on the optimum energy production system, as their efficiencies may differ widely. In order to obtain meaningful results, these calculations will have to be done on a site-specific basis because recommendations on the type and intensity of soil preparation (Smith et al. 2000, 2001; Zwolinski et al. 2002), fertilization (du Toit et al. 2010; Kotze and du Toit 2012) and weed control (Little and Rolando 2008) differ strongly across site types.

The figures most commonly used to evaluate the suitability of bio-energy systems are:

- The energy ratio (energy produced per unit of energy input)
- Net energy yield (energy output minus energy input per hectare)
- The carbon emission coefficient (inclusive of the greenhouse gas emissions expressed as CO_2 equivalents)

Despite the strong dependence of the carbon and energy balance on individual cultural practices when applied to specific site types, a few useful generalisations can be made: Energy ratio's of forestry crops (in temperate climates) typically vary between 10 and 25, compared with annual crops that vary between 1 and 5 (Mead and Pimentel 2006). Energy ratio's as high as 55 have been reported for temperate climate forest crop systems under intensive management and fertilization (Heller et al. 2003). Energy ratio's of 42 have been calculated for warm climate eucalypt crops in semi-arid areas (Wu et al. 2007) and ratio's in excess of 60 have been estimated for a hypothetical *Pinus taeda* and *Eucalyptus grandis* systems that included site preparation and fertilization inputs (Mead and Pimentel 2006). The three warm-climate case studies have all shown considerable room for improvement of the energy ratio if specific vegetation management, harvesting and/or transport regimes are adopted. New research should focus on the ability of warm climate tree crops across different site qualities to produce energy efficient biomass through appropriate silvicultural management strategies.

5.8 Harvesting Larger Percentages of Biomass from the Stand Than Conventional Practices

Biomass procurement per unit of land area can obviously be increase if whole tree harvesting is practised, rather than stem wood harvesting only. However, this relatively small increase in harvestable biomass per hectare comes with a relatively large export of nutrients. Several authors have constructed equations whereby bark, branch, and leaf mass and nutrient contents can be estimated from stem mass or volume, for example Dovey (2009). These tools are useful when calculating the additional nutrient export when the harvesting of additional biomass is considered. Increasingly intensive silviculture and frequent harvesting of ultra-short rotations may result in a net loss of nutrients in many plantations. Recent work in intensively managed short-rotation plantations has shown that fairly large nutrient losses of several elements may occur, which are seldom compensated for by increased inputs (through atmospheric deposition or fertilization). Nutrient losses and gains, as well as the long-term nutritional sustainability of short rotation forestry systems are discussed in detail with some case study data sets in Chap. 10.

References

Ackerman PA, Pulkki RE (2004) "Shorthaul" pulpwood transport in South Africa. A network analysis case study. South Afr For J 201:43–51

Barros NF, Novais RF, Neves JCL, Leal PGL (1992) Fertilizing eucalypt plantations on the Brazilian savannah soils. South Afr For J 160:7–12

Barros NF, Neves JCL, Novais RF (2004) Mineral fertilizer recommendations for eucalypt plantations. In: Gonçalves JLM, Benedetti V (eds) Forest nutrition and fertilization. Institute of Forest Research and Study (IPEF), Piracicaba, pp 269–284

Beard JS (1952) The use of fertiliser in wattle culture. J S Afr For Assoc 22:7–33

Bennett LT, Weston CJ, Attiwill PM (1997) Biomass, nutrient content and growth response to fertilisers of six-year-old *Eucalyptus globulus* plantations at three contrasting sites in Gippsland, Victoria. Aust J Bot 45:103–121

Boreham GR, Pallett RN (2007) The influence of tree improvement and cultural practices associated with mid-rotation *Eucalyptus* productivity in the temperate region of South Africa. In: Eucalypts and diversity: balancing productivity and sustainability. Proceedings of the IUFRO working group 20803 conference, Durban, 22–26 Oct 2007 (CD-ROM)

Carlson CA (2000) Impact of fertilization at first thinning on *Pinus patula* basal area increment in Mpumalanga. South Afr For J 189:35–45

Carlson CA, Allan R, Morris AR (2004) Effects of temperature on *Pinus patula* seedlings growing in pots in a controlled environment. South Afr For J 200:27–38

Carlson CA, Fox TR, Colbert SR, Kelting DL, Allen HL, Albaugh TJ (2006) Growth and survival of *Pinus taeda* in response to surface and subsurface tillage in the Southeastern United States. For Ecol Manage 234(1–3):209–217

Chen ZJ (2010) Molecular mechanisms of polyploidy and hybrid vigour. Trends Plant Sci 15(2):57–71

Coetzee J (1999) Growth prediction for short rotation *E. grandis* crops, ICFR bulletin 9/99. Institute for Commercial Forestry Research, Pietermaritzburg

Coetzee J, Naicker S (1998) The final results for age two to eleven years in the case of Kia-ora spacing trial, ICFR bulletin 3/98. Institute for Commercial Forestry Research, Pietermaritzburg

Coetzee J, Chiswell K, Storey P, Arbuthnot AL (1996) The final results of the *E. grandis* spacing trial, Kwambonambi, for age two to ten years, ICFR bulletin 10/96. Institute for Commercial Forestry Research, Pietermaritzburg

Crous JW, Morris AR, Scholes MC (2007) The significance of residual phosphorus and potassium fertiliser in countering yield decline in a fourth rotation of *Pinus patula* in Swaziland. South Hemisphere For J 69(1):1–8

Crous JW, Morris AR, Scholes MC (2008) Growth and foliar nutrient response to recent applications of phosphorus (P) and potassium (K) and to residual P and K fertiliser applied to the previous rotation of *Pinus patula* at Usutu, Swaziland. For Ecol Manage 256(4):712–721

Deleporte P, Laclau JP, Nzila JD, Kazotti JG, Marien JN, Bouillet JP, Szwarc M, D'Annunzio R, Ranger J (2008) Effects of slash and litter management practices on soil chemical properties and growth in second rotation eucalypts in the Congo In: Nambiar EKS (ed) Site management and productivity in tropical plantation forests: proceedings of workshops in Piracicaba (Brazil), 22–26 Nov 2004 and Bogor (Indonesia) 6–9 Nov 2006, CIFOR, Bogor, pp 5–22

Donald DGM (1987) The application of fertilizer to pines following second thinning. S Afr For J 142:13–16

Dovey SB (2005) Above-ground allomerty, biomass and nutrient content of *Acacia mearnsii* across four ages and three sites in the KwaZulu-Natal. Midlands M.Sc. thesis, University of Kwa-ZuluNatal, Pietermaritzburg, 106pp

Dovey SB (2009) Estimating biomass and macronutrient content of some commercially important plantation species in South Africa. South For 71(3):245–251

Dovey SB (2012) Effects of clear felling and residue management on nutrient pools, productivity and sustainability in a clonal eucalypt stand in South Africa. Ph.D. thesis, University of Stellenbosch, Stellenbosch

Durán A, Gryzenhout M, Slippers B, Ahumada R, Rotella A, Flores F, Wingfield BD, Wingfield MJ (2008) *Phytophthora pinifolia* sp. nov associated with a serious needle disease of *Pinus radiata* in Chile. Plant Pathol 57(4):715–727

du Toit B (2002) Wattle nutrition. In: Dunlop RW, MacLennan L (eds) Black wattle: the South African research experience. Institute for Commercial Forestry Research, Pietermaritzburg, pp 47–64

du Toit B (2008) Effects of site management on growth, biomass partitioning and light use efficiency in a young stand of *Eucalyptus grandis* in South Africa. For Ecol Manage 255(7):2324–2336

du Toit B (2012) Chapter 23: Matching site, species and silvicultural regime to optimise the productivity of commercial softwood species in southern Africa. In: Bredenkamp BV, Upfold S (eds) South African forestry handbook, 5th edn. Southern African Institute of Forestry, Pretoria, pp 53–59

du Toit B, Dovey SB (2005) Effect of site management on leaf area, biomass development and stand growth efficiency of a young stand of *Eucalyptus grandis* in South Africa. Can J For Res 35(4):891–900

du Toit B, Arbuthnot A, Oscroft D, Job RA (2001) The effects of remedial fertilizer treatments on growth and pulp properties of *Eucalyptus grandis* stands established on infertile soils of the Zululand Coastal Plain. South Afr For J 192:9–18

du Toit B, Dovey SB, Smith CW (2008) Effects of slash and site management treatments on soil properties, nutrition and growth of a *Eucalyptus grandis* plantation in South Africa. In: Nambiar EKS (ed) Site management and productivity in tropical plantation forests: proceedings of workshops in Piracicaba (Brazil), 22–26 Nov 2004 and Bogor (Indonesia), 2–9 Nov 2006, Center for International Forestry Research (CIFOR), Bogor, pp 63–78

du Toit B, Smith CW, Little KM, Boreham G, Pallett RN (2010) Intensive, site-specific silviculture: manipulating resource availability at establishment for improved stand productivity. For Ecol Manage 259:1836–1845

Dvorak WS, Hodge GR, Payn KG (2008) The conservation and breeding of *Eucalyptus urophylla*: a case study to better protect important populations and improve productivity. South For 70(2):77–85

Eldridge K, Davidson J, Harwood C, Van Wyk G (1994) Eucalypt domestication and breeding. Oxford University Press, Oxford, 288pp

Fox TR, Allen HL, Albaugh TJ, Rubilar R, Carlson CA (2007a) Tree nutrition and forest fertilization of pine plantations in the Southern United States. South J Appl For 31(1):5–11

Fox TR, Jokela EJ, Allen HL (2007b) The development of pine plantation silviculture in the Southern United States. J For 105(7):337–347

Gardner RAW, Swain TL (1996) Snow damage to timber plantations in KwaZulu-Natal during July 1996 and subsequent species recommendations, ICFR bulletin 8/96. Institute for Commercial Forestry Research, Pietermaritzburg, South Africa, 12p

Gava JL (1997) Efeito da adubação potassica em plantios de *E. grandis* conduzidos em segunda rotação em solos comdiferentes teores de potasssio trocavel. Serie Tecnica, Piracicaba, IPEF 11, pp 89–94

Giardina CP, Ryan MG, Binkley D (2003) Primary production and carbon allocation in relation to nutrient supply in a tropical experimental forest. Glob Change Biol 9:1438–1450

Gonçalves JLM, Stape JL, Benedetti V, Fessel VAG, Gava JL (2004) An evaluation of minimum and intensive soil preparation regarding fertility and tree nutrition. In: Gonçalves JLM, Benedetti V (eds) Forest nutrition and fertilization. Institute of Forest Research and Study (IPEF), Piracicaba, pp 13–64

Gonçalves JLM, Wichert MCP, Gava JL, Masetto AV, Arthur JC Jr, Serrano MIP, Mello SLM (2007) Soil fertility and growth of *Eucalyptus grandis* in Brazil under different residue management practices. South Hemisphere For J 69(2):95–102

Gonçalves JLM, Stape JL, Laclau J, Bouillet J, Ranger J (2008) Assessing the effects of early silvicultural management on long-term site productivity of fast-growing eucalypt plantations: the Brazilian experience. South For 70(2):105–118

Heller MC, Keoleian GA, Volk TA (2003) Life cycle assessment of a willow bioenergy cropping system. Biomass Bioenergy 25:147–165

Herbert MA (1996) Fertilisers and eucalypt plantations in South Africa. In: Attiwill PM, Adams MA (eds) Nutrition of eucalypts. CSIRO Publishing, Collingwood, pp 303–325

Herbert MA (2012) Chapter 24: Site requirements of commercial *Eucalyptus* and *Corymbia* species in Southern Africa. In: Bredenkamp BV, Upfold S (eds) South African forestry handbook, 5th edn. Southern African Institute of Forestry, Pretoria, pp 51–58

Herbert MA, Schönau APG (1989) Fertilizing commercial forest species in Southern Africa: research progress and problems (part I). S Afr For J 151:58–70

Hodge GR, Dvorak WS (2007) Variation in pitch canker resistance among provenances of *Pinus patula* and *Pinus tecunumanii* from Mexico and Central America. New For 33(2):193–206

Hurley BP, Slippers J, Wingfield MJ, Dyer C, Slippers B (2012) Perception and knowledge of the *Sirex* woodwasp and other forest pests in South Africa. Agric For Entomol 14(3):306–316

Kanzler A, Hagedorn SF, Hodge GR, Dvorak WS (2003) Genotype by environment interaction for volume growth at 6 years of age in a series of five *Pinus patula* progeny trials in Southern Africa. South Afr For J 198:3–15

Kotze H, du Toit B (2012) Chapter 211: Silviculture of industrial pine plantations in Southern Africa. In: Bredenkamp BV, Upfold S (eds) South African forestry handbook, 5th edn. Southern African Institute of Forestry, Pretoria, pp 133–149

Kyle KH, Andrews LJ, Fox TR, Aust WM, Burger JA, Hansen GH (2005) Long-term effects of drainage, bedding, and fertilization on growth of loblolly pine (*Pinus taeda* L.) in the coastal plain of Virginia. South J Appl For 29(4):205–214

Lacey ST, Brennan PD, Parekh J (2001) Deep may not be meaningful: cost and effectiveness of various ripping tine configurations in a plantation cultivation trial in eastern Australia. New For 21(3):231–248

Laclau J-P, Deleporte P, Range J, Bouillet J-P, Kazotti G (2003) Nutrient dynamics throughout the rotation of *Eucalyptus* clonal stands in Congo. Ann Bot 91:879–892

Landsberg JJ, Waring RH (1997) A generalised model of forest productivity using simplified concepts of radiation-use efficiency, carbon balance and partitioning. For Ecol Manage 95:209–228

Lincoln MC, Will RE, Morris LA, Carter EA, Markewitz D, Britt JR, Cazell B, Ford V (2007) Soil change and loblolly pine (*Pinus taeda*) seedling growth following site preparation tillage in the upper coastal plain of the Southeastern United States. For Ecol Manage 242(2–3):558–568

Linder S (1985) Potential and actual production in Australian forest stands. In: Landsberg JJ, Parsons W (eds) Research for forest management. CSIRO, Melbourne, pp 11–35

Little KM, Rolando CA (2008) Regional vegetation management standards for commercial *Eucalyptus* plantations in South Africa. South For 70(2):87–97

Little KM, van Staden J (2003) Interspecific competition affects early growth of a *Eucalyptus grandis* × *E. camaldulensis* hybrid clone in Zululand, South Africa. S Afr J Bot 69(4):505–513

Little KM, Smith CW, Norris CH (2001) The influence of various methods of plantation residue management on replanted *Acacia mearnsii* growth. Aust For 63(3):226–234

Little KM, van Staden J, Clarke JPY (2003) *Eucalyptus grandis* × *E. camaldulensis* variability and intra-genotypic competition as a function of different vegetation management treatments. New For 25:227–242

Little KM, Rolando CA, Morris CD (2007) An integrated analysis of 33 *Eucalyptus* trials linking the onset of competition-induced tree growth suppression with management, physiographic and climatic factors. Ann For Sci 64:585–591

Lopez GA, Potts BM, Tilyard PA (2000) F1 hybrid inviability in *Eucalyptus*: the case of *E. ovata* × *E. globulus*. Heredity 85(3):242–250

Louw JH, Scholes M (2002) Forest site classification and evaluation: a South African perspective. For Ecol Manage 171(1–2):153–168

Louw, Smith (2012) Chapter 22: Forest site classification. In: Bredenkamp BV, Upfold S (eds) South African forestry handbook, 5th edn. Southern African Institute of Forestry, Pretoria, pp 27–41

Louw JH, Germishuizen I, Smith CW (2011) A stratification of the South African forestry landscape based on climatic parameters. South For 73(1):51–62

Malan FS, Verryn SD (1996) Effect of genotype-by-environment interaction on the wood properties and qualities of four-year-old *Eucalyptus grandis* and *E grandis* hybrids. S Afr For J 176:47–53

Maree B, Kotze H, du Toit B (2012) Chapter 212: Silviculture of industrial *Eucalyptus* and *Acacia* plantations in South Africa. In: Bredenkamp BV, Upfold S (eds) South African forestry handbook, 5th edn. Southern African Institute of Forestry, Pretoria, pp 151–165

Matthews RW (2001) Modelling of energy and carbon budgets of wood fuel coppice systems. Biomass Bioenergy 21:1–19

Mead DJ, Pimentel D (2006) Use of energy analyses in silvicultural decision-making. Biomass Bioenergy 30:357–362

Mendham DS, Grove TS, O'Connell AM, Rance SJ (2008) Impacts of inter-rotation site management on soil nutrients and plantation productivity in *Eucalyptus globulus* plantations in South-western Australia. In Nambiar EKS (ed) Site management and productivity in tropical plantation forests: proceedings of workshops in Piracicaba (Brazil), 22–26 Nov 2004 and Bogor (Indonesia), 6–9 Nov 2006, CIFOR, Bogor, pp 79–92

Mitchell RG, Wingfield MJ, Hodge GR, Dvorak WS, Coutinho TA (2013) Susceptibility of provenances and families of *Pinus maximinoi* and *Pinus tecunumanii* to frost in South Africa. New For 44:135–146

Nambiar EKS (ed) (2008) Site management and productivity in tropical plantation forests. Center for International Forestry Research, Bogor, 236pp

Nambiar EKS, Sands R (1990) Effects of compaction and simulated root channels in the subsoil on root development, water uptake and growth of radiata pine. Tree Physiol 10:297–306

Noble AD, Herbert MA (1991) Influence of soil organic matter content on the responsiveness of Eucalyptus grandis to nitrogen fertiliser. S Afr For J 156:23–27

Owens SJ, Miller R (2009) Cross- and self-fertilization of plants – Darwin's experiments and what we know now. Bot J Linn Soc 161(4):357–395

Pallett RN (2005) Precision forestry for pulpwood re-establishment silviculture. South Afr For J 203:33–40

Pallett RN, Sale G (2002) Operational gain: measuring the capture of genetic gain. South Afr For J 195:33–38

Payn TW, de Ronde C, Grey DC (1988) Phosphate fertilization of mature *Pinus radiata* stands. S Afr For J 147:26–31

Rubilar R, Blevins L, Toro J, Vita A, Muñoz F (2008) Early response of *Pinus radiata* plantations to weed control and fertilization on metamorphic soils of the Coastal Range, Maule Region, Chile. Bosque 29(1):74–84

Schlamadinger B, Apps M, Bohlin F, Gustavsson L, Jungmeier G, Marland G, Pingoud K, Savolaine I (1997) Towards a standard methodology for greenhouse gas balances of bioenergy systems in comparison with fossil energy systems. Biomass Bioenergy 13(6):359–375

Schönau APG (1983) Fertilisation in South African forestry. S Afr For J 125:1–19

Schönau APG (1984) Silvicultural considerations for high productivity of *Eucalyptus grandis*. For Ecol Manage 9:295–314

Smith CW, du Toit B (2005) The effect of harvesting operations, slash management and fertilisation on the growth of a *Eucalyptus* clonal hybrid on a sandy soil. South Afr For J 203:15–26

Smith CW, Norris CH, Esprey LJ (2000) The effect of land preparation at establishment on the productivity of fast growing hardwoods in the summer rainfall regions of South Africa, ICFR bulletin series 06/2000. Institute for Commercial Forestry Research, Pietermaritzburg

Smith CW, Little KM, Norris CH (2001) The effect of land preparation at re-establishment on the productivity of fast growing hard woods in South Africa. Aust For 64:165–174

Smith CW, Pallett RN, Kunz R, Gardner RAW, du Plessis M (2005) A strategic forestry site classification for the summer rainfall region of Southern Africa based on climate, geology and soils, Institute for commercial forestry research bulletin series 03/2005. Institute for Commercial Forestry Research, Pietermaritzburg

Snowdon P (2002) Modeling Type 1 and Type 2 growth responses in plantations after application of fertiliser or other silvicultural treatments. For Ecol Manage 163:229–244

Snowdon P, Waring HD (1984) Long-term nature of growth responses obtained to fertiliser and weed control applied at planting and their consequences of forest management. In: Grey DC, Schonau APG, Schutz CJ, van Laar A (eds) IUFRO symposium on site and site productivity of fast growing plantations, vol 2. South African Forest Research Institute (SAFRI) and Department of Environmental Affairs, Pretoria, pp 701–711

Sochacki SJ, Harper RJ, Smettem KRJ (2007) Estimation of woody biomass production from a short-rotation bio-energy system in semi-arid Australia. Biomass Bioenergy 31:608–616

Stape JL (2002) Production ecology of clonal Eucalyptus plantations in northeastern Brazil. Ph.D. dissertation, Colorado State University, Ft Collins

Stape JL, Binkley D, Ryan MG (2008) Production and carbon allocation in a clonal *Eucalyptus* plantation with water and nutrient manipulations. For Ecol Manage 255(3-4):920–930

Stape JL, Binkley D, Ryan MG, Fonseca S, Loos RA, Takahashi EN, Azevedo MR (2010) The Brazil eucalyptus potential productivity project: influence of water, nutrients and stand uniformity on wood production. For Ecol Manage 259(9):1684–1694

Swart WJ, Wingfield MJ, Knox-Davies PS (1988) Relative susceptibility to *Sphaeropsis sapinea* of six *Pinus* spp. cultivated in South Africa. Eur J For Pathol 18(3–4):184–189

Tribe GD, Cillié JJ (2004) The spread of *Sirex noctilio* fabricius (Hymenoptera: Siricidae) in South African pine plantations and the introduction and establishment of its biological control agents. Afr Entomol 12(1):9–17

Turnbull CRA, Beadle CL, Bird T, McLeod DE (1988) Volume production in intensively managed eucalypt plantations. Appita J 41:447–450

Turner J, Knott JH, Lambert M (1996) Fertilization of *Pinus radiata* after thinning I Productivity gains. Aust For 59(1):7–21

Verryn SD (2002) Harvesting genetics for productive plantations. South Afr For J 195:83–87

Verryn SD (2008) Breeding for wood quality – a perspective for the future. N Z J For Sci 38(1):5–13

Viero PWM, Little KM (2006) A comparison of different planting methods including hydrogels, and their effect on eucalypt survival and initial growth in South Africa. South Hemisphere For J 208:5–13

Wagner RG, Little KM, Richardson B, McNabb K (2006) The role of vegetation management for enhancing productivity of the world's forests. Forestry 79:57–79

White DA, Crombie DS, Kinal J, Battaglia M, McGrath JF, Mendham DS, Walker SN (2009) Managing productivity and drought risk in eucalyptus globulus plantations in south-western Australia. For Ecol Manage 259(1):33–44

Williams CO (1928) Manuring of wattles. Farming S Africa 3:1013–1019

Wingfield MJ, Slippers B, Hurley BP, Coutinho TA, Wingfield BD, Roux J (2008) Eucalypt pests and diseases: growing threats to plantation productivity. South For 70(2):139–144

Wu HX, Eldridge KG, Colin Matheson A, Powell MB, McRae TA, Butcher TB, Johnson IG (2007) Achievements in forest tree improvement in Australia and New Zealand 8. Successful introduction and breeding of radiata pine in Australia. Aust For 70(4):215–225

Xu ZH, Simpson JA, Osborne DO (1995a) Mineral nutrition of slash pine in subtropical Australia. I. Stand growth response to fertilization. Fertil Res 41(2):93–100

Xu ZH, Simpson JA, Osborne DO (1995b) Mineral nutrition of slash pine in subtropical Australia. III. Relationships between foliar P concentration and stand growth. Fertil Res 41(2):109–115

Zobel B, Talbert J (1984) Applied forest tree improvement. Wiley, New York

Zwolinski J, Johnston M, Kotze H (2002) Soil treatments for improved timber production in *Pinus radiata* plantations in South Africa. For Ecol Manage 171(1–2):199–207

Chapter 6
Biomass Harvesting and Logistics

Pierre Ackerman, Bruce Talbot, and Bo Dahlin

6.1 Introduction

As with conventional timber harvesting and transport, the selection of machine systems for biomass production is often based on local availability, traditional harvesting methods and systems and the innovative spirit of entrepreneurs. However, piecing together an optimal biomass harvesting and transport systems to fulfil sustainable biomass supply requires substantial knowledge and insight into and of the whole biomass supply chain. When considering the number of potential options available at any decision point in the chain, it becomes apparent that biomass supply chains are in fact unidirectional supply networks and not single or unique chains. The best employment of production factors represents the minimum cost flow through the network, from standing tree to boiler grate. Knowledge of the options available and the consequence of employing each of these is therefore important in plotting the best way forward through the network.

Biomass production networks are characterized by a number of state and form combinations. The required state or form of the biomass, e.g., Full-tree (FT – felled trees with branches and top intact), Tree-length (TL – trees felled, debranched and top removed), Cut-to-length (CTL – log assortments), and comminuted material,

P. Ackerman (✉)
Department of Forest and Wood Science, Faculty of AgriSciences, Stellenbosch University, Stellenbosch, South Africa
e-mail: packer@sun.ac.za

B. Talbot
Forest Technology and Economics, Norwegian Forest and Landscape Institute, Ås, Norway
e-mail: bta@skogoglandskap.no

B. Dahlin
Department of Forest Sciences, Faculty of Agriculture and Forestry, University of Helsinki, Helsinki, Finland
e-mail: bo.dahlin@helsinki.fi

T. Seifert (ed.), *Bioenergy from Wood: Sustainable Production in the Tropics*, Managing Forest Ecosystems 26, DOI 10.1007/978-94-007-7448-3_6,

at each stage in the network (e.g., at stump, roadside landing, terminal, plant) determines, or is determined by, the production methods. Some of these can be directly linked in function and time, while others can be totally detached. A full year may pass between extraction and processing of stumps while in some cases hardwood trees can be felled, chipped and combusted on the same day.

Almost all final consumption plants, whether for combustion or as a raw material in further processing to; e.g., briquettes or pellets, requires biomass in a chipped or crushed form. This process of conversion is called comminution. A challenge for the operations manager is determining at what stage in the network comminution should happen. Every alternative has consequences for the choice of harvesting, extracting, processing and transport equipment. In the following overview, examples of a supply network in which comminution takes place at each cardinal point; in-field, at roadside, at a terminal, and at a conversion plant are provided. This chapter provides the reader with broad insight into making these comminution decisions through discussion of the positive and negative aspects at each of these cardinal points.

The proliferation of publications, trade fairs, seminars and internet sites (e.g., www.forestenergy.org) providing information on biomass production equipment and machinery, and the rapid technical developments that are being undertaken limit the relevance of a detailed technical description. In the following section, the working principles and intentions behind the main categories of equipment and machinery are presented and the reader is urged to keep abreast of developments through other media.

6.2 Biomass Felling and Extraction Harvesting Equipment

Felling is a prerequisite of any wood based biomass supply, whether it takes place as an integrated part of traditional roundwood harvesting or as a specific biomass harvest. In this section, distinction is made on the most important types of felling technology. Felling methods range from motor-manual (chainsaws and brush-cutters) to fully mechanized systems, each with rational areas of application. Although mechanised harvesting systems have been practiced for at least three decades, motor-manual methods have been a traditional and historic part of biomass and timber procurement since the early 1950s.

The use of chainsaws and brush-cutters are at times (depending on technological level of the organisation involved with bioenergy production, terrain, extraction system and biomass type and dimension) the preferred means of changing the state of the standing biomass. The felling of spiny or thorny biomass places limitations on the use of motor-manual felling systems, as it is difficult to approach the biomass. Multi-stemmed biomass also negatively impacts motor-manual productivity, particularly when individual stems are of small diameter. Typically this situation requires the use of mechanised multi-stem felling systems to overcome piece size challenges.

In developing countries the use of chainsaws in conventional FT, TL and CTL operations from which biomass residues are retrieved remains an integral part of the harvesting supply chain; i.e., felling, debranching, cross-cutting and topping, where applicable. However, productivity, worker safety and biomass product quality are marginal and modern mechanised systems for felling, debranching and cross-cutting are becoming more the rule than the exception (Pulkki 1992, 2000).

Mechanised felling (and extraction) systems are based on agricultural tractor units, excavators, or purpose-built forest machines, such as feller-bunchers, harvesters or forwarders. What distinguishes these machines from each other is their stability in the terrain, operator safety (ROPS, FOPS and OPS) and working capacity at the boom tip. But caution should be exercised when applying agricultural tractor/trailer systems in forwarding of biomass due to these units operating outside their intended design specifications.

In mechanised systems the felling, handling and eventual processing of the biomass is done using one of a number of specialized harvesting and/or processing heads categorized below. It is generally not necessary to use a sophisticated head for biomass harvesting as specifications on dimensions or quality are low or non-existent. However, some type of multi-stem capability is preferable. Mechanised felling equipment can be categorised as follows:

- Felling head: a felling head grasps a tree and fells it using one of a number of cutting technologies (i.e., chain saw, disc saw, shear, auger) (Fig. 6.1). The felling head is lighter and cheaper than other heads but cannot process a tree, i.e., usually is not fitted with feed rollers or debranching knives.
- Harvesting head: this head has the ability to grasp a tree, fell it, lower it in a controlled fashion/direction, and process it (i.e., debranch, measure the length and cross-cut).
- Accumulating head: this can be either a harvesting head or a felling head that has been fitted with accumulating arms which can hold multiple stems on a plate on which to rest the butt ends.
- Processing head: this head is typically fitted to an excavator or loader boom and is used for processing (i.e., debranching, cross-cutting, and in some cases debarking) FT that have been felled and gathered (e.g., at a landing). These heads do not have the capability to fell trees.

The size and type of the base machine, the crane type, forest conditions and operator skill all influence productivity. A larger base machine would be more stable and powerful when working a bigger head, or accumulating more trees, at a greater distance from the striproad (boom reach). However, smaller scale systems (e.g., agricultural tractor with crane and lightweight felling head) do and will continue to fill an important role in biomass procurement (Russell and Mortimer 2005).

Fig. 6.1 Felling head using a disc saw with a high tolerance of dirt, stones or carbonized bark (Photo: Talbot)

6.3 Collection and Extraction Equipment and Machinery

The extraction of loose material that is not chipped and which weighs less per load than solid wood is less damaging to the site than infield chipping and extraction of heavier loads on the same site (Stupak et al. 2008). This is due to the lower mass and bulk density of the material and reduced impact of the total mass on the soil surface. It also results in inefficiency; however, as less tonnage is extracted due to the loads being volume-restricted. Residual biomass harvesting system selection decisions therefore need to match biomass type with specific machines to result in optimally productive and cost efficient harvesting systems (Stupak et al. 2008). Conventional harvesting systems, which are presently in use form the basis of residual biomass harvesting systems selection, and are dependent on the location of comminution:

- Terrain chipping
- Chipping at roadside or landing
- Terminal chipping
- Chipping at processing plant

However, conventional forwarders or agricultural tractor/trailer units fitted with forestry trailers (i.e., with crane and bogie axle), encompassing CTL, TL and FT systems remain the preferred means of extracting small trees, bundles, tree parts or harvesting residues. This is due to their design and availability to the sector.

These come in various adaptations, all of which aim to maximize the payload of a bulky, low mass product and selection remains specific to particular situations. A FT system incorporating cable and/or grapple skidders is an option as an integrated roundwood/biomass harvesting system provided sufficient space is available on the landing to cater for storage of the resultant biomass. From this point decisions can be made to either comminute at roadside or transport the loose biomass further towards the final consumption point.

6.4 Chipping Equipment and Machinery

Chipping is the most common method of comminuting biomass in preparation for combustion or other form of energy conversion. The two predominant chipping types are disc chippers and drum chippers (Fig. 6.2).

The working principle of the disc chipper is that 2–4 bevelled knives are fixed radially in a fast rotating disc. The knives, which can be adjusted for desired chip size (measured in the fibre direction) cut the biomass perpendicular or slightly offset to the feeding direction, and run up against an anvil to ensure the material is severed. A fan blade mounted on the rear of the disc creates a pneumatic force that blows the chips out of the spout and into a container or onto the ground. In larger chippers (>40 cm intake), the disc can have a diameter of over 120 cm and weigh more than 1,000 kg. Because the disc always cuts at a constant angle to the material, the disc chipper can produce very uniform chips. Disc chippers produce more 'stickers' than others (long slivers which cause stoppages in conveyor systems) as these are pulled into a parallel orientation to the knives. Various solutions have been found to reduce that problem which is more pronounced in small material (small trees, tops and branches).

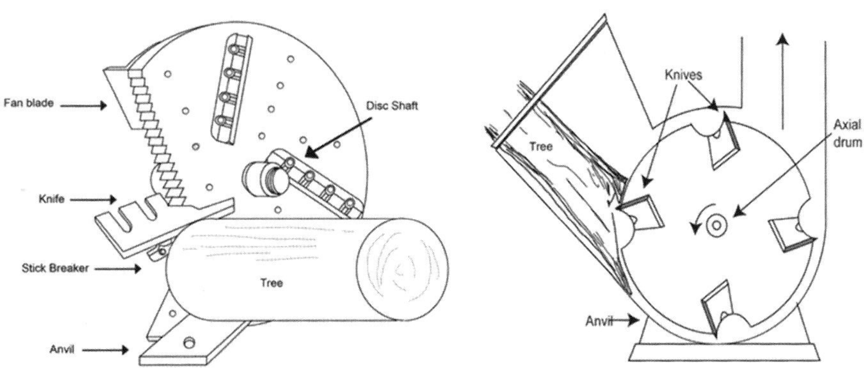

Fig. 6.2 Illustration of the working principles behind the disc and drum chipper (Danish Centre for Biomass Technology)

The drum chipper consists of a number of knives mounted along the longitudinal axis of a steel cylinder, with a smaller diameter than the disc chipper. It is therefore more compact and can be built into smaller spaces (e.g., on chipper trucks). By nature of its design, the knives on a drum chipper cut into the material at different angles, depending on the size of the log or branches. This produces slightly more heterogeneous chips. The drum chipper can generally be built for larger diameter logs, or larger bunches of smaller material, as the disc chipper intake has to be limited to less than the radius of the disc. Provided enough power can be delivered to the drum, it is possible to build drum chippers with larger intake capacities. The length of the knives reduces the negative consequences of hitting dirt or a stone as this would represent a smaller proportion of the knife than the same damage on a shorter disc knife. The knives on both disc and drum chippers have to be maintained (sharpened or reversed) at least once a day, and even more frequently when working with material that has been contaminated with soil, sand or stones, or cutting carbonized bark.

Auger or conical screw chippers are robust and produce homogenous chips of good quality. However, they require much higher power drivers, due to the high forces required in severing the material that is fed in the same direction as the axis of rotation. Chip size is adjusted by exchanging the screw for one with a different pitch, while the whole screw needs to be sharpened in place, or exchanged for a newly sharpened one. An advantage of screw chippers is that large material (chunks – up to 150 mm) can be made for e.g., thermal-gasification. Irrespective of the working principle, chippers are deployed in many sizes and configurations. Some are built onto terrain going base machines, others on trucks or trailers for mobility, while others are located centrally at terminals or conversion plants.

6.5 Biomass Sources and Harvesting Systems

This section discusses the utilization of the most economically accessible biomass resources arising from plantation forestry that include:

- Early thinnings, or dedicated energy roundwood.
- Harvesting residues (i.e., branches, tops and off-cuts).
- FT or salvaging from calamities (i.e., insects, wind, forest fires).
- Stumps and other sources.

Probably the most fundamental issue of biomass supply systems is that they consist of a number of stage-state steps between the standing tree and the boiler grate. Each step requires an action or selection of a method or machine that has implications for the whole downstream chain, and which cannot be reversed. The choice of felling method reduces options for processing or extraction. A decision to do in-field chipping eliminates the option of conventional forwarding and transport. There are a larger number of stage-state combinations, alternating between change of location and change of form, and only the most predominant are discussed in this section.

6.5.1 Harvesting Biomass from Early Thinnings

Unlike forests managed under more natural conditions (e.g., natural forests and woodlands selective cutting and the retention of seed trees for regeneration), plantation forests are typically re-established through the planting of seedlings. The important connotations of this are that spacing is controlled and that trees are typically planted in a geometric pattern which promotes efficient harvesting.

Early thinning is a term often used in conjunction with pre-commercial or sub-economic thinning and this well justified term is continually verified in research. It refers to an operation in which re-spacing is required to be carried out for the benefit of stand development, but the economic results of doing so do not necessarily justify the operations in themselves. Ahtikoski et al. (2008) use a complex calculation showing that energy wood thinnings could be financially viable if the extracted volume at least 42 m^3 ha^{-1} for an average stem volume larger than 0.015 m^3 and the unit price delivered exceeded US$12.00 MW h^{-1}. Plantation managers can partly avoid this cost by expanding initial planting espacement and accepting the consequences of later canopy closure, but the debilitating relationship between productivity and tree size cannot be totally avoided.

However, in cases where a market for smaller roundwood has fallen away (e.g., loss of contract or closure of plant) the demand for biomass could promote early thinnings. Also, the development of a bioenergy conversion facility should stimulate denser establishment (higher number of stems per unit area) on areas that have been managed more extensively (refer Chap. 5). To promote efficiency through mechanized operations, variable row spacing can be used (e.g., closer spaced double rows which will be removed in the first thinning and more widely spaced rows that will not be thinned).

Small trees can be felled in a number of ways, largely determined by the extraction system to be used. Motor-manual felling with a chainsaw is often the most cost effective way of felling trees when they do not need to be processed in whole-tree harvesting (i.e., tree parts and crown intact). This is especially true if the trees will be chipped as they lie in the stand or if they will be extracted with a cable system (e.g., high lead or monocable). If pre-bunching or processing is required, then mechanized felling is preferable. However, opening up of strip roads can be costly if the bunches have to be laid perpendicularly into the stand, as placing them in parallel requires wider striproads, and if driven on they can be contaminated with mineral soil and stones, resulting in increased ash content of the eventual product.

An agricultural tractor fitted with a boom and a multi-tree (or accumulating) felling head can provide a low investment alternative to a feller-buncher or harvester, but versatility is dependent on terrain conditions (Russell and Mortimer 2005). In a study in Scandinavian forests, productivity levels of 3–6 m^3 h^{-1} were obtained for felling and loading small trees onto a trailer for extraction (Belbo 2010). Small tree size thinning harvesters fitted with accumulating felling or harvesting heads are popular because of their greater stability and terrain going capability. However, all harvesting systems are highly sensitive to tree volume (Fig. 6.3).

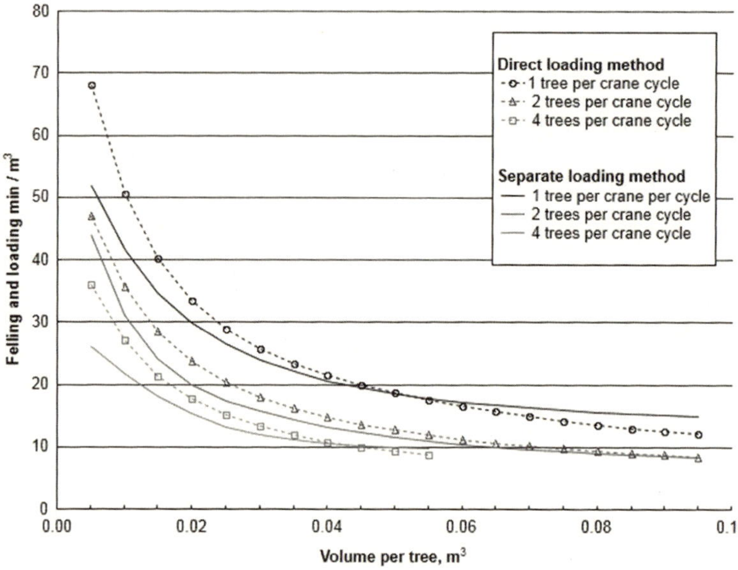

Fig. 6.3 Influence of direct loading and the number of trees per crane cycle on the combined felling and loading time using a Nisula 280 felling head in small trees, by increasing tree size (Belbo 2010)

The extraction of small FT or energy roundwood to roadside can be done using forestry equipped agricultural tractor/trailer units, a grapple skidder or a forwarder (Fig. 6.4). FT have a low bulk density and the load can potentially be compacted with the crane and grab. If the trees are to be extracted individually (e.g., using a monocable or a chute) or if they are to be chipped with a terrain going chipper machine they can simply be left as they fall for transpiration drying and subsequent extraction.

6.5.2 *In-Field Chipping of Full Trees/Tops*

When FT are harvested in a structured manner, and terrain is easily accessible, it is possible to use terrain-going chippers. These are commonly built on a standard forwarder chassis or adapted agricultural tractor fitted with a crane to feed the chipper and a large bin or skip to hold the chips. The orientation of the chipper intake (forward or sideways) has important connotations. Row thinnings imply a largely linear method of working, and a forward oriented chipper easily receives the butt-ends of the FT lying in the row, allowing the machine to move forward at approximately the same rate as the tree is being fed into the chipper. A forward oriented chipper is mounted in front of or under the cabin, leaving the entire loadbed available for a bin, which typically accommodates around 15 m^3 (~3,500 kg).

Fig. 6.4 Extraction of a large load of black wattle (*Acacia mearnsii*) using an agricultural tractor and tipping trailer (Photo: Talbot)

Side oriented chippers are placed behind the cabin, and take up storage space, but subject the cabin to less vibration and noise and have better mass distribution. Side oriented chippers require that the young trees have been laid into the stand perpendicularly to the strip road. A further disadvantage is that the machine cannot move forward before the tops of the trees being chipped clear the closest residual trees in the stand. The side oriented chipper is better suited to chipping tops from later thinnings where a harvester has been used and where more space is available. Maximising bin size is crucial to improving machine utilisation. Large, bulky bins can; however, cause damage to the residual stand, especially if the machine needs to reverse out of the striproad (Fig. 6.5).

In-field chipping requires some form of chip storage at the roadside landing. Tipping the chips onto the ground for later collection decouples production from transport, but results in some losses into the ground, and potential soil contamination, as well as the need for a wheeled loader or self-loading trucks (cranes with buckets) in order to move the chips to the conversion plant. Tipping the chips into a container requires firstly that the bin can be raised to sufficient height (~2.5 m) and that the logistics around the supply and exchange of containers is well managed.

6.5.3 Harvesting Biomass from Harvesting Residues

When applying mechanised processing, residues (i.e., branches and tops) should be dropped in piles that can be collected easily and efficiently. Although over consid-

Fig. 6.5 In-field operation using terrain going chipper which feeds directly into a high tipping bin (Photo: Linddana AS)

eration of this can decrease harvester productivity as trees have to be turned and positioned over the pile (Nurmi 2007). Fewer larger piles also reduces the degree of contamination. While piles can be left on site for a winter (i.e., summer rainfall zone) to promote nutrient recycling through foliage loss, it is more rational to extract the residues to roadside landing while the forwarder is on site. For guidelines on the potential impact on the nutrient status of sites through this practice and the potential nutrient content of the various portions of forest residues, refer to Chap. 5.

Efficient extraction is highly dependent on load density, which has led to the development of extendable load beds on forwarders. Even so, achievable loads are under 50 % of mass pay-load capacity of the forwarder and longer extraction distances make collection infeasible. Laitila et al. (2005) showed a cost reduction of 10 % using an innovative combination of simultaneous residue recovery and site preparation by utilizing a forwarder fitted with disk scarifiers.

The compression of harvesting residues into bundles (e.g., slash bundling) in the stand remains in use to a limited degree but hasn't realized the expected economic benefits (Kärhä and Vartiamäki 2006). In-field bundling does require a specialized base machine (e.g., forwarder) on which the bundling unit (e.g., the John Deere B380) is mounted. In an Australian study, slash was windrowed with an excavator, which resulted in good bundler productivity (21 bundles of 570 kg per productive machine hour), but was expensive and resulted in a high level of soil contamination (8.9 %) (Ghaffariyan et al. 2011). In-field bundling also requires forwarding to roadside, implying that it needs to be carried out in conjunction with the extraction of roundwood, or a forwarder must return to the site. Bundling units can also be mounted on trucks, giving greater mobility but requiring residues to be brought to roadside. When FT harvesting is done using skidders or cable yarders, this material is simultaneously extracted to roadside. Mobile truck mounted bundlers show good potential in serving numerous production points (Spinelli and Magagnotti 2009). These units are however restricted to roadside operations.

6.5.4 Harvesting Biomass from Salvage Operations

The direct combustion of primary forest fuels in larger boilers provides an opportunity to utilize material damaged by fire, drought, wind or insects in a robust and efficient way not offered by any other industry. Burnt trees can be harvested with disc saws or with chainsaw based heads using specially hardened chains. Chipper knives would need to be switched more frequently or the material could be crushed. Most southern hemisphere exotic pine plantations have experienced losses from the *Sirex* woodwasp (*Sirex noctilio F.*) where harvesting and combustion of the trees simultaneously contributes to feedstocks and potentially counteracts the spread of the *Sirex* population. Apart from the normal complexities and dangers of harvesting windblown timber, the high level of contamination with mineral soil and stones (from upturned root mats) make the material less sought after. While volumes can be substantial, additional care needs to be taken while comminuting the material. The end user should be informed of expected higher ash concentrations.

6.5.5 Harvesting Biomass from Stumps

Stumps after harvesting represent a significant potential for increased utilization of bioenergy. But the utilization of stumps for energy in countries like Sweden and Finland is mostly constrained by ecological concerns of the physical, chemical and biological impacts on the soil (Lindholm et al. 2010). In Finland some 850,000 m^3 of stump wood was utilized in 2009, and the production is increasing (METLA 2011). The mass of the stump and coarse roots may be some 25 % of the utilized stem (Marklund 1988). For example, Fig. 6.6 demonstrates the close correlation between mass and stump diameter for Spruce stumps in Europe.

Using specially designed stump pulling and splitting heads (Fig. 6.7), harvesting in good conditions can produce 2–4 dry tonnes per productive machine hour, equating to roughly 100 stumps per productive machine hour (Athanassiadis et al. 2011). The split material is bulky and forwarding productivity rates of 7–9 m^3 per productive machine hour on extraction distances of 50–500 m were found (Laitila et al. 2008). Productivity was lower than for any other forest product, with 27 % of the time being used on unloading alone.

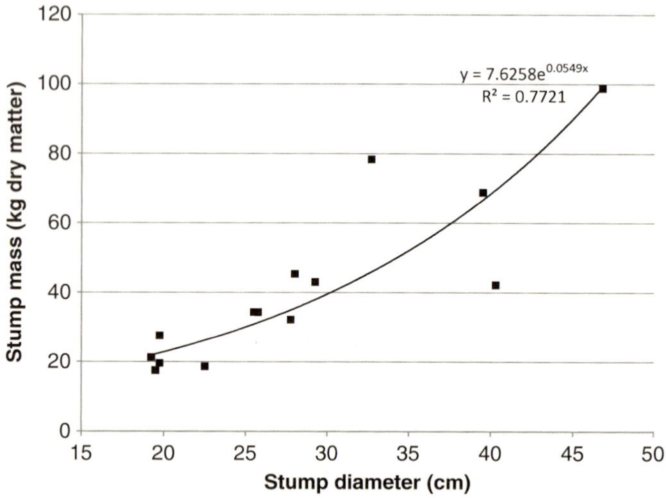

Fig. 6.6 Dry matter content of *Spruce* stumps as a function of the stump diameter at felling cut (Talbot 2010 unpublished data)

Fig. 6.7 Stump lifting device with splitting knife (Photo: Dahlin)

Stumps are normally "seasoned" by leaving them in roadside piles for some time (e.g., one year) mainly to allow precipitation and wind to erode the worst of the soil and stone contamination, and are invariably crushed with tub-grinders before or after transportation. Studies in Sweden indicate that pre-grinding and screening of stumps at landing reduces contaminant levels and loading time, and saves 15–20 % on transport costs (Thorsén et al. 2011). Apart from the volumes and generally good quality of the fuel, a considerable benefit of stump utilization from a cost perspective is that they represent an additional resource within the same procurement area. However, stumps from pine and eucalyptus dominated industrial plantations are not as easily lifted as those of spruce, and larger and more robust lifting heads would be required, depending on soil type, rooting pattern, and stump size.

6.6 Activities at Roadside Landing, Terminal or Plant

The previous section dealt with the harvesting and extraction of biomass in various forms to the roadside landing, which is almost always a discrete point between the primary supply phase (extraction) and the secondary phase of moving the material to the plant. However, the duration of time the material spends at the roadside landing varies from minutes if a "hot" container system is being used, to days, months or even years in the case of stumps. Most of the research looking at roadside storage and moisture management of biomass comes from boreal countries, where autumn and winter are characterized by large amounts of precipitation in the form of rain and snow, and where ice clumping is a problem. Few industrial plantations are located in these climate zones and local knowledge should be developed on good moisture management strategies.

6.6.1 Storage of Trees, Tree Parts or Bundles

Trees, tree parts and bundles should be stored at roadside landings in stacks that are stacked as high as possible while maintaining stability. High stacking minimizes the surface area exposed to rain (i.e., only the top is exposed) and promotes a more uniform material in terms of bulk density and moisture content. Ground contact should be broken by stacking on a simple rack of logs. Stacking butt-ends facing the landing not only promotes the run-off of rain water away from the landing, but makes for easier crane operation when chipping or transporting. These resources are stable and can be stored for long periods of time. The options from this point are roadside chipping and transport of loose chips, or transport of the material "as is" to conversion sites.

6.6.1.1 Storage of Harvesting Residues

Harvesting residues are stacked in the same way as FT or tree-parts, but given the nature of the material (no primary orientation); these stacks do not have the same natural 'roofing' tendency. In wetter climates it is common to cover the stack with a 4 m wide heavy duty paper from a dispenser attached to the forwarder crane. The effect of doing so varies with the time of harvest in relation to the season with only limited differences seen if the material is harvested just prior to the rainy season, whereas there are significant differences in moisture content of up to 15 %, between covered and uncovered material that are relatively dry before going into the wetter period (Filbakk et al. 2011).

6.6.1.2 Chipping at Roadside Landing

An advantage of chipping at the landing is that the harvesting/extraction and the chipping operations are not directly interlinked, and can be separated by hours or even months. This allows for a large feedstock to be built up, facilitating the use of high capacity chippers (>100 m^3h^{-1}) capable of filling a waiting truck within an acceptable time and thereby eliminating chip storage problems. Chipping material at roadside landing is the most common production method in biomass to energy chains in Europe. It can also involve chipping onto the ground or into containers. For chipping onto the ground, suitable preparation of the landing should be carried out beforehand (i.e., a clean and level site), while chipping into containers requires detailed logistics planning that synchronises container arrivals.

For chipping into waiting trucks, the challenge lies in balancing chipper productivity with truck waiting time. Chip transport trucks have loose volume capacities of 85–120 m^3 and should be filled quickly. High performance chippers capable of doing so represent large capital investments that incur expensive waiting time between truck arrivals, while low performance chippers shift the waiting time to the trucks, which can result in queuing at the landing or poor truck utilisation.

Chippers also need to be relocated from site to site. A solution to the challenges of getting this balance right is the use of chipper-trucks, with on board chippers that both chip and transport the material to the plant. The obvious benefit being that they are self-contained and highly mobile, with the drawback being the loss of payload both in terms of mass and volume due the presence of the chipper (Björheden 2008).

Recent research findings (Thorsén et al. 2011) show that self-contained chipping trucks perform well, especially in situations where their high mobility can be utilised to the full.

While gains are made in chipper productivity, the extraction of uncomminuted material (FT, tops, branches) is the least robust link in the chain. Efficient extraction of smaller trees or tree sections requires that they are pre-bunched and well presented for grapple-skidding or forwarding. Pre-bunching almost always implies mechanised felling while grapple-skidding results in higher levels of contamination with mineral soil (high ash levels) and forwarding requires that the trees have been laid in the stand and not in the strip row. FT or tree sections need to be compacted on the forwarder loadbed in order to improve the payload and maximize returns on the time cost of driving in and out of the stand.

6.6.1.3 Storage of Chips at Roadside Landing

Irrespective of whether the material was chipped in the stand and extracted to the landing, or chipped at landing, the storage of chips at roadside landing is normally a short term process but with numerous implications. As a result of the chipping and/or extracting process, the material can either be stored on the ground or in some form of bin container.

- Chips stored on the ground
  The benefits of storing chips temporarily on the ground are that there is no direct coupling with transport and that there is a large space/volume capacity. This option is good for high performing chipping production systems with transport constraints. The immediate disadvantages of chipping onto the ground are that the loading of chips requires specialized equipment or additional machinery and that some of the volume must be forfeited in ensuring that chips contaminated with soil and stones are left *in situ*. It is therefore not a suitable method when harvesting biomass from small, dispersed stands.
- Chips stored in containers
  The chips that have either been extracted from the stand and transloaded into a container or they have been chipped directly into a container at the landing. In the first instance, the availability of containers has to match the performance of the production system or a very high 'interference' penalty will be paid (Talbot and Suadicani 2005). Irrespective of production system, the assumption underlying this storage method is that transport is imminent. Chipping into containers with buffer capacity requires a lower performance (i.e., cheaper) chipper. Full containers left for a weekend for example, should be covered or fitted with sufficient drainage if there is a possibility of rain.

6.6.2 Storage and Handling

For all production systems, biomass needs to be stored and handled a number of times between the stump and the conversion plant. Good supply chain theory suggests that raw materials be kept in their original form as far down the chain as possible. This is to minimize early investments in the form of processing costs, and to allow the "manufacturer" more freedom in utilizing the resource right up until final conversion. The same idea holds true for biomass, though also for biological reasons. Comminuting biomass into chips radically increases the surface area, which together with high moisture content, provides ideal conditions for microbial activity. Exothermic respiration heats up the chip pile and results in dry matter loss due to the breakdown of cellulose and hemi-cellulose and even spontaneous combustion. Dry matter loss transforms directly to a loss in calorific value, and economic erosion. In addition to this, there is a growing awareness of the risks to human health of the clouds of fungal spores that emanate from stored chip piles. People in close contact with these, e.g., truck drivers, should wear respiratory masks when handling chips.

Roundwood is stable and dry matter loss is minimal in the first year after felling. Storage can take place at the point of felling, in bundles on the strip road, in piles at the landing or at the conversion plant. Initially, storage equates to drying, and freshly felled timber can dry to around 40 % moisture content (wet basis) within a number of weeks, depending on the ambient climate. FT felled and left on the ground have a steep drying profile, accelerated by transpiration from the leaves or needles. In spruce, transpirational summer drying is enough to allow the needles and fine fractions to fall to the ground during chipping or handling. This reduces the off take of nutrients from the site and reduces concentrations of corrosive elements (e.g., chlorine) in the fuel (Chap. 5).

FT, tree sections, tops or stemwood for energy can be stored in piles with or without cover. In Finland it is common practice to cover biomass piles with heavy duty paper sheeting as mentioned above. The benefit of doing this is dependent on the time of year the biomass is harvested, and for how long it will be stored. For a single summer, the drying profiles for covered and uncovered stacks are very similar, while biomass that is harvested late in the season will dry substantially faster under cover during autumn and winter (Filbakk et al. 2011).

Handling of biomass is accomplished with conventional forestry equipment as far as possible. Round wood for energy is no different from e.g., pulpwood. However, loose tops and branches are characterized by low densities and benefits can be gained from using adapted grapples that can handle high bulk loads. A residue grapple is made up of four separate grapple arms (tines) that are sharpened and easily penetrate a residue pile. In collecting harvesting residues, it is common to use forwarders with extendable loadbeds, or trailers with the capability to compress the load. For handling chips outside of a specialized terminal, either a front-end loader fitted with a large bucket or a bucket-grapple on a crane is commonly used. Due to the low bulk density, buckets can be over-dimensioned without the risk of exceeding the working capacity of the crane. At the conversion plant, a bunker below ground level allows for trucks to quickly tip a load that is subsequently evenly distributed or mixed with other forms of biomass with an over-head gantry, capable of operating continuously in two dimensions. In modern plants, these gantries operate autonomously, and also serve to feed chips into the boiler in-feed.

Chipping at the conversion plant offers considerable advantages, in that a large and powerful chipper can run consistently, and is well maintained by maintenance staff at the plant, resulting in very little downtime. The stationary chipper can be used to chip material of any size (bundles, stemwood, off-cuts, FT), and is fed and monitored by sophisticated systems, allowing it to operate around the clock. Also, the feedstock is stored in a natural and stable form, and only chipped on demand, reducing the need for covered or paved chip storage areas, and eliminating the risk of fire. If the plant is located near an urban area, noise pollution from centralised chippers can be experienced. Another disadvantage is the fact that all loose material needs to be transported to the plant for comminution. This can have implications on the potential extent of the procurement area.

6.7 Secondary Transport of Biomass

Secondary transport of biomass aims to move units of energy from source to conversion facility at the lowest cost. As a large part of transport cost is made up of fuel costs, this is synonymous with minimizing energy consumption in supply. Biomass ready for transport to the conversion plant exists in many forms with varying degrees of moisture content and differing assumptions on costs and efficiencies can give different suggestions on transport form (Tahvanainen and Anttila 2011). Biomass has a varying but generally low bulk density. For wood chips roughly 40 %, for trees and tree sections about 35 % and for harvesting residues, only 15–20 % of the load volume is solid matter.

6.7.1 Biomass Transport Truck Types

Road transport vehicles are the predominant mode of transport. Permissible loads are governed by the legal gross vehicle (or gross combination) mass and the allowable axle or axle unit mass/es. In South Africa for example, the maximum allowable mass on a single axle (non-steering) is 9,000 kg (7,700 kg on a steering axle), 18,000 kg for a two axle unit and 24,000 kg for a three-axle unit. The least of either the sum of the axle unit masses or the maximum legal gross combination mass (GCM) represents the gross legal allowable mass. The maximum permissible gross vehicle mass for South Africa is 56,000 kg. The maximum width for a vehicle exceeding a gross vehicle mass of 16,000 kg is 2.6 m. Below this gross vehicle mass the width is limited to 2.5 m. The maximum height for all vehicles is 4.3 m (FleetWatch 2012). Container trucks are popular due their versatility while truck-tractor and semitrailer configurations allow larger payloads.

A study examining 58 trucks showed a mean payload of 23,500 kg for container trucks and one of 29,164 kg for articulated trucks. The tare mass of the container trucks was 24,246 kg and for the semitrailer 16,180 kg meaning that the semitrailer is 8,000 kg lighter and has a larger load space (Fig. 6.8). Unloading times were 20.95 min on average for the former and 45.36 min for the latter (Talbot and Suadicani 2006).

Even though the bulk density of forest fuels is low, an increasing moisture content implies that less energy is transported per truck cycle (calculated as the lower heating value), i.e., at the same transport cost. However, at some stage, the total mass limit of the truck is exceeded and the load volume has to be reduced as well, drastically reducing the amount of energy being transported (Fig. 6.9). At 50 % moisture content, the semitrailer (upper line) carries 16 % more energy than the container truck (lower line) while at 55 % MC, it carries 30 % more. This is an important relationship to understand as chipped residues have a higher density than chipped stemwood (shown here) due to the heterogeneity of the material and the higher density of the branches.

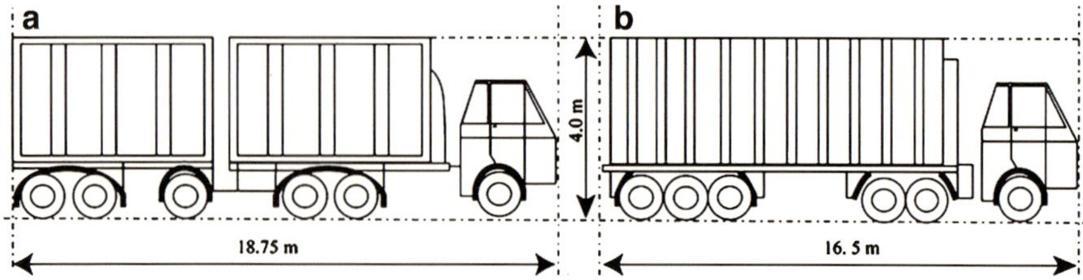

Fig. 6.8 Generic container truck carrying roughly 85 m^3 (**a**) and a semi-trailer, with a capacity of approximately 108 m^3 (**b**)

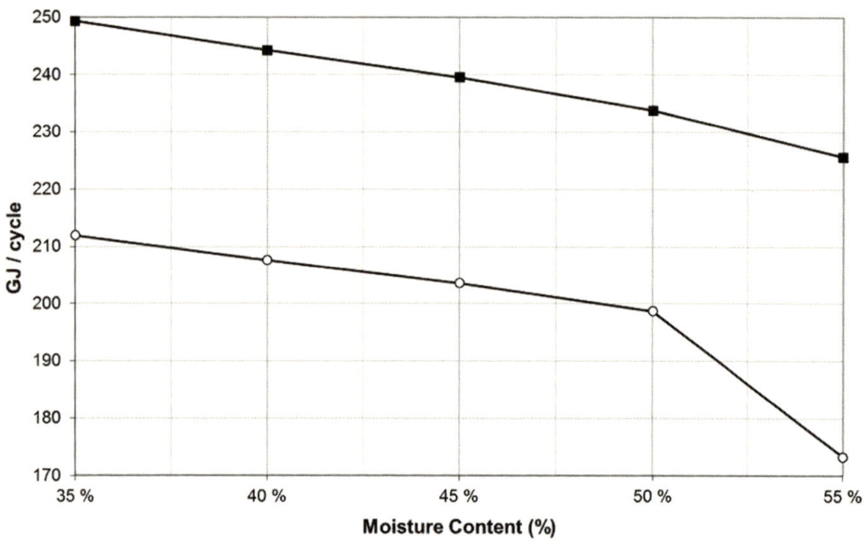

Fig. 6.9 Energy content per load transported against increasing moisture content. For the container truck (*lower line*) the gross mass is met at 50 % and load volume must be reduced to meet legislation (Talbot and Suadicani 2006)

In recognizing that increasing volumes of biomass will be transported over increasing distances in the future, efforts are being made to develop vehicles with increased volumetric and mass carrying capacities. EU directive 96/53/EC allows member states to test and adopt the European Modular System (EMS) which allows for vehicles up to 25.25 m long with a gross mass of 60 tonnes. Countries like Sweden and Finland have benefitted greatly from applying these trucks in transporting forest fuels. In South African Performance Based Standard (PBS) type rigid truck and drawbar trailer combinations are now allowed to operate, with special permits, at lengths of 27 m and 70,000 kg GCM and a payload of up to 49 tonnes (Fig. 6.10). Solely used as roundwood pulpwood vehicles currently (and normally mass limited), these PBS trucks will fulfil a specific role in biomass transport for the same reasons mentioned above.

Fig. 6.10 Loaded Performance Based Standard (PBS) truck on N2 highway, KwaZulu/Natal (Photo: RailRoad Association of South Africa)

6.7.2 Economic Considerations of Transport

The choice of an 'optimal' form of transport for a planned bioenergy plant is not always straightforward, and will depend on whether the enterprise already transports, e.g., pulp chips, whether all transport is to be outsourced, and whether a small or large scale operation is planned. However, some consistencies in making a rudimentary evaluation do exist. If the conversion plant is to be located alongside an existing sawmill in a plantation area, it could be acceptable to use an off-road agricultural tractor/trailer system where the typical distance is under 15 km, while if the intention is to supply a central heating plant in the next city, sophisticated haulers are required. Transport distance is not the only important factor in making the decision on how to invest. In the case below, we will show how factors like loading and unloading (terminal time), load capacity, moisture content, delivered energy price, and investment costs bear influence on a good economic solution. One solution can never be optimal for the diversity of sites encountered in forestry; however, a good solution should perform well on average.

6.7.2.1 Loose Chips

Transporting loose chips calls for as large a loading space and as low a tare weight as possible. Only if the chips are made from very fresh material will the load approach the mass restriction of the truck. The most common solutions seen for transporting loose chips are container systems (from 85 to 120 m^3), i.e. two containers loaded

Fig. 6.11 Container truck exchanging chip containers at landing (Photo: Talbot)

on a rigid truck and drawbar trailer separately, or on semi-trailers drawn by a truck-tractor (Fig. 6.11). The reason that containers remain popular despite their high tare weight is that they can be exchanged quickly at the chipping site, reducing waiting time on both the truck and the chipper. The use of standard freight containers (40 m^3) can be justified by the advantage they offer of being utilisable by a large number of transport contractors; enabling a more flexible local supply chain, where the number of containers or trucks can be varied to meet individual operations. They can also be used for transporting other material during periods of lower demand.

6.7.2.2 Loading and Unloading

Bulk trailers take on a number of forms. Rigid trucks with a tipping bed can be used without a trailer, but transport a limited volume. These trucks are compelled to wait during the loading process, but are agile on poor forest roads and have a short unloading time. For longer chip transport, semi-trailers fitted with reciprocating slatted floors, such as the Walking Floor™ trailer, or side tilting designs, can be used successfully. Articulated trucks can have problems on winding or steep roads, but benefit greatly from their superior load capacity. Semi-trailers and drawbar trailers can be exchanged at the landing, or the truck can wait to be filled. This requires matching with a high production chipper. Side-tipping trailers have minimal offloading times, but require specialized receiving bunkers that run the entire length of the trailer. End-loading trailers can unload into more commonly available bunkers. Chipper-trucks (fitted with own chipper) have reduced transport capacity but have high mobility in that they are not dependent on other machines and work well in areas with a higher number of smaller landings.

6.8 Managing Biomass Trade and Supply

The development of a good business process model for biomass supply is something that many enterprises see as a challenge. More specifically, the measurement of the biomass, the calculation of the heating value, and the varying lead times (from hours to years) are elements that many supply chains wrestle with. In the following section, some fundamental business process structures are presented.

6.8.1 Owner Supplies Directly to Plant

For a small forest owner, this is the simplest form as the forest owner internalises all costs and delays in preparing the biomass for delivery. The basis for measurement is weighbridge mass, corrected for moisture content (see generic description of this below). If the forest owner has utilised contractors for harvesting or transport, the bill for this can be settled in conjunction with conventional harvesting on a volume basis (m^3 – solid) costs using a suitable biomass expansion factor (BEF) which is multiplied against the roundwood harvest. The transport operator can be remunerated directly from the weighbridge bill. The forest owner incurs the costs at the time of each operation.

For a forestry company delivering to its own energy plant (e.g., CHP plant at saw-mill), the model would depend on the accounting processes between internal business units (e.g., harvesting business unit is separate from CHP business unit). From the energy conversion facility's perspective, dealing with a large number of small forest owners (and other suppliers) is complex given that supply is almost impossible to schedule, there is little information on what is in the pipeline, variation in material and its properties can be large, and guarantee of delivery uncertain. These (considerable) disadvantages will be reflected in the price.

6.8.2 Owner Supplies Through Cooperative (Forest Owners Association)

Many forest owners are already members of cooperative organizations that provide services (forest management, harvesting and logistics) and access to market channels. Also, some bioenergy conversion plants create specialized cooperative units which manage the longer term procurement of biomass on their behalf (or a group of forest owners invest in a bioenergy plant). Irrespective of the structure or origin, this business process model is characterized by the undertaking to maximize the interests of their members.

In this case, the forest owner can either supply directly (as in case above) or can choose to leave the procurement process to the cooperative. In the latter case, the cooperative carries the costs to the contractors in the supply chain. Advantages of cooperative supply are: (1) that the interests of the forest owner are seen to; (2) that the cooperative can incubate good contractors and streamline business processes; (3) economies of scale and benefits (e.g., speed of payment) of being a large and consistent supplier to an energy plant; and (4) that dividends, bonuses, or revenue adjustments can be made retrospectively.

6.8.3 Third Party Supply

In this model, a contractor or broker (or other 3rd party) purchases the biomass and undertakes the necessary activity in feeding it into the downstream supply chain. Volumes/energy content must be determined on site at, or after, harvesting. Once again, established BEF can be utilised or the volumes can be estimated from pile dimensions and local guidelines. The advantage for the forest owner is one of a small (but somewhat uncertain) income on site and no responsibility for costs to contractors and uncertainties of supply. The benefit to the energy conversion plant is a longer term relationship with a single supplier representing many forest owners.

Standard models for determining energy content in the supply chain are given below:

6.8.3.1 At the plant

Ultimately the bioenergy plant determines and pays for the energy content of biomass being delivered though its gates. This is done for chips as follows:

1. The (registered) truck driver notes the origin of the biomass on the incoming weighbridge bill, which includes a timestamp. The registration of trucks includes the details of the supplier for whom they are transporting.
2. The driver (or staff member) takes a bucket sample of the chips while they are being poured into the chip bunker – ensuring it is as representative as possible. Some plants are equipped with fully automated sampling devices that bore into the load while still on the truck.
3. The sample is placed in a drying oven together with the weighbridge bill
4. After 24 h at 105° C, the dry matter content of the load is calculated and remuneration to the supplier is affected.

When it comes to other forms of biomass (FT or residues), it is not possible to make accurate estimates of energy content before the material is chipped and dried. With well-developed delivery systems, e.g., from uniform plantation forestry, conversion figures including norms and standard deviations can be developed rapidly and improved on with time. Here, the supplier is paid a value derived from e.g., mean moisture content per specie and season.

6.8.3.2 At the landing or terminal

Solid content conversion rates for stacked harvesting residues, FT, or roughly debranched stems should be developed regionally and according to species. Apart from errors arising from edge effect, a good relationship can be obtained between running metre of stacked volume and solid volume. A simple way of developing such conversion rates is by chipping the stack into a container of known volume. The dry mass of the chips arising from this process is then compared with the dry mass of the tree species in question in determining solid volume equivalents. It should be remembered that high branch content will yield a dry mass higher than the mean for stem-wood for the same species.

6.8.3.3 On Site (pre- or postharvest)

On site estimation of biomass involves an enumeration of standing trees or residues/stumps after harvesting. Methods of ensuring reliable estimations are given in Chap. 3. Some adjustment needs to be made as varying percentages of the measured volumes are utilisable. For harvesting residues after the CTL method, about 70 % of volume is typically recovered.

6.9 Managing Feedstock Supply and Supply Cost Curves

Whether planning the location or capacity of a new plant, or supplying an existing plant, the procurement manager needs to have a good idea of the cost profile for feedstock supply. The location of the resource in relation to the conversion plant is more or less fixed, with annual supply deviations depending on the harvesting or thinning plans. A relatively simple and visually cognitive way of representing the economic availability of the resource is to develop a marginal supply cost curve. This implies deriving and plotting the cost of the last most expensive resource against the cumulative volume acquired up until that point. The cost is a composite figure, which includes harvesting, storage, handling and transport of each biomass type, and from each geographic source point, to a given destination, commonly the plant gate. This means that each resource point (stand) is handled individually in terms of harvesting method, yields, sequencing etc. Such an overview can easily

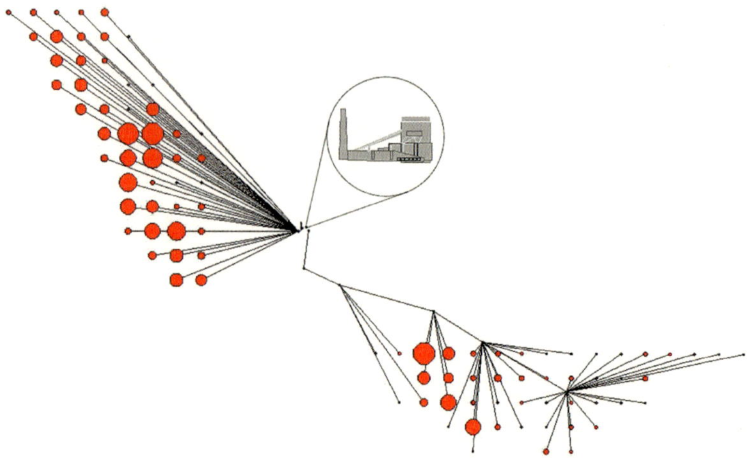

Fig. 6.12 Schematic oversight of the relationship between biomass volumes and transport distances in an extensive woodland

be constructed and maintained in a spreadsheet, although the accuracy is fully dependent on the cost estimations made underway. These can be based on rough estimates e.g., Fig. 6.12 (where each circle represents the volume available in a 1 km^2 grid cell and the lines represent distance to the plant) or modelled in detail in GIS (Möller and Nielsen 2007). The following section provides a step by step guide on how to develop such a curve.

Developing a supply cost curve for managing biomass feedstock

- Step 1. Using a spreadsheet, list all the sources, including either the net energy content or the tonnes of dry matter available as the primary unit.
- Step 2. Estimate a harvesting and transport cost for each source. This can vary considerably depending on the kind of operation (early thinning vs. clearfelling), the terrain, the anticipated transport method and distance.
- Step 3. Sum these costs in a new column and rank the spreadsheet according to increasing delivered cost.
- Step 4. Generate a new column showing the cumulative quantity of energy (sum of all preceding energy quantities). You now have the necessary data for plotting the marginal supply cost curve. However, it is important to know the mean cost as well as the marginal cost.
- Step 5. The mean cost is more complex to calculate as it requires the sum-product of all preceding energy quantities and their respective prices to be divided by the cumulative volume. We use a simpler method to get there: calculate a total cost column (GJ * unit cost) and then sum this up in a cumulative cost column. The mean cost is then the cumulative cost divided by the cumulative volume. Now the supply cost curve can be plotted as in Fig. 6.13).

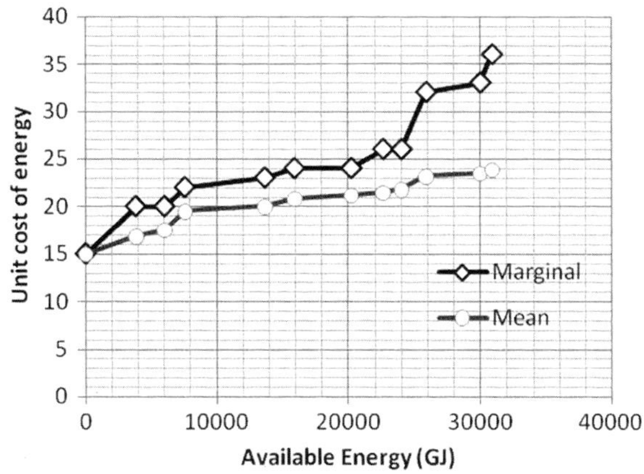

Fig. 6.13 An example of marginal and average value supply cost curves

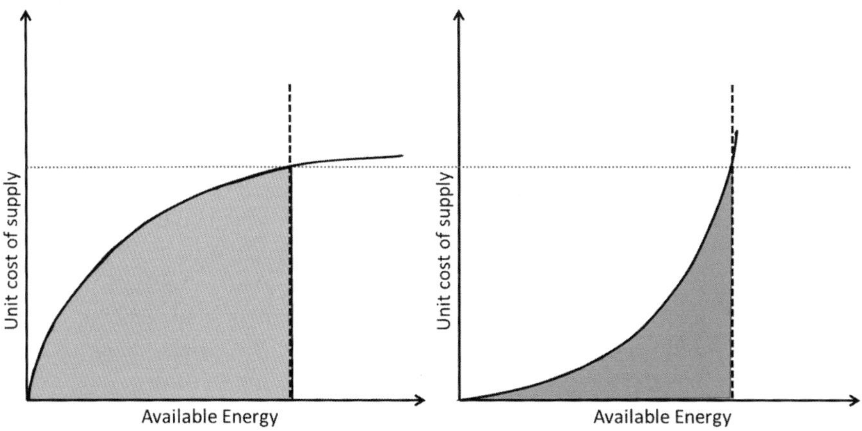

Fig. 6.14 Mean (and marginal) cost curve comparison. The *greyed* area represents total cost and marginal cost is provided by the curved line. The mean cost for the procurement area on the *left* is significantly higher than that on the *right*, despite their having equal marginal costs

The resulting plot from step 4 provides a marginal cost of supply curve. The procurement manager can read what volume is available at what price directly from the curve. However, what is more interesting for procurement management is the mean cost of supply, as managers will normally be working from an operational budget. The manager might have settled on a maximum marginal cost that he/she is willing to pay a supplier, but this might not be the best course of action as is explained below.

The shape of the cost curve function (whether convex or concave) in arriving at the same marginal cost for the same quantity of energy, could imply two vastly different mean costs (Fig. 6.14). The shaded area represents the total cost of supply, i.e. incremental cost multiplied with the incremental quantity. In this example it is easy to see that two very different mean costs are arrived at when the total cost is divided by the total volume. Even if the marginal cost is the same, the mean cost can be quite different as the shapes of these two cost curves illustrate. The mean cost is represented by the shaded area (Fig. 6.14).

There are a number of ways of extending the utility of these curves. Including a code for the present state of the biomass (e.g., loose residues, bundled residues, chipped residues) and/or the stage of where the biomass is in the supply chain (planned harvest, harvested, at landing, at terminal) gives the procurement manager insight into the dynamics of the supply pipeline within a given time horizon, whether it be a week or a year. This means that the manager is able to negotiate prices, or incur heavy costs, in procuring biomass from specific suppliers or sources without compromising the budget.

6.10 Conclusion

This chapter described the possible sources of bioenergy from forests, early thinning, harvesting residues, salvage operations and stumps. Furthermore, the options of collection, extraction, haulage and comminution have been discussed. It has been shown that the place of comminution within the supply chain is decisive for the design and cost-efficiency of this bulky, low value commodity with limited potential for transport efficiency gains. One rule of thumb is that the shorter the transport distance, the later comminution should be performed in the supply chain. On the other hand, for long transport distances and a bulky assortment, comminution could be carried out in an earlier stage of the supply chain in order to decrease transport costs. In most cases the supply is not made up of one chain, but consists of a network of supply chains, where the challenge is to utilize machinery where it is best suited and to minimize costs. Utilising supply cost curves can provide insight into the most suitable supply chain for a particular situation.

References

Ahtikoski A, Heikkila J, Alenius V, Siren M (2008) Economic viability of utilizing biomass energy from young stands – the case of Finland. Biomass Bioenergy 32(11):988–996

Athanassiadis D, Lindroos O, Nordfjell T (2011) Pine and spruce stump harvesting productivity and costs using a Pallari KH 160 stump-lifting tool. Scand J For Res 26(5):437–445

Belbo H (2010) Comparison of two working methods for small tree harvesting with a multi tree felling head mounted on farm tractor. Silva Fenni 44(3):11

Björheden R (2008) Optimal point of comminution in the biomass supply chain. In: Suadicani K, Talbot B (eds) The Nordic-Baltic conference on forest operations. Forest & Landscape, Copenhagen, p 30

Filbakk T, Høibø O, Nurmi J (2011) Modelling natural drying efficiency in covered and uncovered piles of whole broadleaf trees for energy use. Biomass Bioenergy 35(1):454–463

FleetWatch (2012) Maximum mass and dimensions. A FleetWatch Publication. Revised edition 2012:77

Ghaffariyan MR, Andonovski V, Brown M (2011) Slash-bundler in clear felled eucalyptus plantations of Australia. In: Ackerman P, Ham H, Gleasure E (eds) Innovation in forest engineering – adapting to structural change. Stellenbosch University, White River, p 504
Kärhä K, Vartiamäki T (2006) Productivity and costs of slash bundling in Nordic conditions. Biomass Bioenergy 30(12):1043–1052
Laitila J, Asikainen A, Hotari S (2005) Residue recovery and site preparation in a single operation in regeneration areas. Biomass Bioenergy 28(2):161–169
Laitila J, Ranta T, Asikainen A (2008) Productivity of stump harvesting for fuel. Int J For Eng 19(2):37–47
Lindholm EL, Berg S, Hansson PA (2010) Energy efficiency and the environmental impact of harvesting stumps and logging residues. Eur J For Res 129(6):1223–1235
Marklund LG (1988) Biomassafunktioner för tall, gran och björk i Sverige. Rapport 45. Sveriges lantbruksuniversitet, Institutionen för skogstaxering
METLA (2011) Finnish Statistical Yearbook of Forestry 2011. Forest Statistical Bulletins. E. Ylitalo, Finnish Forest Research Institute, Vantaa
Möller B, Nielsen PS (2007) Analysing transport costs of Danish forest wood chip resources by means of continuous cost surfaces. Biomass Bioenergy 31(5):291–298
Nurmi J (2007) Recovery of logging residues for energy from spruce (*Picea abies*) dominated stands. Biomass Bioenergy 31(6):375–380
Pulkki RE (1992) Wood quality considerations in wood procurement decision-making. Pulp Pap Can 93(4):29–34
Pulkki RE (2000) Forest harvesting operations in South Africa. Cent Woodl 3(2):41–45
Russell F, Mortimer D (2005) A review of small-scale harvesting systems in use worldwide and their potential application in Irish forestry. COFORD, Dublin, p 56
Spinelli R, Magagnotti N (2009) Logging residue bundling at the roadside in mountain operations. Scand J For Res 24(2):173–181
Stupak I, Asikainen A, Röser D, Pasanen K (2008) Review of recommendations for forest energy harvesting and wood ash recycling. In: Röser D, Asikainen A, Raulund-Rasmussen K (eds) Sustainable use of forest biomass for energy. Springer, Dordrecht, pp 181–191
Tahvanainen T, Anttila P (2011) Supply chain cost analysis of long-distance transportation of energy wood in Finland. Biomass Bioenergy 35(8):3360–3375
Talbot B, Suadicani K (2005) Analysis of two simulated in-field chipping and extraction systems in spruce thinnings. Biosyst Eng 91(3):283–292
Talbot B, Suadicani K (2006) Road transport of forest chips: containers vs. bulk trailers. Forestry Studies – Metsanduslikud Uurimused 45:11–22
Thorsén Å, Björheden R, Eliasson L (eds.) (2011) Efficient forest fuel supply systems – composite report from a four year R&D programme: 2007–2010. Skogforsk

Chapter 7
Biomass Conversion to Bioenergy Products

Johann F. Görgens, Marion Carrier, and Maria P. García-Aparicio

7.1 Introduction

The rendering of bioenergy products such as heat, fuel and electricity requires the conversion of sustainably produced biomass feedstock by means of thermochemical and biological processes. Such processes convert feedstocks into higher energy-value products amenable to industrial and domestic applications. This chapter deals with the nature of the conversion processes, the biomass feedstock requirements for these processes and the resulting quality of bioenergy products. In addition, the present chapter will also consider the application potential of different conversion technologies to both industrial and rural areas in the Southern Hemisphere.

Conversion of biomass feedstocks is a key step in bioenergy production. The value of bioenergy products is related to their suitability for particular energy applications, which is determined by the interaction between characteristics of the feedstock and the conversion process applied. Thermochemical conversion processes are primarily combustion, pyrolysis, gasification and direct liquefaction, while biochemical conversion of biomass involves hydrolysis to monomeric sugars and organic acids, followed by fermentation/digestion to liquid and gaseous biofuels.

Technology selection for biomass conversion represents a key decision in the formulation and selection of a bioenergy production process. These decisions are driven to a large extent by the availability of feedstocks and market demands. The present chapter will consider technologies applicable to tree-based biomass produced in tropical regions, as discussed in Chaps. 2, 3, 4, 5 and 6, while serving as an introduction to Chap. 8, where the impact of tree-quality on these

J.F. Görgens (✉) • M. Carrier • M.P. García-Aparicio
Department of Process Engineering, Stellenbosch University, South Africa

T. Seifert (ed.), *Bioenergy from Wood: Sustainable Production in the Tropics*, Managing Forest Ecosystems 26, DOI 10.1007/978-94-007-7448-3_7,

conversion processes is investigated. Aspects of technology selection and feedstock are included in the global bioenergy analysis in Chaps. 9, 10 and 11 as well as case studies presented in Chap. 12.

7.2 Types of Bioenergy Products

The term *bioenergy* refers to all types of energy derived from plant biomass such as the lignocellulose feedstocks under consideration here. The following bioenergy types can be obtained by application of different transformation technologies to the conversion of lignocellulose biomass:

1. *Thermal energy* is one of the most commonly used products of woody biomass transformation and provides heat required for cooking and heating through direct combustion. In addition, steam produced by combustion can be used for both domestic and industrial processes (e.g. drying, boiling, ceramic oven heating/baking, etc.). Heating needs differ and can be distinguished between urban and rural, domestic and industrial applications.
2. *Electric energy* can be obtained through several transformation technologies, the choice of which is mostly determined by the type and amount of biomass available. For instance, the steam generated in a combustion process can be used to produce electricity (cogeneration). Additionally, other conversion technologies such as gasification, pyrolysis and anaerobic digestion all produce gases (synthesis gas, bio-gas) suitable for electricity production.
3. *Transportation fuel* is the energy obtained in self-propulsion motors from biofuels. These so-called 'second generation biofuels' originate from lignocellulosic biomass and can be obtained through several thermochemical and biochemical transformation technologies, as described in the sections below. Second generation biofuels particularly refers to those that employ lignocellulose biomass resources as feedstock without competing with food production. Lignocellulose biomass can be collected as residues from various activities in different sectors, although it is also possible to specifically produce ethically suitable "energy crops" as feedstocks for bio-energy production (Sánchez and Cardona 2008).

Woody biomass is a renewable feedstock for the production of bioenergy that is available in relatively large amounts in many parts of the world. It can be collected as by-products from lumber, pulp and paper production or from dedicated energy crops such as short rotation woody crops (see Chap. 6). Lignocellulosic biomasses are characterized by their heterogeneous composition and structure, multiplying the possible approaches for conversion into bioenergy products. Among the different feedstocks, this chapter will focus on the following genera: *Pinus* (pines), *Eucalyptus* (gums) and *Acacia* (wattles), given their wide and ready availability in the Southern hemisphere. Bioenergy transformation technologies are generally classified as being either thermochemical (combustion, pyrolysis, gasification, liquefaction) or biochemical (anaerobic digestion, microbial fermentation) in nature.

7.3 Thermochemical Conversion Technologies

The various thermochemical conversion technologies that may be applied to processing of lignocellulosic biomass are presented in Fig. 7.1.

7.3.1 *Combustion*

The production of thermal heat and electricity from lignocellulose, as well as the production of intermediate bioenergy products such as pellets, charcoal, gases and liquid fuels derived from lignocellulose, all proceed via combustion. Combustion processes combine three elements: a feedstock as fuel, air as oxidant of the feedstock and the application of a specifically required temperature from a heat source. The carbon and hydrogen components of a feedstock are totally or partially oxidized and converted into heat. Normally, combustion precedes pyrolysis. Burning of woody material proceeds in four steps: the temperature of the starting material is increased by application of heat, which leads to the evaporation of volatile species and char formation, followed by combustion of volatiles species (primary combustion) and finally the combustion of char (Gonzalez et al. 2005).

The production of steam in a boiler, facilitated by energy derived from combustion, is used either directly as thermal heat or to drive steam turbines for

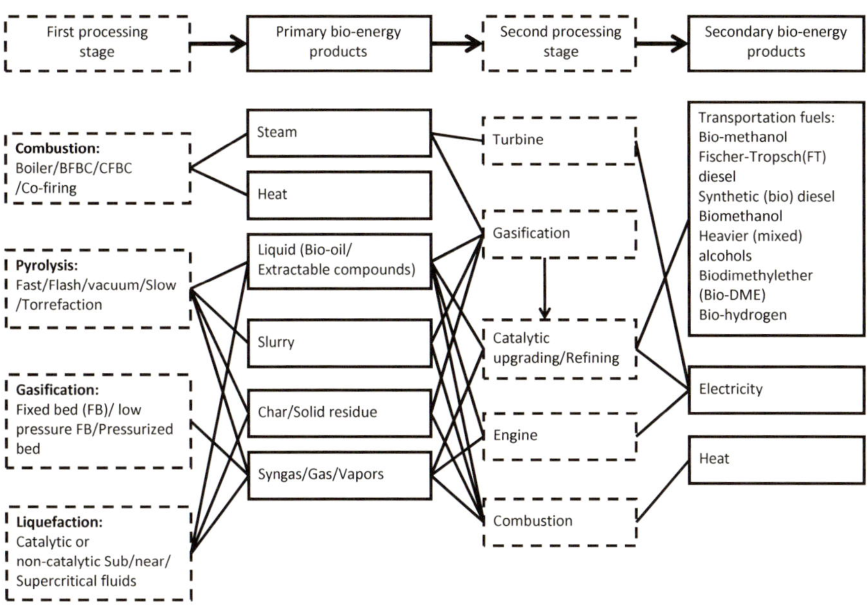

Fig. 7.1 Overview of thermochemical processes for conversion of woody biomass into bioenergy products

Table 7.1 Experimental conditions for the different pyrolysis processes

	Experimental conditions		
Pyrolysis process	Temperature (K)	Vapour residence time	Heating rate (K s^{-1})
Torrefaction (Mild)	450–575	15–30 min	0.1–1
Slow (Conventional)	550–950	5–30 min/45–550 s	0.1–1
Vacuum	600–700	2–30 s	0.1–1
Fast	850–1,250	0.5–10 s	10–200
Flash	1,050–1,300	<0.5 s	>1,000
Pressurized	<750	<10 s	

electricity production, with high boiler pressures combined with multistage turbines providing the highest overall process efficiency. Similarly, the gases produced through gasification, anaerobic digestion and pyrolysis can be combusted directly in gas turbines or gas engines, with the biomass integrated gasifier/combined cycle (BIG/CC) systems delivering significantly higher conversion efficiencies than the boiler-steam turbine process (Laser et al. 2009). The recovery of waste heat from boilers, gas engines/turbines, steam turbines and other combustion applications can provide a useful form of low quality heat (e.g. warm water), which increases conversion efficiency.

The heat energy and steam delivered by gasification systems can be used for heating purposes and electrical power generation, respectively. Known and applied for many decades, many biomass combustion technologies can be found on the market and mainly categorized in two types, fixed-bed (Water-cooled vibrating grate (VG)) and fluidised-bed systems (Bubbling fluidised bed combustion (BFBC), Circulating fluidised bed combustion (CFBC)). Fluidised-bed boilers present a series of advantages such as limited emissions and relatively complete combustion which improves overall efficiency and makes it suitable for processing of a wide range of feedstocks (Wright et al. 2010; IEA ETSAP 2010; Demirbas 2005).

7.3.2 Pyrolysis

Pyrolysis is defined as the thermochemical decomposition of organic materials in the absence of oxygen or other reactants executed under specific parameter conditions (Table 7.1). Several types of pyrolysis have been developed: vacuum pyrolysis (Roy 2000), pressurized pyrolysis (Tomeczek and Stanislaw 2003), fast pyrolysis (Bridgwater and Peacocke 2000), flash pyrolysis (Demirbas and Arin 2002), torrefaction pyrolysis (Prins et al. 2006) and slow pyrolysis (Antal and Grønli 2003). The processes differ from each other according to the conditions maintained in the pyrolysis reactor (Table 7.1). Fast pyrolysis leads to a high yield of bio-oil, while vacuum and slow pyrolysis offer a good compromise for the production of char and bio-oil, providing relatively high yields of both as well as providing superior quality char products (Bridgewater 2011; Bridgwater and Peacocke 2000; Chen et al. 2003). Bio-oil represents a valuable liquid fuel for boilers, while chemicals, nutritional supplements and/or pharmaceutical products may be isolated from it, provided that the challenging separation of these compounds can be achieved efficiently. The char represents a good feedstock both for boiler fuel and for the production of activated carbon. In more recent years char has also become an option for soil improvement and carbon sequestration (Antal and Grønli 2003).

Fast pyrolysis thermal decomposition of a feedstock using a relatively high heating rate can yield liquids (collectively termed bio-oil) of up to70–75 % of the weight of the starting material (Bridgwater 2011; Butler et al. 2011). One of the main advantages of fast pyrolysis lies in the fact that it is an effective method for densification of voluminous biomass. Different fast pyrolysis technologies exist, namely ablative, cyclonic, rotating cone, entrained flow, bubbling fluidised bed (BFB), auger, circulating fluid bed (CFB), transported bed, screw and auger kiln and wire-mesh reactor fast pyrolysis (Hoekstra et al. 2012; Bridgwater 2011; Butler et al. 2011). Currently, BFB and screw kiln reactors can be used for commercial-scale production of biofuel. Because of its properties relating to aging, instability, corrosion and viscosity, fast pyrolysis bio-oil may be upgraded physically, chemically or catalytically (Bridgwater 2011). Bio-oil can be a substitute for fuel oil used for electricity generation and biorefinery.

Slow and vacuum pyrolysis processes deliver higher char and gas yields, limiting their energy applications to electricity and heat production. Slow pyrolysis follows the conventional carbonisation process of charcoal or biochar production, with recent improvements in recycling of gaseous/liquid products providing much of the heat required by the process, increasing overall energetic efficiency. The main difference between vacuum and slow pyrolysis lies in the method of removing gaseous vapours from the reaction zone. In vacuum pyrolysis, a vacuum is created which serves the same function as purge gas used in slow pyrolysis. Because of the lower pressures applied, aerosols tend to evaporate more easily. This removes them from the reaction zone and results in a significantly reduced vapour residence time of 2–3 s for vacuum pyrolysis, relative to the 165–170 s required for slow pyrolysis. Vacuum pyrolysis is therefore a modified slow pyrolysis resulting in improved quality of both liquid and char products (Carrier et al. 2011).

Torrefaction is classified as a mild pyrolysis technique, because it takes place in an inert atmosphere at relatively low temperatures (between 200 and 300 °C (Uslu et al. 2008)). The technology is less sophisticated than fast, flash, slow and vacuum pyrolysis technologies and can be seen as something in-between the combustion of dried biomass and of pyrolysis products. From a chemical point of view, torrefaction removes oxygen from the original biomass resulting in a solid product which has a lower O/C molar ratio (Van der Stelt et al. 2011). Torrefied biomass has potential for application in various industries: as raw minerals for pellet production, as reducer for smelters in the steel industry, for the manufacturing of charcoal or activated carbon, for use in gasification and for co-firing during boiler operation.

7.3.3 *Gasification and Synthesis*

Gasification involves the conversion of biomass into a combustible, non-condensable gas mixture by partial oxidation of biomass at high temperature (800–1,300 °C). The resulting gaseous mixtures consist mostly of carbon monoxide (CO), hydrogen (H_2) and traces of methane (CH_4). Unlike bio-oil, incondensable biogas cannot be stored easily, safely or economically. Therefore, this low heating value gas fuel must be used immediately. Different gasification technologies exist using a broad range of gasifiers, namely: fixed bed gasifiers; including updraft, downdraft and crossdraft gasifiers, fluidised bed gasifiers; including bubbling fluidised bed (BFB), circulating fluidised bed and entrained flow gasifiers, multi-bed gasifiers; including indirect heating, pyrolysis/char, cyclone and plasma gasifiers (Bridgwater 1995; Siedlecki et al. 2011; Gabra et al. 2001; Rutberg et al. 2011). Gasification conditions in various types of gasifiers have been thoroughly reviewed by Pfeifer et al. (2011).

The gasification products of lignocellulose can be applied to various channels of bioenergy production, such as thermal heat generation (steam, hot water), electrical power production by a steam or gas turbine/engine and as synthesis gas (or syngas), which can be used for the production of liquid fuels (biodiesel), hydrogen, methane, mixed alcohols and other chemicals. From raw syngas (the main product of gasification and a by-product of pyrolysis) a series of liquid hydrocarbons can be produced, such as bio-synthetic natural gas (SNG), bio-hydrogen, biomethanol, ethanol, dimethylether and Fischer-Tropsch fuels, which are synthesised via different catalytic processes (Swain et al. 2011; Zinoviev et al. 2010). Fischer-Tropsch (FT) fuels are mainly aliphatic straight chain and branched hydrocarbons and primary alcohols. The product distribution obtained from FT fuels include light hydrocarbon methane (CH_4), ethylene (C_2H_4), ethane (C_2H_5), LPG (C_3–C_4), propane (C_3), butane (C_4), gasoline (C_5–C_{12}), diesel fuel (C_{13}–C_{22}) and wax (C_{23}–C_{33}) (Naik et al. 2010).

7.3.4 *Direct Liquefaction*

Thermal conversion of biomass to liquids can proceed via non-pyrolytic processes, in which the feedstock is directly heated in a liquid medium that may or may not interact with the biomass (Cheng et al. 2010; Klass 1998; Titirici et al. 2007). Low temperature and atmospheric pressure in the presence of a solvent with acidic or basic catalysts can be used, but these solvents have to be recycled, rendering the process energetically and financially costly. The use of higher temperatures and pressures in water have shown promising results and recently an increase of interest regarding hot-compressed or sub-/supercritical water technologies (hydrothermal technologies) for biomass conversion has appeared (Peterson et al. 2008). Hydrothermal processing can be divided into three main regions, namely: liquefaction, catalytic gasification and high-temperature gasification, depending on the

Table 7.2 Product yields resulting from thermochemical conversion of *Pinus*, *Eucalyptus*, and *Acacia* species

Lignocellulose source	Conversion technologies	Yields (wt.%)	Reference
Eucalyptus species	Gasification	S: 6–15	Pindoria et al. (1998)
		G: 86–88	
	Fast pyrolysis	L: 53–76	Kumar et al. (2010)
		S: 13–20	Oasmaa et al. (2010)
		G: 11–25	
	Slow pyrolysis	L: 46–50	Pimenta et al. (1998)
		S: 30–38	
		G: 16–20	Kumar et al. (2010)
	Combustion	S: 24–44	Suarez et al. (2010)
	Liquefaction	L: 50–90	Zhang et al. (2012)
	Liquefaction	L: 51 daf, wt.%	Sugano et al. (2008)
		S: 41 daf, wt.%	
		G: 8 daf, wt.%	
Acacia species	Vacuum pyrolysis	L: 27	Uras et al. (2012)
		S: 38	
		G: 35	
	Slow pyrolysis	C: 30–34	El-Juhany et al. (2003)
Pinus species	Gasification	L: 40–66	Sun et al. (2011)
		G: 14–40	
		S: $\sim$ 20	
	Flash pyrolysis	L: 65–75	Wagenaar et al. (1993)
		S: 5–18	
		G: 5–20	
	Slow pyrolysis	L: 21–30	Sensoz and Can (2002)
		S: 23–36	
		G: 11–23	
	Supercritical	G: $\sim$40	Kersten et al. (2006)
	Flash pyrolysis	L: 50	Bhattacharya et al. (2009)
		S: 19	
		G: 31	

L liquid, *S* solid, *G* gas

temperature and pressure conditions (Klingler and Vogel 2010). The production of biocrude, aqueous organics, combustible gases (H_2, CO, CO_2, CH_4) and light hydrocarbons is expected.

7.3.5 Overall Comparison of Thermochemical Conversion Technologies

A comparison of typical process yields obtained with thermochemical conversion of woody biomasses available for bioenergy production in the Southern hemisphere is presented in Table 7.2. The product yields depend on the process used and the nature of the feedstocks involved. The relative portions of cellulose, hemicelluloses and lignin in biomass feedstocks differ between hardwoods (Eucalyptus, Acacia, etc.) and softwoods (Pine, etc.) which have a significant impact on the quality of bioenergy products, further aspects of which are addressed in Chap. 8.

7.4 Biochemical Conversion Technologies

Lignocellulosic biomass can be biochemically transformed into different bioenergy products which are either liquid (bioethanol, biobutanol, 1,2-biobutanol, branched liquid alcohol mixtures) or gaseous (biogas, biohydrogen) and can be used as replacement for conventional combustible fuels used for electricity generation, transport and various heating applications. Table 7.3 lists the lignocellulose source (energy crops or residues generated by the forestry industry and pulp mills), the technology for biological transformation and associated products of the transformation.

Table 7.3 Second generation biofuels obtained by biochemical transformation of lignocellulose materials and industrial residues

Woody biomass	Fraction of interest	Production process		Biofuel type
Woody energy crops	Sugars (C5,C6)	Gasification and fermentation		Bioethanol (cellulosic ethanol)
	Syngas	Pretreatment	Hydrolysis and fermentation	
	Sugars		Hydrolysis and ABE fermentation	Biobutanol
			Hydrolysis and synthetic non fermentative pathway	1,2-biobutanol
Industrial forest residues	Glucose		Hydrolysis and butanediol fermentation	Branched alcohol mixtures (high isobutanol content)
Paper sludge	Organic acid		Hydrolysis and fermentation (dark fermentation/ photofermentation)	Biohydrogen
	Glucose			
	Polymers		Anaerobic digestion	Biogas (upgraded biogas)
	Proteins			
	Carbohydrates			
	Lipids			

Lignocellulose conversion into biofuels through biochemical transformation involves the following steps:

- Pretreatment of biomass for enhanced accessibility and digestibility by enzymes
- Hydrolysis of cellulose and hemicellulose to sugars and/or organic acids
- Microbial transformation of sugars/acids into designated biofuels
- Product recovery

From the abovementioned bioenergy products, bioethanol and biogas have been the most extensively studied and are nearest to commercialization, while butanol is receiving increased attention. The conversion technologies involved as well as biomass specifications are described in more detail in the following sections.

7.4.1 Pretreatment of Woody Biomass for Hydrolysis

The biochemical production of different biofuels occurs through the hydrolysis of polysaccharides contained in lignocellulosic biomass into fermentable sugars and organic acids. Hydrolytic catalysis of biomass is achieved using either acid-catalysed hydrolysis or enzyme-catalysed hydrolysis. While acid hydrolysis is applicable to certain instances of biofuel production, enzymatic hydrolysis is a more economically and technologically advantageous option given the specificity of enzymes, the mild conditions required for the process and the high product yields gained.

The complex structure of lignocellulose calls for a pretreatment step in order to render the substrate more amenable to hydrolysis. Knowledge of the composition and structure of the raw materials is imperative to determine the most suitable pretreatment, to choose which enzymes to use as catalysts and to select the microorganisms most suited for optimal biofuel production. In fact, identifying the most suitable pretreatments and conditions for the fractionation of different lignocellulosic materials into its main structural components is one of the most important goals of research and development (http://www.eng.auburn.edu/cafi/index.htm). Furthermore, the pretreatment step is currently considered the second largest contributor to the cost of second generation biofuels after feedstock acquisition (Stephen et al. 2010). The effectiveness of the pretreatment will determine the yield in each successive step and therefore also the final product yield.

The heterogeneity of lignocellulose feedstocks has driven the investigation of numerous pretreatments but these can be broadly categorized into physical, chemical and biochemical pretreatments or can be a combination of these different classes of pretreatment (Alvira et al. 2010; Agbor et al. 2011). Among the different pretreatment options, organosolv, dilute acid prehydrolysis, acid-catalysed steam explosion and the novel SPORL technique (Sulphite Pretreatment to Overcome Recalcitrance of Lignocellulose) are the most promising technologies to facilitate the commercialization of biorefining of woody materials (Wang et al. 2009).

However, other pretreatments could be of interest for forestry residues that are more easily digestible. Conventional as well as some novel pretreatments and their mode of action are briefly described next with reference to specific studies on *Pinus*, *Eucalyptus* and *Acacia*.

7.4.1.1 Physical Pretreatments

Mechanical comminution (see Chap. 8) is a requirement for the biochemical conversion of biomass to alter the recalcitrance of lignocelluloses by reducing particle sizes and the degree of polymerization of the celluloses (Vidal et al. 2011). The reduction of particle size favours heat and mass transfer during pretreatment, increasing the susceptibility of biomass to hydrolysis. The ultimate particle size is determined by the feedstock type and the pretreatment method applied. In the case of woody biomass, the reduction of particle size to less than 3 mm (based on screen openings used for biomass fractionation) does not impact the digestibility further. Furthermore, the excessive reduction of particle size can have a detrimental effect on the digestibility of softwoods (Cullis et al. 2004). Different pretreatments tolerate different particle sizes: steam explosion allows greater particle sizes (large, >10 mm) followed by liquid hot water (intermediate, 1–15 mm), in turn followed by dilute acid and base pretreatments (low, <3 mm). In the case of biogas production, particle reduction increases digestibility by 5–25 % as well as reducing hydrolysis time by 23–59 % (Kratky and Jirout 2011).

During extrusion, materials are exposed to friction, heat, mixing and shearing, which results in chemical and physical modifications as the material moves through the extruding device. Moreover, the fact that extrusion is a continuous treatment supports its industrial application and subsequent commercialisation. Extrusion induces depolymerisation of cellulose, hemicellulose, lignin and protein which enhances lignocellulose conversion and the yields of biogas or alcohols (Hjorth et al. 2011). This pretreatment has been tested using pine wood chips, resulting in similar sugar recovery to conventional pretreatments but with no by-product formation (Karunanithy et al. 2012).

7.4.1.2 Chemical Pretreatments

Alkali pretreatment uses bases (sodium, potassium, calcium or ammonia hydroxide) as catalysts. Lime [$Ca(OH)_2$] treatment is most beneficial from an economic and health and safety point of view and lime can be easily recovered for re-use (Yang and Wyman 2008). Alkali treatment induces swelling of celluloses and selective removal of hemicelluloses coupled with lignin solubilisation, which renders the lignocellulose more accessible to enzymes and bacteria. Although the amount of inhibitors generated is lower than in acid pretreatments, hemicelluloses can undergo peeling reactions and be degraded into furans that, together with the solubilised lignin, could negatively impact the functioning of microorganisms. This technology can increase the methane yield of residual materials such as newsprint (Fox et al. 2003) but is less attractive for pretreatment of woody materials given the negative effects of high lignin content on the process.

Acid hydrolysis pretreatment is mostly performed using sulphuric acid catalysts, but other mineral acids such as hydrochloric, nitric and trifluoroacetic acids have also been applied. Processes utilising acid catalysts can either be carried out by concentrated-acid/low temperature or dilute acid/high temperature hydrolysis. Concentrated acid allows the hydrolysis of both cellulose and hemicellulose under moderate temperatures, but requires high acid concentrations (72 % H_2SO_4, 41 % HCl or 100 % TFA), which makes recovery costly and can lead to equipment corrosion (Gírio et al. 2010). Lower concentrations (30 %) of acid have been reported to retrieve 41 % of the total theoretical glucose from pine sawdust (Miller and Hester 2007). Dilute acid selectively hydrolyses the hemicellulose fraction, while the cellulose remains in a solid fraction that can be hydrolysed by enzymes or by a second dilute acid step. Dilute acid has been applied to species of *Eucalyptus* (McInstoch et al. 2012; Silva et al. 2011; Wei et al. 2012), *Acacia* (Ferreira et al. 2011) and *Pinus* (Huang and Ragauskas 2012). Although dilute acid treatment implies lower acid consumption by the substrate when compared to concentrated acid treatments, the higher temperatures of operation can lead to greater equipment corrosion and higher levels of hemicellulose degradation. Both acid schemes entail a neutralization step prior to biochemical transformation.

Ozonolysis is achieved by the treatment of lignocellulose with oxidizing agents such as ozone, which mainly reduces lignin content, slightly affects hemicellulose and increases the sugar yield of enzymatic hydrolysis. It can be conducted at normal pressure and room temperature, so that it does not generate compounds that are toxic to further hydrolysis and fermentation. Ozonolysis could be effective for the pretreatment of lignocellulose-rich residues such as sawdust (Ncibi 2010), but the amounts of ozone required makes this process costly.

The *organosolv* process employs organic or aqueous organic solvent mixtures with ethanol, methanol, acetone, ethylene glycol and tetrahydrofurfuryl alcohol as potential components, which can be supplemented with an acid catalyst (HCl, H_2SO_4, oxalic or salicylic acid) to disrupt the link between hemicellulose and lignin. This improves the susceptibility of cellulose to enzymatic hydrolysis by increasing enzyme accessibility to cellulose in both hardwoods (Romaní et al. 2011) and softwoods (Park et al. 2010). An additional advantage of this process is the recovery of relatively pure lignin.

The *novel SPORL* (*Sulphite Pretreatment to Overcome Recalcitrance of Lignocellulose*) *approach* to pretreatment is based on the pulping of biomass in the presence of sulphites and was developed to enhance the biochemical conversion of softwoods with large particle sizes (Zhu et al. 2009). This technology consists of sulphite/bisulphite treatment of wood chips under acidic conditions and, contrary to conventional pretreatment technologies, is followed by a reduction of particle size by means of disk milling. The removal of hemicelluloses (pulping spent liquor) and sulfonation of lignin is considered to be critical for enhanced cellulose conversion. Moreover, this technology reduces the energy consumption required for size-reduction to values equivalent to agricultural biomass (Zhu et al. 2010).

Ionic liquids (*ILs*), also named "green solvents", are organic salts composed mainly of organic cations with small amounts of either organic or inorganic anions, with the ability to dissolve a wide range of organic and inorganic compounds. IL solvents have several valuable properties including a low melting point, chemical and thermal stability, negligible vapour pressure and relatively low toxicity (Liu et al. 2012). Additionally, its solvent properties can be adapted for a particle substrate by adjusting the ratio of cations to anions. The synergy of ILs with other compounds such as acids (Diedericks at al. 2012) or solid super acids (Brønsted superacids or Lewis acids) have also been investigated (Gírio et al. 2010). The most common ILs can be classified according to their cations in four groups: quaternary ammonium ILs, N-alkylpyridinium ILs, N-alkyl-isoquinolinium ILs, and 1-alkyl-3-methylimidazolium ILs. Most of these solvents remove lignin and alter cellulose structure, which increases the accessibility of cellulolytic enzymes. Imidazolium-based ionic liquids have been used to dissolve hardwoods and softwoods (Mora-Pale et al. 2011). The ionic liquid 1-ethyl-3-methyl imidazolium acetate ([C2mim][OAc]) has been shown to increase cellulose digestibility of species of *Eucalyptus* (Çetinkol et al. 2010) and *Pinus* (Torr et al. 2012).

Although the technology based on ILs would require less equipment and energy input compared to conventional pretreatments, efficient methods for both recovery of the different fractions and the recycling of ILs should be developed for large scale application. The development of such processes would circumvent the negative impact that some ILs have on enzyme activity and effective microorganism functioning (Wang et al. 2011).

7.4.1.3 Physicochemical Pretreatments

Steam explosion is one of the most studied pretreatments since its development for commercial scale application (Masonite technology) and has been applied successfully on a variety of lignocellulosic materials (Ballesteros et al. 2004; Taherzadeh and Karimi 2007). This process combines thermal (high temperature), mechanical (sudden vaporization of the water) and chemical (hydrolysis of hemicelluloses) alteration of biomass. During steam explosion the biomass is exposed to saturated steam at high pressure for a period of time (seconds to several minutes) after which it is suddenly depressurised. The water penetrates into the lignocellulose structure and hydrolyses acetyl residues from the hemicelluloses, which in turn promotes partial hydrolysis of the hemicelluloses. The sudden depressurization induces the mechanical rupturing of fibres and redistribution of lignin. As a result, lignocellulose is more accessible to enzymes and the solubilised hemicelluloses can be easily recovered by filtration. However, solubilised lignin and sugars can be further degraded into compounds that can be inhibitory to further bioprocessing.

The concentration of degradation products, and therefore the extent of toxicity, depends on the source of biomass, the severity of the pretreatment (often measured as the combination of temperature and residence time) and properties of the specific enzymes/microorganisms involved in the subsequent biochemical conversion.

Despite its feedstock versatility, steam explosion is not very effective on softwood materials owing to its lower acetyl content. The addition of an acid catalyst (H_2SO_4, SO_2) has been recommended as an option to improve the performance of steam explosion on softwoods and other woody materials, which will result in reduction of pretreatment severity and maximise sugar yields (Ewanick et al. 2007; García-Aparicio et al. 2011). Similarly, the application of CO_2 with organic acids has been suggested to reduce the severity of pretreatment, in turn reducing hemicellulose degradation into inhibitory compounds (Gírio et al. 2010).

Liquid hot water is a hydrothermal treatment that employs compressed hot water (pressurised and above saturation point) at high temperatures to disrupt the lignocellulose structure. Compared to steam explosion, this treatment is very effective for hemicellulose solubilisation, leading to reduced inhibitor formation when the reaction pH is kept between 4 and 7. However, the concentration of hemicellulose-derived sugars is reduced due to the higher water input.

Ammonia fibre explosion (*AFEX*) combines the use of liquid anhydrous ammonia at high temperature (60–100 °C) and pressure. After a variable period of heating, sudden decompression provokes the expansion of ammonia gas altering lignocellulose structure with limited inhibitor formation. Contrary to other thermochemical pretreatments, AFEX treated material is a solid with a similar carbohydrate composition to the raw material but is more digestible by cellulases and hemicellulases. Moreover, the residual ammonia after recycling can reduce nutritional requirements in the following fermentation step. Although it has been suggested that AFEX alters lignin, reducing its ability to bind enzymes (Kumar and Wyman 2009), this pretreatment is not very effective when applied to woody biomass (Kumar et al. 2009).

In *ammonia recycle percolation* (*ARP*) the biomass is subjected to aqueous ammonia (5–15 %, w) in a flow-through system (approximately 5 ml/min) at high temperatures (normally 170 °C). ARP solubilises hemicelluloses and lignin providing cellulose-enriched residues with high digestibility. ARP has shown to be efficient in increasing digestibility of hardwoods, waste paper and softwood pulp mill sludges (Gírio et al. 2010). Sugar degradation is minimal during ARP, but the solubilised lignin can be toxic for microorganisms.

Wet oxidation involves the exposure of biomass to oxygen or air at high temperatures (170–200 °C) and pressures (10–12 bars) for short periods of time (10–15 min). It solubilizes hemicelluloses (mainly in polymeric form) and lignin (Hendricks and Zeeman 2008). The phenolic compounds are further degraded to carboxylic acids but furans formation is lower in comparison with steam explosion or liquid hot water.

The *choice of pretreatment* depends on the type of feedstock and the desired biofuel output. The majority of pretreatments generate a material known as slurry which consists of a solid fraction enriched in cellulose (and lignin, depending on the pretreatment) designated as the water insoluble fraction (WIS) and a liquid fraction or prehydrolysate containing the sugars solubilised during the pretreatment (mainly hemicellulose-derived sugars). Depending on the severity of the pretreatment, the sugars can be further degraded into furans that, coupled with the solubilised lignin and acetic acid released from the hemicelluloses, impact negatively on biochemical transformation. For this reason, most studies separate slurry into the separate liquid and solid fractions to optimise the conversion of each fraction into biofuels. Current research focuses not only on the development of detoxification processes but also on the cultivation of robust microorganisms that are able to effectively convert hexoses and pentoses in slurry into biofuels. This aspect seems to have less impact for biogas production given the higher tolerance of methanogenic bacteria to inhibitors (Hendriks and Zeeman 2008).

7.4.2 Enzymatic Hydrolysis of Pretreated Lignocellulose

Following pretreatment, hydrolysis and transformation of sugars into biofuels is carried out, generally by fermentation. As mentioned before, enzyme-based processes are preferred due to their high specificity under milder conditions with no formation of toxic compounds. The main enzymes involved in the hydrolysis of polysaccharides are cellulases and hemicellulases, which belong to the glycosyl hydrolase family (http://www.cazy.org/).

Cellulases, mainly derived from fungi such as *Trichoderma reesei*, consist of a mixture of several enzymes that act synergistically to facilitate cellulose degradation. The core enzymes of cellulases are endoglucanase (EG, endo-1,4-β-glucanohydrolase, or EC 3.2.1.4), exoglucanase (CBH, 1,4-β-glucan cellobiohydrolase) and β-glucosidase (EC 3.2.1.21) and decompose the substrate in a stepwise manner. EG acts on amorphous regions of cellulose, creating free chain-ends that are further hydrolysed by CBH releasing cellobiose units, which are cleaved into glucose by the β-glucosidase.

Hemicelluloses are normally solubilised during pretreatment into monomers and oligomers with different degrees of polymerization. In other pretreatments such as AFEX, both celluloses and hemicelluloses remain in the pretreated material. Therefore, hemicellulases are necessary for conversion into monomeric sugars. Furthermore, these enzymes enhance cellulose hydrolysis by removing residual hemicelluloses from the fibres and reducing the inhibitory effect of the xylo-oligosaccharides (Qing et al. 2010). The complex structure of hemicellulose requires several enzymes to complete its hydrolysis, but they can be broadly classified as xylanases and mannanases (Gírio et al. 2010).

The addition of other enzymes that affect lignin (peroxidases, oxidases, and laccases) have been also recommended as part of biochemical pretreatment to reduce lignin content (Wang et al. 2012) or to facilitate detoxification of pretreated materials prior to enzymatic hydrolysis and fermentation (Moreno et al. 2012).

Enzyme production and application has been estimated to be one of the main contributors to the cost of second generation ethanol production (Klein-Marcuschamer et al. 2012). Cellulases have lower specific activity than amylases used in first generation ethanol production, thus increasing enzyme dosage requirements. The enzyme loading for optimal conversion is determined by the type of feedstock used, pretreatment technology applied, pretreatment severity and solid concentration. Apart from optimization of pretreatment, different strategies are implemented to reduce enzyme loadings. These strategies include the development of robust enzyme producers, the improvement of existing enzyme systems and bioprospecting of new enzymes (Banerjee et al. 2010; Huang et al. 2011). Regarding the enzymatic process, the construction of tailor-made enzyme combinations adapted to feedstock and pretreatment and/or the addition of additives such as surfactants enhances the yield of enzymatic hydrolysis while reducing cellulase requirements. Moreover, surfactant addition could favour enzyme recycling (Tu and Saddler 2010).

7.4.3 Fermentation of Sugars for the Production of Cellulosic Ethanol

Alcoholic fermentation is a biochemical transformation where monomeric hexoses (glucose and mannose) and pentoses (xylose, arabinose) are transformed into ethanol and carbon dioxide. There are several ethanologenic microorganisms with the ability to ferment sugars to ethanol, with *Saccharomyces cerevisiae* being the most commonly used due to its high tolerance to ethanol and toxic compounds, delivering conversion yields close to the theoretical maximum (0.51 g ethanol/g sugar). However, the main limitation of *S. cerevisiae* is that it lacks the ability to carry out pentose fermentation.

Currently there is no natural microorganism that meets all the traits desirable for an efficient and complete fermentation of lignocellulosic ethanol (e.g. cofermentation of hexoses and pentoses, high yields and productivity, tolerance to ethanol and other inhibitors). Since the conversion of pentoses would contribute to cost reduction, microorganisms are being modified by metabolic engineering to introduce the pentose utilization pathway into their metabolome (Madhavan et al. 2012; Kuhad et al. 2011). Apart from *S. cerevisiae*, the microorganisms *Escherichia coli*, *Zymomonas mobilis* and *Pichia stipitis* are projected to be the most beneficial species for second generation ethanol production (Gírio et al. 2010).

High solid loading (15–20 %) during fermentation is crucial to obtain ethanol concentrations of at least 4 % (v/v), which is considered to be the benchmark to reduce the energy consumption of the distillation step. However, the maximum loading of solids is limited by end product inhibition of enzymes, as inhibitors are generated during pretreatment and mass transfer limitations. The application of fed-batch processes and the development of bioreactors with improved mixing capacity and low energy consumption have been proposed as strategies for increasing ethanol concentration.

Enzymatic hydrolysis (EH) and fermentation can be performed with different levels of integration. Separate hydrolysis and fermentation (SHF) is a two-stage process in which enzymatic hydrolysis and fermentation is carried out in two different reactors. During EH, the polysaccharides of the pretreated material are hydrolysed into monomers by the action of enzymes, mainly cellulases. This hydrolysate is transferred to a second reactor where the sugars are transformed to ethanol by fermenting microorganisms (bacteria or yeasts). The advantage of this process is that the two different steps can be optimised individually. However, glucose and xylo-oligomers inhibit cellulolytic enzymes. End-product inhibition can be minimised by the application of a simultaneous saccharification and fermentation (SSF) or simultaneous saccharification and cofermentation (SSCF), which results in improved overall yields. In fact, economic analysis has indicated that high-solid SSF process will significantly reduce the operating cost of ethanol production (Rudolf et al. 2005). The major difficulty faced when trying to combine EH and fermentation into a single step lies in the difference in optimal conditions of the two processes, mainly embodied by differences in preferred reaction temperature. Since the optimum temperature of enzymes involved in EH is close to 50 °C, the use of thermotolerant microorganisms would be recommended when combining EH with fermentation. The maximum level of integration is reached in consolidated bioprocessing (CBP), where enzyme production, enzymatic hydrolysis and fermentation are carried out in the same reactor by one or several microorganisms (La Grange et al. 2010; van Rensburg et al. 2012). Bacteria from the genera *Clostridium* are the most studied given their ability to hydrolyse cellulose and coferment glucose and xylose into ethanol and other products (Li et al. 2012).

7.4.4 Fermentation of Sugars to Butanol, Butanediol and Alcohol Mixtures

Butanol can be generated biologically by acetone-butanol-ethanol (ABE) fermentation. This metabolic route is present in bacterial species in the genus *Clostridium* such as *C. acetobutylicum* and *C. beijerinckii*. ABE fermentation has been industrially applied using starchy biomass as an alternative to chemical synthesis from fossil-fuels. Similarly, the sugars present in lignocellulose hydrolysates can be fermented into 1.2-biobutanol, also known as 2,3-butylene glycol (2,3-BD), by mixed acid-butanediol fermentation (Menon and Rao 2012). Several microorganisms have the 2,3-BD pathway but the most commonly used is *Klebsiella oxytoca*, given its wide sugar spectrum and adaptive potential. Branched alcohol mixtures with high iso-butanol content can also be formed from glucose through synthetic pathways present in bacteria such as *Escherichia coli* (Rodríguez and Atsumi 2012).

Although these technologies are not as mature as those involved in the production of bioethanol, there is a growing interest in these kinds of biofuels, reflected by the investment in research for their commercial application by companies such as

Dupont, BP, Gevo and Green Biologics. The economic feasibility of production of these biofuels depends on the use of cheaper feedstocks to increase yields and productivities as well as on the development of more efficient recovery processes (Jin et al. 2011). In this context, the use of lignocellulose residues, selection of bacterial strains as well as process development is expected to reduce butanol production costs (Kumar et al. 2012).

Typical process yields for biochemical conversion of woody biomass available in the Southern hemisphere to bio-alcohols are presented in Table 7.4. The impact of biomass properties on biochemical conversion processes is addressed in Chap. 8.

7.4.5 Anaerobic Digestion for Biogas Production

Anaerobic digestion is a complex process involving various chemical reactions such as hydrolysis, acidogenesis and methanogenesis by means of several bacterial strains industrially applied for bio-gas production. This process can be applied to a wide range of biomass types, especially those with high moisture content. There is a growing interest in the application of this technology to transformation of lignocellulosic biomass. Pure lignocellulosic biomass represents an under-utilised source for biogas and -ethanol production (Hendriks and Zeeman 2008), primarily due to the recalcitrance of lignocellulose to biological degradation. Hydrolysis of lignocellulosic materials is the first step for either digestion to biogas or fermentation to ethanol, but it is considered to be the most rate-limiting step (Lissens et al. 2004; Sanders et al. 2000). Pretreatments developed for alcoholic fermentation (see Sect. 7.3.1) can improve the efficiency of lignocellulose hydrolysis during anaerobic digestion, which leads to increased yields and productivity while decreasing residence times for lignocellulose digestion (Taherzadeh and Karimi 2008). Among physicochemical processes, steam pretreatment, lime pretreatment, liquid hot water pretreatment and ammonia based pretreatments have high potential for application to biogas production (Hendriks and Zeeman 2008). Steam explosion and thermal pretreatments are widely investigated for improving biogas production from different materials such as forest residues (Hooper and Li 1996). Liu et al. (2002) investigated and developed steam pressure disruption as a treatment step to render lignocellulose-rich, solid municipal waste more digestible, which would result in increased biogas yields. During most steam pretreatment processes small amounts of sugar degradation products are formed, with varying amounts of furfural, acetic acid, HMF and soluble phenolic compounds present in the pretreated lignocellulose. These by-products may be inhibitory to fermentation by methanogenic bacteria and thus methane production. The consortium of microorganisms involved in anaerobic digestion is however capable of adapting to overcome the effects of such compounds, even though there are limits to the adaptive potential. Bacterial adaptation to fermentation in the presence of inhibitory compounds was demonstrated by Benjamin et al. (1984), Fox et al. (2003), over several bacterial generations.

Table 7.4 Product yields from the biochemical conversion of *Pinus*, *Eucalyptus*, and *Acacia* species to bio-alcohols as liquid biofuels

Lignocellulose source	Conversion technology	Ethanol concentration (g/l)/cellulose conversion/sugar recovery	Reference
Eucalyptus feedstock			
Eucalyptus thinnings	Dilute acid	CSF = 1–1.6 93 % xylose recovery	McIntosh et al. (2012)
		CSF = 2.48 73 % glucan conversion and 18 g/l ethanol	
Eucalyptus bark	CO_2 hydrothermal treatment	80 % glucose yield	Matsushita et al. (2010)
Eucalyptus chips	Autohydrolysis at Severity = 4.67	15.1 g/l ethanol	Romaní et al. (2010)
	Dilute acid (160 °C, 0.75 % acid concentration for 10 min)	82 % of total sugars	Wei et al. (2012)
	Acid hydrolysis (1.2 % H_2SO_4, 121 °C for 45 min)	28.7 g/l ethanol	Silva et al. (2011)
	Delignification (60 % ethanol) at H = 125,001	35 g/l ethanol	Muñoz et al. (2011)
	Autohydrolysis at So = 4.67	67.4 g/l ethanol	Romaní et al. (2012)
	Autohydrolysis and organosolv So = 3.64, TD = 198, C = 60 kg ethanol/100 kg liquor	18.1 kg of saccharides (as monomers and sugars) from first step, 17.9 kg of soluble lignin and 41.9 kg glucose from second step	Romaní et al. (2011)
	Hot water (160 °C for 30 min) and ball milling (BM) (20 min)	70 % yield of total sugar with a cellulase loading of 4 FPU/g substrate (10 times reduction)	Inoue et al. (2008)
	Two step liquid hot water (180 °C for 20 min followed by 200 °C for 20 min)	total sugar recovery of 96.63 %	Yu et al. (2010)
	Steam explosion (25 atm for 3 min) and anaerobic fermentation	80 % conversion of cellulose into biogas	Nakamura and Mtui (2003)
	Steam explosion (210 °C for 4 min)	17 g/l ethanol (62.5 % yield)	Ballesteros et al. (2004)
Acacia feedstock			
	Dilute acid 180 °C, 0.8 % H_2SO_4 for 15 min	10.3 g/l ethanol	Ferreira et al. (2011)
	Hydrothermal followed by alkaline treatment (NaOH concentration = 4.5 %, 130 °C for 3 h),	47.3 g glucose/100 g solids from autohydrolysis	Kaida et al. (2009)
	Autohydrolysis	77–99 % xylose solubilization	Yáñez et al. (2009)

Pinus feedstock			
	Acid-Catalysed Steam Pretreatment 200 °C, 5 min, and 4 % SO_2	SHF overall ethanol yield of 72 %, SSF 63 % of theoretical total (22 g/l)	Ewanick et al. (2007)
	Supercritical water (425 °C) 220 atm for 60 s with 0.05 % hydrochloric acid	25.1 % of monomeric sugars	Kim et al. (2010)
	Extrusion: screw speed of 150 rpm and a barrel temperature of 180 °C with a moisture content of 25 %	Total sugar recovery of 66.1 %	Karunanithy et al. (2012)
Pinus sawdust	Concentrated acid at 30 wt.%, 130 °C for 3 min 130 ° C, 5 wt.% acid, 25 min	41 % recovery of total theoretical glucose 50 % of the theoretical glucose	Miller and Hester (2007)
	SPORL sodium bisulphite solution at 180 °C for 20 min	270 l/t wood	Tian et al. (2010)
	Sodium bisulphite and sulphuric acid charges on oven-dried wood were 8 % w/w and 2.21 % w/w respectively		
	Organosolv with 1 % sulphuric acid	55–60 % digestibility providing 70 % of total theoretical yield	Park et al. (2010)

7.5 Integration of Different Conversion Technologies

The cost disadvantages of bioenergy conversion from lignocellulose relative to fossil-based energy sources can be addressed through innovative methods of process integration, the goal of which is to minimize capital investment, maximize energy efficiency and thus improve overall economics. Such optimisation of overall process performance and energy efficiency will also increase the environmental benefits that can be derived from bioenergy production. Heat integration within biochemical and thermochemical routes of lignocellulose conversion has the potential to increase overall energy efficiency by as much as 15 % and can reduce capital and operational costs substantially (Van Zyl et al. 2011; van der Drift et al. 2004). Similarly, the integration of energy cycles for biomass conversion processes with adjacent/associated industrial processes can address both energy efficiency and production costs of the lignocellulose conversion process. Process integration with adjacent industrial processes can be broadly classified as (i) integration with electricity production from biomass or fossil fuels, (ii) integration with biomass processing for pulp or sugar production, (iii) integration of first and second-generation biofuel production by biochemical processing, (iv) integration of second-generation biofuel production by thermochemical processing with petrochemical processing and (v) integration of biochemical and thermochemical processing of lignocellulose to second-generation biofuels. The economics of process integration carries scale-dependent economic benefits, whereby more expensive, high efficiency equipment becomes affordable at larger production scales. Several of the conversion technologies presented in this chapter may be combined to form a value chain, in particular through the production of bioenergy intermediates such as wood chips, pellets and briquettes and liquid products such as bio-oil. These intermediates have higher bulk density than harvested lignocellulose which significantly reduces biomass feedstock transportation costs (Stephen et al. 2010; see Chap. 6). Further examples of integration of more than one conversion technology are presented below.

Combustion and gasification: Gasification combined with pre-combustion carbon capture can be used to produce either biofuels or electricity and improve the efficiency of Integrated Gasification Combined Cycle processes (IGCC) (Prins et al. 2012). Pre-treating biomass with hydrothermal carbonization (HTC) produces a coal-like substance (biocoal) which is potentially better suited for entrained flow gasification than raw biomass (Erlach et al. 2012).

Gasification and combustion: Biomass downdraft reactors coupled with reciprocating internal combustion engines (RICEs) are a viable technology for small scale heat and power generation (Martínez et al. 2012). Dry gasification oxy-combustion (DGOC) is a process best described as a hybrid between gasification and oxy-combustion systems (Walker et al. 2011).

Torrefaction and gasification: The main idea behind combining biomass torrefaction and gasification is that the heat produced during gasification in the form of steam is recovered for application to torrefaction (Van der Stelt et al. 2011;

Prins et al. 2006). Gasification using torrefied biomass allows for improved flow properties of the feedstock, increases levels of H_2 and CO in the resulting syngas and improves overall process efficiencies.

Torrefaction and combustion: Combustion reactivity of torrefied biomass has been evaluated and shows promise for biomass co-firing in existing coal-fired power stations (Bergman et al. 2005; Bridgeman et al. 2008).

Torrefaction and fast-pyrolysis: Recent development of torrefaction as a pretreatment technology for fast pyrolysis results in enhancement of bio-oil properties by reducing oxygen-to-carbon ratios and water content (Meng et al. 2012).

Fast-pyrolysis and gasification: The adaptation of distributed fast pyrolysis biomass processing systems to central gasification systems in order to facilitate the production of hydrocarbon transport fuels is currently being developed by KIT (Dahmen et al. 2012). Fast pyrolysis bio-oil can also be gasified through a catalytic steam reformer (Czernik et al. 2002). Bio-oil gasification in entrained flow, oxygen blown pressurised gasifier systems is also feasible with applications currently in use by Texaco and Shell (Bridgwater 2011).

Fast-pyrolysis and combustion: The integration of fast pyrolysis and combustion technologies have been extensively studied (Czernik and Bridgwater 2004) and commercially applied (VTT, Dynamotive, Ensyn, btg-btl), while also being under further development (Khodier et al. 2009).

Direct liquefaction and gasification: Pre-treating biomass with hydrothermal carbonization (HTC) produces biocoal which is potentially better suited for entrained flow gasification than raw biomass (Erlach et al. 2012).

7.6 Technology Maturity and Economic Considerations for Biomass Conversion

The degree of technological maturity, economic cost considerations and conversion efficiency strongly influence technology selection decisions when conversion of woody biomass into bioenergy products is undertaken. Conversion efficiencies in particular have a key impact on the overall environmental impact of bioenergy value chains and therefore receive in-depth attention in this regard in Chap. 11, while not being discussed further here. Technology maturity and economic considerations are considered here for each of the thermochemical and biochemical conversion methods presented.

7.6.1 Technology Maturity for Commercialisation

Several innovative technologies for the thermochemical or biochemical conversion of woody biomass into useful bioenergy products are presently under development, although few of these have reached commercialization, as discussed below. A complete list of lignocellulose processing facilities for bioenergy production, including types of feedstocks used, processing scale and operational status, is available at http://www.bioenergy2020.eu/files/publications/pdf/2010-bericht-demoplants.pdf.

Combustion: Systems that employ direct combustion to convert biomass and charcoal into energy for heat, power, and CHP (Combined heat & Power) are widely utilized and commercially available for small-, medium- and large-scale applications. Large scale co-firing of bio-oil has been carried out at Manitowac and Red Arrow, but few other cases of application exist.

Pyrolysis: Different pyrolytic technologies, including torrefaction (Brownsort 2009), pressurized pyrolytic reactor processing (Antal and Grønli 2003), slow pyrolysis (Brownsort 2009), vacuum pyrolysis (Bridgwater and Peacocke 2000) and fast pyrolysis (Dahmen et al. 2012; Bridgwater and Peacocke 2000) are all considered to be mature technologies. Slow pyrolysis technologies for charcoal and biochar production are commercially available (Bioenergy Ltd., Yury Yudkevitch, Biogreen Energy, Enecon, Pty Ltd, ICM Inc., Pacific Pyrolysis (formerly BEST Energies)), while fast pyrolysis is on the verge of commercialisation (Dynamotive, Ensyn, BTG, Biomass Eng., KIT/Lurgi, Pytec, ARBI-Tech, ROI, Agri-Therm, Anhui Yineng, Metso Consortium).

Gasification-synthesis: Biomass gasification technologies have been sufficiently developed to be considered as a significant contributor to global sustainable energy production. Nevertheless, there are still some issues with biomass processing (pre-treatment, gas cleaning, reforming efficiency, etc.) to be addressed before successful large-scale commercial introduction of biomass gasification takes place (Bridgwater et al. 2002; Tijmensen et al. 2002). Commercial-scale technology for Fischer-Tropsch synthesis using syngas has been in operation for several decades in South Africa, Malaysia and elsewhere.

Direct liquefaction: Commercialisation of hydrothermal technologies suffers from difficulties that arise with the conversion of batch reactors to continuously processing systems, as it is difficult to pump fluids at high pressures and low flow rates (Peterson and Haase 2009). Nevertheless, the supercritical water gasification process appears to be a suitable technique for hydrogen production from biomass at the commercial scale (Calzavara et al. 2005).

Alcoholic fermentation from lignocellulose: The biochemical conversion of lignocellulose into cellulosic ethanol by fermentation has been substantially developed in the past few decades, comparatively more than thermochemical technologies for liquid transportation fuel production (Anex et al. 2010). The first commercial projects for cellulosic ethanol production using woody biomass and/or forest residues as feedstocks are presently under development, including several efforts by Borregaard Industries (Norway), Mascoma Corp (USA), KL Energy Corporation (USA) and SEKAB (Sweden). Butanol fermentation from lignocellulose requires further development prior to commercialisation.

Anaerobic digestion of lignocellulose: Anaerobic digestion (AD) for the production of biogas is a well-established commercial technology. However, the limitations in terms of conversion efficiency and productivity of lignocellulose conversion requires either the co-digestion of limited amounts of lignocellulose with readily digestible, high-nitrogen content substrates (such as sewage sludge) or alternatively further development of pretreatments applied to lignocellulose to improve digestibility. Pretreatment development needs to facilitate a greater degree of pure lignocellulose digestion as well as higher tolerance of AD to inhibitors formed by sugar degradation during pretreatment.

7.6.2 *Economies of Scale*

The potential for small-scale, rural application of bioenergy production processes depends to a large extent on the economic costs associated with small-scale application, which will be considered here on the basis of several examples. The economic aspects presented here should be integrated with the biomass harvesting and transport aspects considered in Chap. 6.

7.6.2.1 Electricity

Electricity production from lignocellulose, as well as retrieval of intermediate lignocellulose-derived bioenergy products can be performed at small-, medium- or large-scales, but smaller scales are invariably more costly. Production costs of combustion, gasification and fast pyrolysis technologies have been compared for biomass-derived electricity production in the range of 1–20 MW (Bridgwater et al. 2002). Although the costs of technologies differed significantly at small scales, production costs between technologies converged at larger scales. The abovementioned study recommended the use of pyrolysis bio-oil in diesel engines and gas turbines for direct production of electricity at small scales, although further development is required to address the corrosive effects of bio-oil on equipment. Community-level projects for the production of electricity from community-managed forests were investigated for the state of Madhya Pradesh, India (Dwivedi and Alavalapati 2008). Gasification technology was preferred for this application, and showed a five-fold reduction in electricity production costs from 5 to 100 MWe. The average cost of electricity at the consumer level produced using the largest capacity 100 kW gasifier was \$0.15/kWh, which was greater than the \$0.08/kWh price of electricity supplied from the grid. The study demonstrated significant differences in the hectare requirements for different types of biomass, based not only on the projected annual biomass increment per hectare, but also on the calorific value (energy content) of the various types of biomass (see Chap. 8). Such changes in biomass yields will also affect the environmental impact of bioenergy supply chains (see Chap. 11). Large-scale electricity production from forests showed an economic optimum in the range of 450–3,150 MW, indicating that a wide range of scales can be considered, rather than only a minimum production scale (Kumar et al. 2003). A similar plateau-effect in the capital investment for electricity production has been observed, showing increased costs at smaller scales (Uslu 2005; Uslu et al. 2008; Bridgwater et al. 2002). Although a minimum cost of electricity production can be achieved by increasing production scale, economic benefits should be balanced against increased biomass feedstock demand, which may increase production costs, as will be considered in the section below. As a result of limitations in biomass supply, a smaller scale of electricity production may need to be employed in certain instances. For large-scale biomass co-firing in coal-fired power plants, a location near a large deep-water harbour to the facilitate shipping of large quantities of feedstocks and bioenergy products is an important advantage for economic competitiveness (IEA ETSAP 2010).

7.6.2.2 Liquid Biofuels

Although there are significant differences in the capital investments for pyrolysis-upgrading, gasification-synthesis and biochemical conversion of lignocellulose to liquid biofuels for transportation, these technologies all require large-scale implementation to make the high capital costs worthwhile (Anex et al. 2010; Kazi et al. 2010; Stephen et al. 2012; Kumar et al. 2012; Brown et al. 2013). Furthermore, it is also expected that the first commercial plants for these technologies will perform at less than design capacity and have capital costs that are higher than anticipated (Anex et al. 2010). Despite the need for bioenergy in transportation, the high capital costs and technology risks associated with newer technologies represent hurdles on the pathway to commercialisation (Stephen et al. 2012; Anex et al. 2010). The wide range of multiple, competing, commercially unproven technologies for the production of liquid transportation fuels from lignocellulose also indicates that standardisation of the conversion process, with associated economies-of-scale in plant component production to reduce capital costs, will not be achieved soon (Stephen et al. 2012).

The large production scales required for cost effective production of liquid transportation fuels from lignocellulose have to be balanced against the cost of feedstock supply. The transport cost of biomass supply generally increases with an increase in production capacity (Gnansounou and Dauriat 2010; Seabra et al. 2010; see Chap. 6). Regional conditions for the supply of biomass will determine optimal plant size, since the reduction of capital investment per unit energy produced as production scale increases is balanced by an increased requirement of feedstocks, the local cost and availability of which determines the maximum scale-dependant cost-benefit that can be achieved (Stephen et al. 2010). Various combinations of lignocellulose feedstock may therefore be used to increase biomass supply in order to reach the desired economies of scale (Seabra et al. 2010). In some scenarios the increase in cost effectiveness with an increase in production scale may outweigh the associated increase in feedstock cost (Amigun et al. 2010). It has been demonstrated that for torrefaction processes there is no benefit in terms of economies of scale for production scale beyond 40MWth (Uslu 2005; Uslu et al. 2008).

7.7 Conclusion

Combustion, pyrolysis and anaerobic digestion of lignocellulose are well-established technologies available for commercial application; while gasification, liquefaction, hydrolysis-fermentation and fractionation hold promise for future implementation.

Overall, it is expected that the production of bioenergy will increase in the Southern hemisphere, based on the overall sustainability of biomass production and cost comparisons with fossil-derived energy.

References

Agbor VB, Cicek N, Sparling R, Berlin A, Levin DB (2011) Biomass pretreatment: fundamentals toward application. Biotechnol Adv 29:675–685

Alvira P, Tomás-Pejo E, Ballesteros M, Negro MJ (2010) Pretreatment technologies for an efficient bioethanol production process on enzymatic hydrolysis: a review. Bioresour Technol 101: 4851–4861

Amigun B, Gorgens J, Knoetze JH (2010) Biomethanol production from gasification of non-woody plant in South Africa: optimum scale and economic performance. Energy Policy 38:312–322

Anex RP, Aden A, Kazi FK, Fortman J, Swanson RM, Wright MM, Satrio JA, Brown RC, Daugaard DE, Platon A, Kothandaraman G, Hsu DD, Dutta A (2010) Techno-economic comparison of biomass-to-transportation fuels via pyrolysis, gasification, and biochemical pathways. Fuel 89:S29–S35

Antal MJ, Grønli M (2003) The art, science and technology of charcoal production. Ind Eng Chem Res 42:1619–1640

Ballesteros M, Oliva JM, Negro MJ, Manzanares P, Ballesteros I (2004) Ethanol from lignocellulosic materials by a simultaneous saccharification and fermentation process (SFS) with *Kluyveromyces marxianus* CECT 10875. Process Biochem 39:1843–1848

Banerjee G, Scott-Craig JS, Walton JD (2010) Improving enzymes for biomass conversion: a basic research perspective. Bioenergy Res 3:82–92

Benjamin MM, Woods SL, Ferguson JF (1984) Anaerobic toxicity and biodegradability of pulp mill waste constituents. Water Res 18:601–607

Bergman PCA, Boersma AR, Zwart RWR, Kiel JHA (2005) Torrefaction for biomass co-firing in existing coal-fired power stations: BIOCOAL. ECN report. ftp://kernenergie.nl/pub/www/library/report/2005/c05013.pdf

Bhattacharya P, Steele PH, Hassan EBM, Mitchell B, Ingram L, Pittman CU Jr (2009) Wood/plastic copyrolysis in an auger reactor: chemical and physical analysis of the products. Fuel 88: 1251–1260

Bridgeman TG, Jones JM, Shield I, Williams PT (2008) Torrefaction of reed canary grass, wheat straw and willow to enhance solid fuel qualities and combustion properties. Fuel 87:844–856

Bridgwater AV (1995) The technical and economic feasibility of biomass gasification for power generation. Fuel 74:631–653

Bridgwater AV (2011) Review of fast pyrolysis of biomass and product upgrading. Biomass bioenergy 38:1–27

Bridgwater AV, Peacocke GVC (2000) Fast pyrolysis processes for biomass. Renew Sustain Energy Rev 4:1–73

Bridgwater AV, Toft AJ, Brammer JG (2002) A techno-economic comparison of power production by biomass fast pyrolysis with gasification and combustion. Renew Sustain Energy Rev 6(833): 181–248

Brown TR, Thilakaratne R, Brown RC, Hu G (2013) Techno-economic analysis of biomass to transportation fuels and electricity via fast pyrolysis and hydroprocessing. Fuel 106:463–469
Brownsort PA (2009) Biomass pyrolysis processes: performance parameters and their influence on biochar system benefits. Dissertation, University of Edinburgh, Edinburgh
Butler E, Devlin G, Meier D, McDonnell K (2011) A review of recent laboratory research and commercial developments in fast pyrolysis and upgrading. Renew Sustain Energy Rev 15:4171–4186
Calzavara Y, Joussot-Dubien C, Boissonnet G, Sarrade S (2005) Evaluation of biomass gasification in supercritical water process for hydrogen production. Energy Convers Manag 46:615–631
Carrier M, Hugo T, Gorgens J, Knoetze JH (2011) Comparison of slow and vacuum pyrolysis of sugar cane bagasse. J Anal Appl Pyrolysis 90:18–26
Çetinkol OP, Dibble DC, Cheng G, Kent MS, Knierim B, Auer M, Wemmer DE, Pelton JG, Melnichenko YB, Ralph J, Simmons BA, Holmes BM (2010) Understanding the impact of ionic liquid pretreatment on eucalyptus. Biofuels 1(1):33–46
Chen G, Andries J, Luo Z, Spliethoff H (2003) Biomass pyrolysis/gasification for product gas production: the overall investigation of parametric effects. Energy Convers Manag 44: 1875–1884
Cheng S, D'cruz I, Wang M, Leitch M, Xu CC (2010) Highly efficient liquefaction of woody biomass in hot-compressed alcohol-water Co-solvents. Energy Fuel 24:4659–4667
Cullis IF, Saddler JN, Mansfield SD (2004) Effect of initial moisture content and chip size on the bioconversion efficiency of softwood lignocellulosics. Biotechnol Bioeng 85(4):413–421
Czernik S, Bridgwater AV (2004) Overview of applications of biomass fast pyrolysis oil. Energy Fuel 18:590–598
Czernik S, French R, Feik C, Chornet E (2002) Hydrogen by catalytic steam reforming of liquid byproducts from biomass thermoconversion processes. Ind Eng Chem Res 41:4209–4215
Dahmen N, Dinjus E, Kolb T, Arnold U, Leibold H, Stahl R (2012) State of the art of the bioliq® process for synthetic biofuels production. Environ Prog Sustain Energy 31:176–181
Demirbas A (2005) Potential applications of renewable energy sources, biomass combustion problems in boiler power systems and combustion related environmental issues. Prog Energy Combust Sci 31:171–192
Demirbas A, Arin G (2002) An overview of biomass pyrolysis. Energy Sources 24:471–482
Diedericks D, van Rensburg E, García-Aparicio MP, Görgens JF (2012) Enhancing the enzymatic digestibility of sugarcane bagasse through the application of an ionic liquid in combination with an acid catalyst. Biotechnol Prog 28(1):76–84
Dwivedi P, Alavalapati JRR (2008) Economic feasibility of electricity production from energy plantations present on community-managed forest lands in Madhya Pradesh, India. Energy Policy 37:352–360
El-Juhany LI, Aref IM, Megahed MM (2003) Properties of charcoal produced from some endemic and exotic acacia species grown in Riyadh, Saudi Arabia. J Adv Agric Res 8:695–704
Erlach B, Harder B, Tsatsaronis G (2012) Combined hydrothermal carbonization and gasification of biomass with carbon capture. Energy 45:329–338
Ewanick SM, Bura R, Saddler JN (2007) Acid-catalyzed steam pretreatment of lodgepole pine and subsequent enzymatic hydrolysis and fermentation to ethanol. Biotechnol Bioeng 98(4): 737–746
Ferreira S, Gil N, Queiroz JA, Duarte AP, Domingues FC (2011) An evaluation of the potential of *Acacia dealbata* as raw material for bioethanol production. Bioresour Technol 102(7): 4766–4773
Fox MH, Noike T, Ohki T (2003) Alkaline subcritical-water and alkaline heat treatment for the increase in biodegradability of newsprint. Water Sci Technol 48:77–84
Gabra M, Pettersson E, Backman R, Kjellstrom B (2001) Evaluation of cyclone gasifier performance for gasification of sugarcane residue – part 1: gasification of bagasse. Biomass Bioenerg 21:351–369

García-Aparicio M, Parawira W, van Rensburg E, Diedericks D, Galbe M, Rosslander C, Zacchi G, Görgens JF (2011) Evaluation of steam-treated giant bamboo for production of fermentable sugars. Biotechnol Progr 27(3):641–649

Gírio FM, Fonseca C, Carvalheiro F, Duarte LC, Marques S, Bogel-Lukasik R (2010) Hemicelluloses for fuel ethanol: a review. Bioresour Technol 101(13):4775–4800

Gnansounou E, Dauriat A (2010) Techno-economic analysis of lignocellulosic ethanol: a review. Bioresour Technol 101:4980–4991

Gonzalez JF, Gonzalez-Garcıa CM, Ramiro A, Ganan J, Gonzalez J, Sabio E, Roman S, Turegano J (2005) Use of almond residues for domestic heating. Study of the combustion parameters in a mural boiler. Fuel Process Technol 86:1351–1368

Hendriks ATWM, Zeeman G (2008) Pretreatments to enhance the digestibility of lignocellulosic biomass. Bioresour Technol 100(1):10–18

Hjorth M, Gränitz K, Adamsen APS, Møller HB (2011) Extrusion as a pretreatment to increase biogas production. Bioresour Technol 102(8):4989–4994

Hoekstra E, Van Swaaij WPM, Kersten SRA, Kees JA, Hogen-doorn KJA (2012) Fast pyrolysis in a novel wire-mesh reactor: decomposition of pine wood and model compounds. Chem Eng J 187:172–184

Hooper RJ, Li J (1996) Summary of the factors critical to the commercial application of bioenergy technologies. Biomass Bioeng 11:469–474

Huang F, Ragauskas AJ (2012) Dilute H_2SO_4 and SO_2 pretreatments of loblolly pine wood residue for bioethanol production. Ind Biotechnol 8(1):22–30

Huang R, Su R, Qi W, He Z (2011) Bioconversion of lignocellulose into bioethanol: process intensification and mechanism research. Bioenerg Res 4(4):225–245

IEA ETSAP (2010) Technology Brief E05. www.etsap.org

Inoue H, Yano S, Endo T, Sakaki T, Sawayama S (2008) Combining hot-compressed water and ball milling pretreatments to improve the efficiency of the enzymatic hydrolysis of eucalyptus. Biotechnol Biofuels 1:2

Jin C, Yao M, Liu H, Lee CFF, Ji J (2011) Progress in the production and application of n-butanol as a biofuel. Renew Sustain Energy Rev 15(8):4080–4106

Kaida R, Kaku T, Baba K, Oyadomari M, Watanabe T, Hartati S, Sudarmonowati E, Hayashi T (2009) Enzymatic saccharification and ethanol production of *Acacia mangium* and *paraserianthes falcataria* wood, and *Elaeis guineensis* trunk. J Wood Sci 55(5):381–386

Karunanithy C, Muthukumarappan K, Gibbons WR (2012) Extrusion pretreatment of pine wood chips. Appl Biochem Biotechnol 167(1):81–99

Kazi FK, Fortman JA, Anex RP, Hsu DD, Aden A, Dutta A, Kothandaraman G (2010) Techno-economic comparison of process technologies for biochemical ethanol production from corn stover. Fuel 89:S20–S28

Kersten SRA, Potic B, Prins W, Swaaij WPMV (2006) Gasification of model compounds and wood in hot compressed water. Ind Eng Chem Res 45:4169–4177

Khodier A, Kilgallon P, Legrave N, Simms N, Oakey J, Bridgwater T (2009) Pilot-scale combustion of fast-pyrolysis bio-oil: ash deposition and gaseous emissions. Environ Prog Sustain Energy 28:397–403

Kim KH, Eom IY, Lee SM, Cho ST, Choi IG, Choi JW (2010) Applicability of sub- and supercritical water hydrolysis of woody biomass to produce monomeric sugars for cellulosic bioethanol fermentation. J Ind Eng Chem 16(6):918–922

Klass DL (1998) Thermal conversion: pyrolysis and liquefaction. In: Klass DL (ed) Biomass for renewable energy, fuels, and chemicals. Entech International, Inc., Barrington

Klein-Marcuschamer D, Oleskowicz-Popiel P, Simmons BA, Blanch HW (2012) The challenge of enzyme cost in the production of lignocellulosic biofuels. Biotechnol Bioeng 109(4): 1083–1087

Klingler D, Vogel H (2010) Influence of process parameters on the hydrothermal decomposition and oxidation of glucose in sub- and supercritical water. J Supercrit Fluids 55:259–270

Kratky L, Jirout T (2011) Biomass size reduction machines for enhancing biogas production. Chem Eng Technol 34(3):391–399
Kuhad RC, Gupta R, Khasa YP, Singh A, Zhang YHP (2011) Bioethanol production from pentose sugars: current status and future prospects. Renew Sustain Energy Rev 15(9):4950–4962
Kumar R, Wyman CE (2009) Access of cellulase to cellulose and lignin for poplar solids produced by leading pretreatment technologies. Biotechnol Prog 25:807–819
Kumar A, Cameron JB, Flynn PC (2003) Biomass power cost and optimum plant size in western Canada. Biomass Bioenerg 24:445–464
Kumar P, Barrett DM, Delwiche MJ, Stroeve P (2009) Methods for pretreatment of lignocellulosic biomass for efficient hydrolysis and biofuel production. Ind Eng Chem Res 48(8):3713–3729
Kumar G, Panda AK, Singh RK (2010) Optimization of process for the production of bio-oil from eucalyptus wood. J Fuel Chem Technol 38:162–167
Kumar M, Goyal Y, Sarkar A, Gayen K (2012) Comparative economic assessment of ABE fermentation based on cellulosic and non-cellulosic feestocks. Appl Energy 93:193–204
La Grange DC, Den Haan R, Van Zyl WH (2010) Engineering cellulolytic ability into bioprocessing organisms. Appl Microbiol Biotechnol 87(4):1195–1208
Laser M, Jin H, Jayawardhana K, Lynd LR (2009) Coproduction of ethanol and power from switchgrass. Biofuels Bioprod Biorefin 3:195–218
Li Y, Tschaplinski TJ, Engle NL, Hamilton CY, Rodríguez M, Liao JC, Schadt CW, Guss AM, Yang Y, Graham DE (2012) Combined inactivation of the *Clostridium cellulolyticum* lactate and malate dehydrogenase genes substantially increases ethanol yield from cellulose and switchgrassf. Biotechnol Biofuels 5:2. doi:10.1186/1754-6834-5-2
Lissens G, Verstraete W, Albrecht T, Brunner G, Creuly C, Seon J, Dussap G, Lasseur C (2004) Advanced anaerobic bioconversion of lignocellulosic waste for bioregenerative life support following thermal water treatment and biodegradationby *Fibrobacter succinogenes*. Biodegradation 15:173–184
Liu HW, Walter HK, Vogt GM, Vogt HS, Hobein BE (2002) Steam pressure disruption of municipal solid waste enhances anaerobic digestion kinetics and biogas yield. Biotechnol Bioeng 77: 121–130
Liu CZ, Wang F, Stiles AR, Guo C (2012) Ionic liquids for biofuel production: opportunities and challenges. Appl Energy 92:406–414
Madhavan A, Srivastava A, Kondo A, Bisaria VS (2012) Bioconversion of lignocellulose-derived sugars to ethanol by engineered *Saccharomyces cerevisiae*. Crc Cr Rev Biotechnol 32(1):22–48
Martínez JD, Mahkamov K, Andrade RV, Silva Lora EE (2012) Syngas production in downdraft biomass gasifiers and its application using internal combustion engines. Renew Energy 38:1–9
Matsushita Y, Yamauchi K, Takabe K, Awano T, Yoshinaga A, Kato M, Kobayashi T, Asada T, Furujyo A, Fukushima K (2010) Enzymatic saccharification of eucalyptus bark using hydrothermal pre-treatment with carbon dioxide. Bioresour Technol 101(13):4936–4939
McIntosh S, Vancov T, Palmer J, Spain M (2012) Ethanol production from eucalyptus plantation thinnings. Bioresour Technol 110:264–272
Meng J, Park J, Tilotta D, Park S (2012) The effect of torrefaction on the chemistry of fast-pyrolysis bio-oil. Bioresour Technol 111:439–446
Menon V, Rao M (2012) Trends in bioconversion of lignocellulose: biofuels, platform chemicals and biorefinery concept. Prog Energy Combust 38(4):522–550
Miller S, Hester R (2007) Concentrated acid conversion of pine sawdust to sugars part II: high-temperature batch reactor kinetics of pretreated pine sawdust. Chem Eng Commun 194(1): 103–116
Mora-Pale M, Meli L, Doherty TV, Linhardt RJ, Dordick JS (2011) Room temperature ionic liquids as emerging solvents for the pretreatment of lignocellulosic biomass. Biotechnol Bioeng 108(6):1229–1245
Moreno AD, Ibarra D, Fernández JL, Ballesteros M (2012) Different laccase detoxification strategies for ethanol production from lignocellulosic biomass by the thermotolerant yeast *Kluyveromyces marxianus* CECT 10875. Bioresour Technol 106:101–109

Muñoz C, Mendonça R, Baeza J, Berlin A, Saddler J, Freer J (2011) Bioethanol production from bio-organosolv pulps of *Pinus radiata* and *Acacia dealbata*. J Chem Technol Biotechnol 82(8):767–774

Naik SN, Goud VV, Rout PK, Dalai AK (2010) Production of first and second generation biofuels: a comprehensive review. Renew Sustain Energy Rev 14:578–597

Nakamura Y, Mtui G (2003) Anaerobic fermentation of woody biomass treated by various methods. Biotechnol Bioprocess E 8(3):179–182

Ncibi MC (2010) Bioconversion of renewable bioresources and agricultural by-products into bioethanol. Recent Pat Chem Eng 3(3):165–179

Oasmaa A, Solantausta Y, Arpiainen V, Kuoppala E, Sipila K (2010) Fast pyrolysis bio-oils from wood and agricultural residues. Energy Fuel 24:1380–1388

Park N, Kim HY, Koo BW, Yeo H, Choi IG (2010) Organosolv pre-treatment with various catalysts for enhancing enzymatic hydrolysis of pitch pine (*Pinus rigida*). Bioresour Technol 101(18):7046–7053

Peterson D, Haase S (2009) Market assessment of biomass gasification and combustion technology for small- and medium-scale applications. Technical report NREL. http://www.nrel.gov/docs/fy09osti/46190.pdf

Peterson AA, Vogel F, Lachance RP, Froling M, Antal MJ Jr, Tester JW (2008) Thermochemical biofuel production in hydrothermal media: a review of sub- and supercritical water technologies. Energy Environ Sci 1:32–65

Pfeifer C, Koppatz S, Hofbauer H (2011) Steam gasification of various feedstocks at a dual fluidised bed gasifier: impacts of operation conditions and bed materials. Biomass Convers Biorefin 1:39–53

Pimenta AS, Vital BR, Bayona JM, Alzaga R (1998) Characterisation of polycyclic aromatic hydrocarbons in liquid products from pyrolysis of eucalyptus grands by supercritical fluid extraction and GC-MS determination. Fuel 77:1133–1139

Pindoria RV, Megaritis A, Messenbhk RC, Dugwell DR, Kandiyoti R (1998) Comparison of the pyrolysis and gasification of biomass: effect of reacting gas atmosphere and pressure on eucalyptus wood. Fuel 77:1247–1251

Prins MJ, Ptasinski KJ, Janssen FJJG (2006) More efficient biomass gasification via torrefaction. Energy 31:3458–3470

Prins M, van der Berg R, van Holthoon E, van Dorst E, Geuzebroek F (2012) Technological developments IGCC for carbon capture. Chem Eng Technol 35:413–419

Qing Q, Yang B, Wyman CE (2010) Xylooligomers are strong inhibitors of cellulose hydrolysis by enzymes. Bioresour Technol 101(24):9624–9630

Rodríguez GM, Atsumi S (2012) Synthetic biology approaches to produce C3–C6 alcohols from microorganisms. Curr Chem Biology 6(1):32–41

Romaní A, Garrote G, Alonso JL, Parajó JC (2010) Bioethanol production from hydrothermally pretreated *Eucalyptus globulus* wood. Bioresour Technol 101(22):8706–8712

Romaní A, Garrote G, López F, Parajó JC (2011) *Eucalyptus globulus* wood fractionation by autohydrolysis and organosolv delignification. Bioresour Technol 102(10):5896–5904

Romaní A, Garrote G, Parajó JC (2012) Bioethanol production from autohydrolyzed *Eucalyptus globulus* by simultaneous saccharification and fermentation operating at high solids loading. Fuel 94:305–312

Roy C (2000) Vacuum pyrolysis breakthrough. PyNe Issue 10. http://www.pyne.co.uk/Resources/user/PyNews%2010.pdf

Rudolf A, Alkasrawi M, Zacchi G, Lidén G (2005) A comparison between batch and fed-batch simultaneous saccharification and fermentation of steam pretreated spruce. Enzyme Microb Technol 37:195–204

Rutberg PG, Bratsev AN, Kuznetsov VA, Popov VE, Ufimtsev AA, Shtengel SV (2011) On efficiency of plasma gasification of wood residues. Biomass Bioenerg 35:495–504

Sánchez OJ, Cardona CA (2008) Trends in biotechnological production of fuel ethanol from different feedstocks. Bioresour Technol 99:5270–5295

Sanders WTM, Geerink M, Zeeman G, Letting G (2000) Anaerobic hydrolysis kinetics of particulate substrates. Water Sci Technol 41:17–24

Seabra JEA, Tao L, Chuma HL, Macedo IC (2010) A techno-economic evaluation of the effects of centralized cellulosic ethanol and co-products refinery options with sugarcane mill clustering. Biomass Bioenerg 34:1065–1078

Sensoz S, Can M (2002) Pyrolysis of pine (*Pinus brutia* ten.) chips: 1. Effect of pyrolysis temperature and heating rate on the product yields. Energy Sour 24:347–355

Siedlecki M, de Jong W, Verkooijen AHM (2011) Fluidized bed gasification as a mature and reliable technology for the production of bio-syngas and applied in the production of liquid transportation fuels-a review. Energies 4:389–434

Silva NLC, Betancur GJV, Vasquez MP, De Barros GE, Pereira N Jr (2011) Ethanol production from residual wood chips of cellulose industry: acid pretreatment investigation, hemicellulosic hydrolysate fermentation, and remaining solid fraction fermentation by SSF process. Appl Biochem Biotechnol 163(7):928–936

Stephen JD, Mabee WE, Saddler JN (2010) Biomass logistics as a determinant of second-generation biofuel facility scale, location and technology selection. Biofuels Bioprod Biorefin 4:503–518

Stephen JD, Mabee WE, Saddler JN (2012) Will second-generation ethanol be able to compete with first-generation ethanol? Opportunities for cost reduction. Biofuels Bioprod Biorefin 6:159–176

Suarez AC, Tancredi N, Pinheiro PCC, Yoshida MI (2010) Thermal analysis of the combustion of charcoals from *Eucalyptus dunnii* obtained at different pyrolysis temperatures. J Therm Anal Calorim 100:1051–1054

Sugano M, Takagi H, Hirano K, Mashimo K (2008) Hydrothermal liquefaction of plantation biomass with two kinds of wastewater from paper industry. J Mater Sci 43:2476–2486

Sun Q, Yu S, Wang F, Wang J (2011) Decomposition and gasification of pyrolysis volatiles from pine wood through bed of hot char. Fuel 90:1041–1048

Swain PK, Das LM, Naik SN (2011) Biomass to liquid: a prospective challenge to research and development in 21st century. Renew Sustain Energy Rev 15:4917–4933

Taherzadeh MJ, Karimi K (2007) Enzyme-based hydrolysis processes for ethanol from lignocellulosic materials: a review. Bioresources 2:707–738

Taherzadeh MJ, Karimi K (2008) Pretreatment of lignocellulosic wastes to improve ethanol and biogas production: a review. Int J Mol Sci 9:1621–1651

Tian S, Luo XL, Yang XS, Zhu JY (2010) Robust cellulosic ethanol production from SPORL-pretreated lodgepole pine using an adapted strain *Saccharomyces cerevisiae* without detoxification. Bioresour Technol 101(22):8678–8685

Tijmensen MJA, Faaij APC, Hamelinck CN, van Hardeveld MRM (2002) Exploration of the possibilities for production of Fischer-Tropsch liquids and power via biomass gasification. Biomass Bioenerg 23:129–152

Titirici MM, Thomas A, Antonietti M (2007) Back in the black: hydrothermal carbonization of plant material as an efficient chemical process to treat the CO_2 problem? New J Chem 31: 787–789

Tomeczek J, Stanisław G (2003) Volatiles release and porosity evolution during high pressure coal pyrolysis. Fuel 82:285–292

Torr KM, Love KT, Etinkol OP, Donaldson LA, George A, Holmes BM, Simmons BA (2012) The impact of ionic liquid pretreatment on the chemistry and enzymatic digestibility of *Pinus radiata* compression wood. Green Chem 14(3):778–787

Tu M, Saddler JN (2010) Potential enzyme cost reduction with the addition ofs during the hydrolysis of pretreated softwood. Appl Biochem Biotechnol 161(1–8):274–287

Uras Ü, Carrier M, Hardie AG, Knoetze JH (2012) Physico-chemical characterization of biochars from vacuum pyrolysis of South African agricultural wastes for application as soil amendments. J Anal Appl Pyrolysis 98:207–213

Uslu A (2005) Pre-treatment technologies, and their effects on the international bioenergy supply chain logistics: techno-economic evaluation of torrefaction, fast pyrolysis and pelletisation. Dissertation, Utrecht University, Utrecht

Uslu A, Faaij APC, Bergman PCA (2008) Pre-treatment technologies, and their effects on the international bioenergy supply chain logistics: techno-economic evaluation of torrefaction, fast pyrolysis and pelletisation. Energy 33:1206–1223

van der Drift A, Boerrigter H, Coda B, Cieplik MK, Hemmes K (2004) Entrained flow gasification of biomass: ash behaviour, feeding issues, and system analyses. Report ECN-C–04-039. http://www.ecn.nl/docs/library/report/2004/c04039.pdf

Van der Stelt MJC, Gerhauser H, Kiel JHA, Ptasinski KJ (2011) Biomass upgrading by torrefaction for the production of biofuels: a review. Biomass Bioenerg 35:3748–3762

Van Rensburg E, Den Haan R, Smith J, Van Zyl WH, Gorgens JF (2012) The metabolic burden of cellulase expression by recombinant *Saccharomyces cerevisiae* Y294 in aerobic batch culture. Appl Microbiol Biotechnol 96(1):197–209

Van Zyl WH, Chimphango AFA, den Haan R, Görgens JF, Chirwa PWC (2011) Next-generation cellulosic ethanol technologies and their contribution to a sustainable Africa. Interface Focus 1:196–211

Vidal BC Jr, Dien BS, Ting KC, Singh V Jr (2011) Influence of feedstock particle size on lignocellulose conversion – a review. Appl Biochem Biotechnol 164(8):1405–1421

Wagenaar BM, Prins W, van Swaaij WPM (1993) Flash pyrolysis kinetics of pine wood. Fuel Process Technol 36:291–298

Walker ME, Abbasian J, Chmielewski DJ, Castaldi MJ (2011) Dry gasification oxy-combustion power cycle. Energy Fuel 25:2258–2266

Wang GS, Pan XJ, Zhu JY, Gleisner R, Rockwood D (2009) Sulfite pretreatment to overcome recalcitrance of lignocellulose (SPORL) for robust enzymatic saccharification of hardwoods. Biotechnol Prog 25(4):1086–1093

Wang Y, Radosevich M, Hayes D, Labbé N (2011) Compatible ionic liquid-cellulases system for hydrolysis of lignocellulosic biomass. Biotechnol Bioeng 108(5):1042–1048

Wang W, Yuan T, Wang K, Cui B, Dai Y (2012) Combination of biological pretreatment with liquid hot water pretreatment to enhance enzymatic hydrolysis of *Populus tomentosa*. Bioresour Technol 107:282–286

Wei W, Wu S, Liu L (2012) Enzymatic saccharification of dilute acid pretreated eucalyptus chips for fermentable sugar production. Bioresour Technol 110:302–307

Wright MM, Satrio JA, Brown RC, Daugaard DE, Hsu DD (2010) Techno-economic analysis of biomass fast pyrolysis to transportation fuels. Technical report NREL/TP-6A20-46586. http://www.nrel.gov/docs/fy11osti/46586.pdf

Yáñez R, Romaní A, Garrote G, Alonso JL, Parajó JC (2009) Experimental evaluation of alkaline treatment as a method for enhancing the enzymatic digestibility of autohydrolysed *Acacia dealbata*. J Chem Technol Biotechnol 84(7):1070–1077

Yang B, Wyman CE (2008) Pretreatment: the key to unlocking low-cost cellulosic ethanol. Biofuel Bioprod Biorefin 2(1):26–40

Yu Q, Zhuang X, Yuan Z, Wang Q, Qi W, Wang W, Zhang Y, Xu J, Xu H (2010) Two-step liquid hot water pretreatment of *Eucalyptus grandis* to enhance sugar recovery and enzymatic digestibility of cellulose. Bioresour Technol 101(13):4895–4899

Zhang H, Pang H, Shi J, Fu T, Liao B (2012) Investigation of liquefied wood residues based on cellulose, hemicellulose, and lignin. J Appl Polymer Sci 123:850–856

Zhu JY, Pan XJ, Wang GS, Gleisner R (2009) Sulfite pretreatment (SPORL) for robust enzymatic saccharification of spruce and red pine. Bioresour Technol 100(8):2411–2418

Zhu W, Zhu JY, Gleisner R, Pan XJ (2010) On energy consumption for size-reduction and yields from subsequent enzymatic saccharification of pretreated lodgepole pine. Bioresour Technol 101(8):2782–2792

Zinoviev S, Muller-Langer F, Das P, Bertero N, Fornasiero P, Kaltschmitt M, Centi G, Miertus S (2010) Next-generation biofuels: survey of emerging technologies and sustainability issues. ChemSusChem 3:1106–1133

Chapter 8
Biomass Quality

Martina Meincken and Luvuyo Tyhoda

8.1 Introduction

Biomass used for energy conversion ranges from wood, especially planted for energy purposes, over harvesting residues, other woody biomass, such as shrubs or bamboo to waste materials, such as sawdust or pulp residues. These types of biomass differ widely in their properties and furthermore there is a variation within each species or type of biomass, due to biological variation. An assessment of the biomass quality is therefore vital to decide on its feasibility for conversion, the most suitable type of conversion and the need for further processing.

Wood, for example, differs in density, moisture content (MC) and chemical composition within one tree in horizontal and vertical direction. The variation between different trees is even larger, as these properties depend on the quality of the site where the tree is grown (water availability, temperature etc.) and of course there is a difference between wood species as well, although the difference due to site may be larger than the differences due to species. Therefore, in order to characterise biomass, care has to be taken, to work with statistically meaningful values. This means that samples should be obtained at the very least from different positions of a tree, but rather also from different trees and that sufficient samples have to be measured in order to obtain a representative average value. Practically, biomass is collected on a much larger scale than just a few trees, so sampling for quality assessment is commonly done by mixing e.g. the wood chips of many trees thoroughly and taking a few randomly chosen chips for sampling.

M. Meincken (✉) • L. Tyhoda
Department of Forest and Wood Science, Stellenbosch University, Private Bag X1, Matiland 7602, South Africa

T. Seifert (ed.), *Bioenergy from Wood: Sustainable Production in the Tropics*, Managing Forest Ecosystems 26, DOI 10.1007/978-94-007-7448-3_8,

The obvious advantage of biomass compared to fossil fuels is its renewable character. Its disadvantages are low density, which makes it bulky and difficult to store, the inhomogeneous form, which could be a problem for industrial equipment, a fairly large content of inorganic substances, which remain as ash and generally a high MC, which leads to an energy loss, because energy is needed to evaporate the water. Due to these characteristics, biomass usually requires additional processing before it can be used as a biofuel. Preparations, such as chipping, dehydration, densification and removal of incombustible material might be necessary.

8.2 Drying and Storage

After harvesting, the biomass is commonly stored and air-dried before it is transported and further processed. The rate of drying depends strongly on environmental factors, such as temperature and humidity, the particle size and the stacking method. Whole logs lose their moisture slowly and reach a MC of about 30 % after drying anything between 4 weeks and 6 months. This time can be decreased, if the logs are sheltered and cut into smaller pieces. Chips of a few cm size can reach a MC of about 12 % after a few weeks in sunny, arid conditions (Sturos et al. 1983).

Good ventilation is vital to good drying results. This becomes especially a problem, when the biomass is first comminuted and then dried. If the chips are piled on top of each other, ventilation is inhibited and fungi and bacteria can attack the particles. Biological and chemical degradation caused by bacteria and other organisms lead to an increase of temperature, which can in some cases lead to self-ignition of the entire pile.

Air drying should ideally happen on a flat area without contact to moisture (e.g. on a concrete slab under shelter) and the pile should be turned over on a regular basis.

All these problems can be avoided by oven or kiln drying the biomass, which on the other hand increases the processing costs drastically and sets free carbon dioxide, which affects the energy balance negatively (see Chap. 10).

8.3 Particle Size

For many applications the particle size plays a vital role. Conversion reactors can only be optimized with regards to their processing parameters if the biofuel is evenly sized. Fixed bed gasifiers, for example, can be fed with chunks or pellets of a few centimetre diameter, while fluidized bed gasifiers require fuel of smaller particle size.

As most biomass is very heterogeneous in size and density, it needs to be comminuted (made smaller) and sometimes compressed into pellets or briquettes.

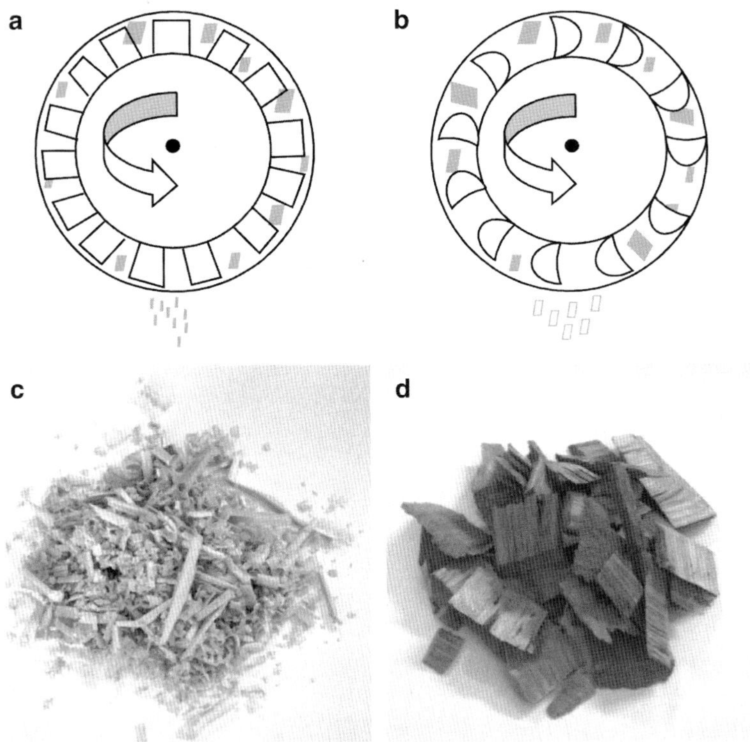

Fig. 8.1 schematic drawing of a (**a**) hammermill and (**b**) chipper and the resulting (**c**) wood particles and (**d**) chips

Pellets are cylindrical particles with a diameter of 5–10 mm and about 1–3 cm length. Briquettes are larger, with a diameter of 5–10 cm and a length of about 15–20 cm.

After felling, trees are typically debarked and cut into logs and the remaining bark, branches, twigs and leaves form the harvest residue. The logs can be further comminuted into chunks or fire logs, with a length of about 30 cm and a diameter of about 10 cm. The harvest residue is too heterogeneous in size to be used directly and needs to be processed further.

8.3.1 Comminution

Wood chunks or residues can be further comminuted with a hammermill (Fig. 8.1a) or a chipper (Fig. 8.1b). A hammermill breaks down particles with blunt metal hammers and the resulting particles have a fairly wide size distribution. The

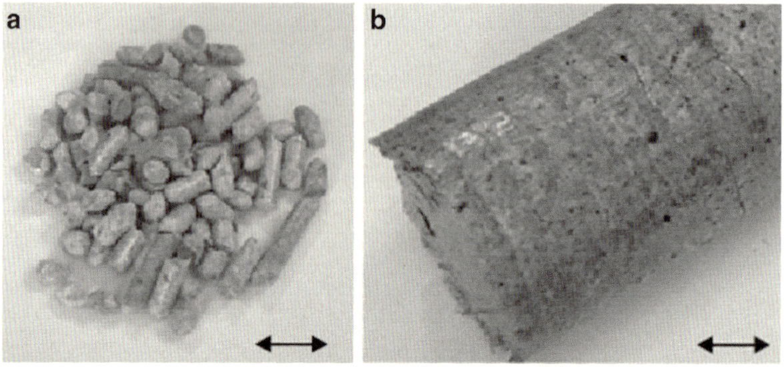

Fig. 8.2 Pellets (**a**) and a briquette (**b**). The bar indicates 1 cm length

maximum particle size can be determined by a sieve through which the particles have to pass before they can leave the mill. In a disk- or drum-chipper the particles are cut along the grain with a knife, which results in fairly even sized chips of a few centimetre side lengths.

Further mechanical comminution of chips can be attained by grinding or milling (0.2–2 mm) with various mills. Ball mills, vibratory mills, hammer mills, knife mills, two-roll mills, colloid mills, attrition mills or extruders can be employed for this purpose. The choice of equipment is largely determined by the moisture content of the biomass (Kratky and Jirout 2011). While colloid mills and extruders are only appropriate for materials with a moisture content of 15–20 %, ball or vibratory ball mills can be employed for dry and wet materials. The energy requirements and hence subsequent costs of the comminution step are dictated by the type of mill, the original and target particle size, and the lignocellulose characteristics.

8.3.2 Compression

If biomass needs to be compressed into pellets or briquettes, they need to be further comminuted into particles of a few millimetre lengths. This happens with refining mills that grind the biomass between serrated disks. These particles are then compressed in a pellet or briquette press to particles of uniform size and density. The heat produced during pressing degrades the lignin present in the biomass and allows it to flow freely between the particles. Lignin is a natural binder and as the particle leaves the press and cool it hardens resulting in a stable biomass particle (Fig. 8.2).

8.4 Moisture Content

The moisture content (MC) is the mass of moisture in biomass and can be either expressed as a percentage/fraction of the ovendry mass (mass with 0 % MC)

$$MC = \frac{m - m_0}{m_0} \times 100\% \tag{8.1}$$

$m = mass\ of\ wet\ wood;\ m_0 = ovendry\ mass$

or as a percentage/fraction of the wet mass

$$MC = \frac{m - m_0}{m} \times 100\% \tag{8.2}$$

$m = mass\ of\ wet\ wood;\ m_0 = ovendry\ mass$

The MC on dry basis is commonly used in the wood processing industry, while the MC on a wet basis is used in the Forestry and Pulp & Paper industry. The MC on wet basis has a maximum value of 100 %, while the MC on dry basis can be larger than 100 %. The MC on a wet basis is mostly used for practical reasons, as all transport costs are based on the weight of the wet biomass. The problem with the MC on wet basis is that it is not well defined, because in contrast to the MC on dry basis the weight of wood is very variable.

Typically, considerable MC variation can be found in all biomass, depending on location, age, season etc. In wood, for example, heartwood generally has a lower MC than sapwood and in softwoods it tends to be larger close to the bark and higher up. Softwoods generally have a larger MC than hardwoods. Care has to be taken therefore to determine a statistically significant average value that describes the entire biomass. Typical values are displayed in Table 8.1.

The MC is the most important property for biomass utilisation as fuel, because it affects the entire supply chain and the related costs, i.e. transport, storage, energy content, conversion methods and end use.

Table 8.1 Typical MC values for different types of biomass

Biomass	MC (%)
Freshly harvested trees	80–180
Trees, 6 month stored	30–60
Freshly chopped wood	60–120
Air-dried chopped wood	15–30
Bark, fresh	60–120
Wood chips, sawmill waste	30–60
Chips, biomass	60–100
Wood pellets/briquettes	8–12

From Marutzky and Seeger (1999)

8.4.1 MC Determination

The MC of partially dried wood can be measured with a resistance meter. In this method two pins are inserted into the wood and the measured resistance can be converted to a MC value. This technique works, however, only for MC values below fibre saturation point (around 30 % MC), on large enough samples and is generally not very accurate.

For bioenergy purposes the MC is typically determined with the ovendry method (ASTM D1762). For this a small sample is cut, weighed when wet, dried at 103 ± 2 °C for at least 24 h and weighed again when dry. MC can then be calculated according to Eqs. 8.1 or 8.2.

8.4.2 Dehydration

Free water in excess of 65 % MC can be extracted by compressing the biomass with a press. In this way the MC can be reduced to about 20 %. To remove the remaining water, the biomass must either be air or oven dried. The drying time depends on the particle size and the composition of the biomass used. Typically wood chips reach a MC of 10–12 % after a few weeks of air drying in the Western Cape in South Africa and this process is faster for smaller particles (Sturos et al. 1983; Wondifraw 2010). Drying in a kiln or oven can remove the water entirely, depending on the time and temperature employed. Torrefaction (see Chap. 7) is an alternative to drying, resulting in products highly suited for thermochemical processing that can be stored for long periods of time.

Any moisture present in biomass reduces its calorific value. For combustion, gasification and pyrolysis the MC should therefore be as low as possible. Acceptable values for most reactor types are between 10 and 20 % (Schuck 2006; McKendry 2002). If too much moisture is present, much of the energy contained in the biomass is used to evaporate the water. Furthermore, biomass with a high MC cannot generate certain compounds when pyrolysed, which also affects the acidity and composition of liquid produced (Guillen and Ibargoitia 1999).

On the other hand, moisture is desirable for biochemical conversion, such as fermentation and anaerobic digestion (see Chap. 7) and the biomass is kept as wet as possible, typically with MC values between 80 and 90 %.

8.5 Density

Density is the mass (m) contained in a volume (V) of material.

$$r = \frac{m}{V} \quad \left[\frac{g}{cm^3}\right] \quad or \quad \left[\frac{kg}{m^3}\right]$$

Table 8.2 Typical density values of South African biofuels as determined in house for various samples

Species	Density (kg/m^3)
Eucalypts	700–800
Acacias	600–900
Pines	400–550
Other woody biomass (shrubs etc.)	350–500
Bark	450–500

When looking at biomass, however, care has to be taken as to *which* mass and volume are regarded, as both parameters depend on the moisture content, wood structure (earlywood/latewood, etc.) and chemical composition.

The only reproducible density values are the "ovendry" density, which is

$$r_0 = \frac{\textit{Ovendry Weight}}{\textit{Ovendry Volume}} \tag{8.3}$$

and the basic density, which is defined as

$$R = \frac{\textit{Ovendry Weight}}{\textit{Fully saturated Volume}} \tag{8.4}$$

Typical average density ranges of woody biomass regarded in this book are given in Table 8.2.

As with the moisture content a large variation of density can be found within one tree. Generally juvenile wood has a lower density than mature wood, heartwood is denser than softwood (this is more pronounced in hardwoods) and in softwoods the density decreases with height (Tsoumis 2009).

For bioenergy purposes the bulk density of chips is determined via shock impact (BS EN 15103). A cylindrical vessel with known volume is filled to the rim with chips and shock exposed (dropped from a certain height) to compact the chips. The vessel is then either refilled to maximum level or the surplus material is removed. The basic bulk density of the wet chips is then given by:

$$BD_w = (m_2 - m_1)/V \tag{8.5}$$

with m_1: weight of vessel, m_2: weight of vessel + biomass, V = inner volume of vessel

The bulk density of the dry chips can be calculated if the MC is known:

$$BD_d = BD_w * (100 - MC)/100 \tag{8.6}$$

The Diana Smith method (Smith 1959) is often used to determine the exact basic density of oddly shaped samples, such as wood chips. This method is more time consuming, but disposes of the volume in the equation. The samples are submerged

Table 8.3 Typical distribution of cellulose, hemicelluloses and lignin in wood

Wood type	Cellulose (%)	Hemicelluloses (%)	Lignin (%)
Hardwood[1]	40–44	15–35	18–25
Softwood[1]	40–44	20–32	25–35
Pine[2]	26.4	44.7	18.6
Eucalyptus[2]	27.7–25.9	49.5–57.3	13.1–16.8
Black Wattle[3]	17.9–21.2	63.9	12.7

From: [1]Walker (2006), [2]Hamelick et al. (2005), [3]Kumar and Gupta (1992)

in water and cyclically exposed to pressure (to get water in) and under-pressure (to get air out). The basic density is then calculated according to:

$$R = \frac{1}{\frac{m_{sat}-m_0}{m_0} + \frac{1}{1.53}} \quad \left[\frac{g}{cm^3}\right] \tag{8.7}$$

$m_{sat} = saturated\ weight;\ m_0 = ovendry\ weight$

The volume to weight ratio of most biomass is generally rather unfavourable, which decreases the possible energy output. The solid content of wood chips is only around 0.4, which is a major reason for densification (e.g. pelletising). This decreases transport and storage costs and increases the energy density at the same time. The calorific value increases linear with density, because more material is available (Kataki and Konwer 2001; Munalula and Meincken 2009). For combustion and gasification the biomass should therefore have a density as high as possible, whereas for fermentation and digestion a low density is more desirable, because this is correlated to a looser wood structure, which can be degraded more easily (see Chap. 7).

8.6 Chemical Composition

Wood contains a significant amount of carbohydrates and consists of about 50 % Carbon, 6 % Hydrogen and 44 % Oxygen and other elements, often grouped together under the name "extractives". Wood is primarily composed of macromolecular substances, which are mainly polysaccharides (cellulose and hemicelluloses) and lignin (Table 8.3).

8.6.1 Cellulose

Cellulose is one of the most abundant, naturally-occurring organic compounds in the world. Approximately 40–45 % of dry substance in most wood species is cellulose, located mainly in the secondary cell wall. It is a linear homopolysaccharide composed exclusively of β-D-anhydro-glucopyranose units, which are linked together by β (1 → 4)-glycosidic bonds. It is the main structural component of plant cell walls. Because of its strong tendency for intra- and intermolecular hydrogen bonding, bundles of cellulose molecules aggregate into microfibrils, which form either highly ordered (crystalline) or less ordered (amorphous) regions. This highly ordered three-dimensional structure confers the mechanical strength of cellulose, and also results in its low susceptibility to chemical and enzymatic attack.

It is often assumed that wood of the same species is identical in all structural and physical characteristics. However, this is not true as different pieces of wood from the same tree are never identical but are similar only within broad limits. Therefore, structural components, such as cellulose, which determine the physical and chemical properties, are also never found in the same quantities throughout a tree or in different trees of the same species (Howard 1973). The cellulose content differs between the roots, stem, branches, normal wood versus reaction wood, juvenile wood versus mature wood, earlywood versus latewood and varies from the pith to the bark (Downes et al. 2000; Haygreen and Bowyer 2007).

Most studies on the variation of cellulose content have been carried out on softwoods, showing minimum values of cellulose of 40 % in earlywood and maximum values of 50 % in the latewood (Downes et al. 2000). The cellulose content in latewood is not only higher but the cellulose also has a higher degree of polymerisation, which is very important for most applications, which use cellulose as a raw material. The latewood cellulose also has a higher packing density and a higher degree of crystallinity than that in the earlywood.

The cellulose content increases from pith to bark correlated with the tracheid length, which increases from juvenile wood to mature wood. It has been found that the cellulose content decreases about 2 % vertically in *Pinus densiflora S* and in *Pinus radiata* (Panshin and De Zeeuw 1980).

Reaction wood also shows a considerable difference in cellulose content. The tracheids in compression wood are about 30 % shorter than in normal wood, leading to a decrease in cellulose content of about 10 %.

The different cellulose content in branches can be ascribed to a larger amount of bark, knots, as well as the presence of reaction wood. Branches also have narrower growth rings, resulting in overall lower cellulose content than in stem wood (Haygreen and Bowyer 2007).

8.6.2 Hemicelluloses

Hemicelluloses are short chains of branched hetero-polysaccharides composed of both hexoses and pentoses. D-xylose and L-arabinose are the main constituents of pentosans (xylans), while D-glucose, D-galactose and D-mannose are the main constituents of the hexosans (mannans). The major hemicelluloses component of softwood is mannan-based whilst the hemicelluloses in hardwoods are xylan-based. Hemicelluloses comprise 20–25 % of the material in hardwood and 7–12 % in softwoods. The close association of hemicelluloses with cellulose and lignin in the fibre cell walls contributes both rigidity and flexibility. The type and amount of hemicelluloses vary widely, depending on plant material, tissue type, growth rate, growth conditions, storage and method of extraction. A study of *Pinus resinosa Ait* has shown that the xylose content in earlywood was about 1–2 % higher than in latewood and the ratios were reversed for mannose (Panshin and De Zeeuw 1980). No difference was found between early wood and late wood for galactose, arabinose, and glucose in young trees. Analysis of successive growth increments within the tree in *Pinus radiata* D. Don showed a 3 % reduction of hemicelluloses from pith to bark and from the top to the butt of a tree. A maximum hemicelluloses content of about 11 % was found near the pith and toward the top of the tree. Glucose and mannose were shown to increases with age and decrease upwards while arabinose, xylose and galactose were shown to decrease with age and increased with height. Compression wood contains about 8–9 % more hemicelluloses than normal wood (Haygreen and Bowyer 2007).

8.6.3 Lignin

Lignin is the second most abundant organic compound after cellulose. It is an integral part of the wood cell wall. In wood, it carries the major part of the methoxyl content, is unhydrolysable by acids, readily oxidisable and is soluble in hot alkali and bisulphite (Schubert 1965). It has a network structure and lacks a defined primary structure (Herman 1987). According to Hatfield and Vermerris (2001), lignin can be described in two ways: either from a chemical point of view i.e. its functional groups and lignin-type sub-structure compositions, or from a functional point of view that stresses what lignin does within a plant. The lignin content and composition (syringyl/guaiacyl ratio) do not change significantly with tree height or diameter (Sykes et al. 2008).

Lignins contain up to 67 % carbon, depending on the method of isolation (Jakab et al. 1995). As such, lignin is a major energy-bearing compound in wood with a calorific value of about 23 MJ/kg (Blunk and Jenkins 2000). Softwoods in general have a higher energy content than hardwoods, due to the proportionally higher lignin content.

Lignin is a natural bonding agent for the cellulose fibres when combined with hemicelluloses and imparts rigidity to the cell wall, generating a composite structure that is outstandingly resistant towards impact, compression and bending. Lignin also protects the polysaccharides from microbial attack. This makes a high lignin content undesirable for anaerobic digestion or hydrolysis-fermentation.

Lignin, due to its structure and high molecular weight also serves as UV protection and flame retardant agent. The latter property is of benefit in forest fires where the lignin-carbohydrate complex is able to protect wood from the effects of fire (Wünning 2001).

Table 8.4 Chemical composition throughout a tree in *Pinus elliottii* (Howard 1973)

Tree part	Cellulose (%)	Hemicelluloses (%)	Lignin (%)	Extractives (%)	Ash (%)
Bark	23.7	24.9	50.0	13.0	0.9
Needles	42.6	22.3	37.7	26.2	2.4
Branches	36.9	33.7	35.0	13.6	1.2
Top	41.5	31.2	32.5	11.0	0.8
Roots	44.6	25.6	31.3	11.7	1.6
Stem	51.1	26.8	27.8	9.1	0.3

8.6.4 *Other Cell Wall Components*

In addition to cellulose, hemicelluloses and lignin, wood cells also contain extraneous substances. Many of these substances are extractable with neutral solvents and are generally referred to as extractives. Extractives (proteins, fats, fatty acids, terpenes, resins, phenols and alcohol, etc.) can represent between 4 and 10 % of lignocelluloses (Fengel and Wegener 2003). There is a considerable variation in the type, distribution and amount of extractives between tree species and even within species. Sugars and other sap-soluble constituents, such as starch are found in the cell lumens of parenchyma cells. Phenolic materials are deposited in the heartwood to protect the wood from insect, fungal or bacterial attack (Harju et al. 2003). Water repellent fats are found in the parenchyma cells, especially in the ray parenchyma, whereas resins are secreted by epithelial cells and tend to form resin ducts. Some extractives are utilized commercially such as vegetable tannins, turpentine and tall oil, fatty acids etc.

Resins have a high CV of >30 MJ/kg (Novaes et al. 2010), which explains why softwoods generally have a higher CV than hardwoods and also leads to a fairly high CV value of bark.

Variation in extractives within the stem is mostly between sap- and heartwood, except for resins. The resin content in softwoods is reported to be highest near the pith and at the butt of the tree, decreasing upward and towards the outside (Fengel and Wegener 2003). The resin content is significantly lower in the sapwood than in the heartwood and slightly lower in earlywood than latewood. Many heartwood extractives, on the other hand, increase from the pith to the outer heartwood boundary (Panshin and De Zeeuw 1980) (Table 8.4).

8.7 Elemental Composition

All biomass contains Carbon (C), Oxygen (O), Hydrogen (H), Nitrogen (N) and various other trace elements. On average wood contains about 49 % C, 6 % H and 44 % O with some variation between wood species (Fengel and Wegener 2003).

A high carbon content is directly related to a high density and therefore a high calorific value and desirable for combustion and gasification.

The major elements contained in any biomass, such as C, N, H, can be determined with elemental analysis and/or x-ray fluorescence (XRF) as described by Perkel (2012).

C, N and H can be determined by combustion in a pure oxygen atmosphere at high temperatures, which converts the sample elements into CO_2, H_2O and N_2. The amount of the product gases is proportional to the amount of C, N and H in the sample.

In the XRF technique a material is exposed to high energy x-rays, which eject electrons from the sample atoms. The transition of outer electrons to the vacant spaces results in photo energy that is characteristic for the element and can be detected by a fluorescence detector.

8.8 Calorific Value

C, O and H can be broken down with the release of energy (heat) to form CO_2, H_2O and other by-products. The calorific value or energy content is a measure for the heat that can be produced and is measured in MJ/kg. The two major breakdown processes are:

$$C + O \rightarrow CO_2 + 32.8\ \mathrm{MJ/kg\ (C)}$$

$$2H_2 + O_2 \rightarrow 2H_2O + 142.2\ \mathrm{MJ/kg\ (H)}$$

The amount of the released energy depends on the ratio of C, H and O in the biomass used and the MC, as part of the released energy will be consumed for the vaporization of present water. Commonly it is differentiated between gross and net calorific value or higher heating value (HHV) and lower heating value (LHV). The gross CV is the energy content of the sample without any moisture and therefore the maximum value, while the net value is measured in a moist sample, as it would be used for conversion.

Typical gross calorific values for different types of biomass that might be used for energy conversion are given in Table 8.5. It can be seen that the values do not differ much for woody biomass – they are generally between 19 and 20 MJ/kg. This suggests that other factors, such as ash content, are more important when deciding which biomass should be used for energy conversion.

The energy content is typically determined with a bomb calorimeter, which is a closed system in which small temperature increases after combustion in a pressurized oxygen atmosphere can be determined with high sensitivity. The calorific value is defined as the amount of thermal energy per weight unit produced by the total combustion of the sample in MJ/kg.

About 0.5 g of biomass are combusted in a pressurized oxygen atmosphere with a pressure of about 3,000 kPa. The measured temperature increase of the vessel is very small, requiring very sensitive temperature sensors in the calorimeter and good calibration before each measurement. Calibration is generally performed with benzoic acid.

8.9 Ash Content

Ash is the inorganic residue such as silica, potassium, calcium, sulphur and chlorine that remains after combustion at high temperatures and constitutes between 0.2 and 20 % of the biomass weight. In general most biomass contains similar levels of C, H and O, but the amounts of Si, Ca, Mg, K, P, S, Cl, Al, Fe and Mn, or heavy metals such as Cu, Zn, Co, Mo, As, Ni, Cr, Pb, Cd, V and Hg can differ widely (Obernberger et al. 1997).

The ash content of different types of biomass used in South Africa for energy conversion is given in Table 8.5.

8.9.1 Determination of the Ash Content

To determine the ash content, ovendry samples are placed in a ceramic crucible and the weight of each crucible and the biomass is noted. The crucibles are then placed and in the furnace at a temperature of 575 °C for 3 h (EN 14775).

After cooling, the ash content is calculated according to Eq. 8.8:

$$Ash\ (\%) = \frac{m_A \times 100}{m_0} \tag{8.8}$$

$m_A = mass\ of\ ash,\ m_0 = mass\ of\ ovendry\ sample$

Table 8.5 Calorific values and ash content of typical South African wood species (Munalula and Meincken 2009; Smit 2010)

Biomass	Gross CV (MJ/kg)	Ash content (%)
Acacia cyclops (Rooikrans)	19.4	2.79
Eucalyptus cladocalyx (Blue Gum)	19.3	2.38
Pinus patula (Pine)	19.0	0.45
Acacia eriloba (Camelthorn)	19.3	2.79
Vitis vinifera (Vine stumps)	19.2	0.34
Acacia saligna (Port Jackson)	19.1	2.47
Acacia mearnsii (Black Wattle)	19.2	1.64
Acacia longifolia (Long Leaved Wattle)	19.1	1.32
Casuarina cunninghamiana (Beefwood)	19.0	1.67
Pinus pinaster (Cluster pine)	19.6	0.78

8.10 Volatile Content

The volatile content is given by the mass loss (excluding moisture) due to thermal degradation when the biomass is heated. The degradation products are gaseous substances, such as CO, CO_2, NO_x etc. The combustion of these volatile components results in the bright flame when wood is combusted and its colour and temperature depend on the chemical composition of the wood.

Because of its low C/H ratio, wood has a rather high volatile content – between 75 and 90 %. A high volatile content is directly proportional to a lower CV, resulting in the low energy density of biofuels compared to, e.g. coal.

To determine the volatile content, ovendry samples are placed in a ceramic crucible and the weight of each crucible and the biomass is noted. The crucibles are then placed and in the furnace at a temperature of 900 °C for 7 min (ASTM E872-82).

After cooling, the volatile content is calculated according to Eq. 8.9:

$$Volatiles\ (\%) = \frac{100 * (m2 - m3)}{m2 - m1} \tag{8.9}$$

m1 = mass of crucible, m2 = mass of ovendry sample and crucible, m3 = mass of contents and crucible after heating

8.11 Biomass Requirements

The conversion techniques described in Chap. 7 have different requirements with regards to the biomass feedstock. The energy yield can be optimised, if the biomass is chosen and prepared correctly, as the pre-treatment of the initial feedstock has a big impact on the product yield and quality. These pre-treatments can be mechanical and/or chemical, such as sieving, water washing, solvent or acid-leaching. Table 8.6 lists some of the requirements for a selection of conversion techniques and reactor types (extracted from van Loo and Koppejan 2008; Stephen et al. 2010):

8.11.1 Thermochemical Conversion: Combustion, Torrefaction, Pyrolysis and Gasification

Ideally the ash content of biomass should be as low as possible for thermochemical conversion, as it poses many practical problems in the conversion process, ranging from slagging to corrosion of the reactor. A high nitrogen content in the biomass will lead to increased NO_x emissions during combustion. Particularly undesirable in ash are Silicon (Si), Chlorine (Cl), Potassium (K) and Sulphur (S), as they form silicates, sulphates and alkali chlorides. For example, high potassium and chlorine

Table 8.6 Biomass requirements of various conversion techniques

		Biomass			
Conversion type	Reactor	Type/size	Density (kg/m^3)	MC (%)	Ash (%)
Combustion	Fixed grate	Chunks, briquettes ∅ < 50 mm	As high as possible	<20	As low as possible
	Suspension boiler	Sawdust, small shavings ∅ < 10 mm	150–200	<15	<5
	FB or CFB boiler	Sawdust, small shavings, low alkali ∅ < 50 mm	400–600	<50	<5
Gasification	Fixed bed	Wood chunks, briquettes etc. not too small ∅ < 100 mm	100–600	<20	<10
	Moving bed	Chips, pellets ∅ < 50 mm	200–300	<15	<20
	CFB	Chips ∅ < 50 mm	200–500	<50	<10
Pyrolysis	Slow	Chips, pellets ∅ < 70 mm	200–500	<15	<5
	Fast	Chips, pellets ∅ < 5 mm	200–500	<10	<20
Alcoholic fermentation	Pretreatment-hydrolysis-fermentation	Wet chips; never dried material			As low as possible
Anaerobic digestion	Single stage	Chips, residues	Low	High > 50 %	N/A

contents can cause slagging and corrosion in the combustion unit (Skrifvars et al. 2004). For an efficient use of biomass for power generation, the amounts of K and Cl should thus be as low as possible. The ash content is typically also inversely related to the calorific value, making biomass with high ash content undesirable for combustion and gasification. For pyrolysis, solvent-leaching can be used as a de-ashing step for either the biomass feedstock or the char product, to increase the char adsorption properties (Carrier et al. 2012). These changes in char adsorption due to de-ashing are directly related to the pyrolysis rate and volatile yield (Raveendran and Ganesh 1998; Carrier et al. 2012).

Thermochemical conversion processes are well suited to uniform, densified feedstock, such as pyrolysis products or pellets (Stephen et al. 2010). Torrefaction is an example of such a pre-treatment step to decrease the MC of the biomass, while at the same time increasing the energy density and CV. Subsequent size reduction, such as grinding, before feeding the reactor is easier with torrefied biomass, as it is more brittle.

The Lignin content has a significant impact on the suitability of biomass for thermochemical conversion. Although the high CV of lignin makes it desirable for combustion and gasification, an increased lignin content also affects the rate of thermal degradation. As a result, biomass with a higher lignin content will pyrolyse more slowly, while wood with high cellulose content can be pyrolysed faster (Gani and Naruse 2007). The positive impact of lignin on the CV, however, generally outweighs the negative effects of reduced reaction rate.

In the case of vacuum pyrolysis of wood, extractives showed an inhibiting effect on the oil yield, as they inhibited the levoglucosan formation. Removal of extractives did modify the hemicelluloses composition significantly, as reflected by the similar acetic acid yield derived from wood and extractive-free wood material (Roy et al. 1990).

8.11.2 Biological Conversion: Anaerobic Digestion and Alcoholic Fermentation

Compared to agricultural biomass (i.e. wheat straw), woody biomass has a higher density, higher cellulose and lignin content and lower hemicelluloses content, which means it is more resistant to biological conversion. Enzymatic hydrolysis and digestibility of lignocellulose are the key steps in biological conversion and they depend strongly on the feedstock composition and structure, pre-processing of the material and dosage and efficiency of the enzymes used for hydrolysis.

Biological conversion of lignocellulose is enhanced by mechanical comminution that reduces the particle size as well as the degree of polymerization of the cellulose. For biochemical conversion the biomass should have a loose structure that can easily be penetrated by enzymes. Therefore additional pre-treatment is needed to improve accessibility of the cellulose and increase the digestibility to above 50 % for enzymatic hydrolysis (Vidal et al. 2011). High crystallinity of the cellulose and a high degree of polymerization limit enzymatic hydrolysis, mainly the initial hydrolysis rate.

For biochemical conversion feedstock with a high moisture content is preferred and drying reduces the accessibility of biomass to chemicals, steam and enzymes (Stephen et al. 2010; Liu et al. 2002). Special care must therefore be taken to avoid drying and the associated "hornification" phenomenon. Freshly harvested wood chips are the optimal woody feedstock for bioconversion of lignocellulose.

The disruption of the lignocellulose structure by pre-treatments can significantly reduce the recalcitrance of lignocellulose to biological degradation. Most of the pretreatments alter not only the chemical composition but also the physical structure of the biomass by increasing the accessible surface area and pore volume, thereby enhancing cellulase attack.

An increase in lignin content reduces the digestibility of biomass, thereby decreasing the rate and extent of enzymatic hydrolysis. Lignin with a high syringyl content (typical for hardwoods) can be easier degraded by pre-treatments (diluted acid, alkali, hydrogen peroxide, etc.). Apart from being a physical barrier to hydrolysis, lignin can adsorb irreversibly to the enzymes, thereby reducing the yield and increasing time required for effective conversion. The adsorption capacity of lignin depends on the type of lignin and pretreatment applied. Unlike in the pulping process, where the objective is to maintain the fiber integrity, these pretreatments aim for maximum digestibility by removing lignin and loosening the fibre structure.

The type of hemicelluloses also affects the choice of pre-treatment and the enzymes used for the fermentation process. For example, the acetyl content in hemicelluloses from hardwoods is involved in autohydrolysis reactions during thermochemical pretreatments, enhancing the cellulose accessibility. On the other hand, residual acetyl groups in pre-treated material constitute a steric hindrance for the enzymatic hydrolysis.

However, the pretreatment of lignocellulose to improve digestibility may also result in the production of sugar and lignin degradation compounds, which may lead to hydrolysis/fermentation inhibition.

Biomass with a low ash content is preferred for biological conversion, not only because it maximizes the availability of carbohydrates and lignin for the conversion process, but also because the buffering capacity of ash may increase the chemicals requirement in acid-catalysed pretreatments for biological conversion.

References

ASTM D1762-84: American Standard, Chemical analysis of wood and charcoal, Philadelphia: ASTM International, 2007

ASTM E872-82: American Standard, Volatile matter of wood fuels

Blunk SL, Jenkins BM (2000) Combustion properties of lignin residue form lignocellulose fermentation. NREL Subcontract ACG-8-180-21-01. National Renewable Energy Laboratory, Davis

BS EN 15103:2009: British Standard, Solid biofuels, determination of bulk density

Carrier M, Hardie AG, Uras Ü, Görgens J, Knoetze JH (2012) Production of char from vacuum pyrolysis of South-African sugar cane bagasse and its characterization as activated carbon and biochar. J Anal Appl Pyrolysis 96:24–32

Downes GM, Evans R, Schimleck LS, Fritts HC (2000) The commercial cambium: understanding the origin of wood property variation. Society for experimental biology annual general meeting, march 1999, Edinburgh, UK. In: Savidge RA, Barnett JR, Napier R (eds) Cell and molecular biology of wood formation. BIOS Science, Oxford, pp 325–336

EN 14775: British Standard, Solid biofuels, determination of ash content

Fengel D, Wegener G (2003) Wood chemistry, ultrastructure and reactions. Kessel Verlag (reprint), Remagen, Germany

Gani A, Naruse I (2007) Effect of cellulose and lignin content on pyrolysis and combustion characteristics for several types of biomass. Renew Energy 32:649–661

Guillen MD, Ibargoitia ML (1999) Influence of the moisture content on the composition of the liquid smoke produced in the pyrolysis process of *Fagus sylvatica* L. Wood J Agric Food Chem 47:4126–4136

Hamelick CN, Suurs RAA, Faaij APC (2005) International bioenergy transport costs and energy balance. Biomass Bioenergy 29:114–134

Harju AM, Venäläinen M, Anttonen S, Viitanen H, Kainulainen P, Saranpää P, Vapaavuori E (2003) Chemical factors affecting the brown-rot decay resistance of Scots pine heartwood. Trees 17:263–268

Hatfield R, Vermerris W (2001) Lignin formation in plants. The dilemma of linkage specificity. Plant Physiol 126:1351–1357

Haygreen JG, Bowyer JL (2007) Forest products and wood science: an introduction. Iowa State Press, Ames

Herman FM (1987) In: Herman FM, Bikales NM, Meyers G, Overberg CG, Kroschwitz JI (eds) Encyclopedia of polymer science and engineering: identification of Lignin, vol 8. Wiley Interscience, New York, pp 795–852

Howard ET (1973) Physical and chemical properties of slash pine parts. Wood Sci Technol 5(4):312–317

Jakab E, Faix O, Till F, Szekely T (1995) Thermogravimetry/mass spectrometry study of six lignins within the scope of an international round robin test. J Anal Appl Pyrolysis 35(2):167–179

Kataki R, Konwer D (2001) Fuelwood characteristics of some indigenous woody species of north-east India. Biomass Bioenergy 20:17–23

Kratky L, Jirout T (2011) Biomass size reduction machines for enhancing biogas production. Chem Eng Technol 34(3):391–399

Kumar M, Gupta RC (1992) Properties of acacia and eucalyptus woods. J Mater Sci Lett 11: 1439–1440

Liu HW, Walter HK, Vogt GM, Vogt HS, Hobein BE (2002) Steam pressure disruption of municipal solid waste enhances anaerobic digestion kinetics and biogas yield. Biotechnol Bioeng 77: 121–130

Marutzky R, Seeger K (1999) Energie aus Holz und anderer Biomasse. DRW Verlag, Leinfelden-Echterdingen

McKendry P (2002) Energy production from biomass (part 1): overview of biomass. Bioresour Technol 83:37–46

Munalula F, Meincken M (2009) An evaluation of south African fuelwood with regards to calorific value and environmental impact. Biomass Bioenergy 33:415–420

Novaes E, Kirst M, Chiang V, Winter-Sederoff H, Sederoff R (2010) Lignin and biomass: a negative correlation for wood formation and lignin content in trees. Plant Physiol 154: 555–561

Obernberger I, Biedermann F, Widmann W, Riedl R (1997) Concentrations of inorganic elements in biomass fuels and recovery in the different ash fractions. Biomass Bioenergy 2:211–224

Panshin AJ, De Zeeuw C (1980) Textbook of wood technology: structure, identification, properties, and uses of the commercial woods of the United States and Canada. McGraw-Hill, New York

Perkel JM (2012) New elemental analysis techniques for the analytical chemistry laboratory. Application note, American Laboratory. http://www.americanlaboratory.com/914-Application-Notes/38746-New-Elemental-Analysis-Techniques-for-the-Analytical-Chemistry-Laboratory/. Accessed on 18 Sep 2013

Raveendran K, Ganesh A (1998) Adsorption characteristics and pore-development of biomass-pyrolysis char. Fuel 77:769–781

Roy C, Pakdel H, Brouillard D (1990) The role of extractives during vacuum pyrolysis of wood. J Appl Polymer Sci 41:337–348

Schubert WJ (1965) Lignin biochemistry. Academic, New York

Schuck S (2006) Biomass as an energy source. Int J Environ Stud 63:823–835

Skrifvars BJ, Lauren T, Hupa M, Korbee R, Ljung P (2004) Ash behaviour in a pulverised wood fired boiler. Fuel 83:1371–1379

Smit HC (2010) Evaluation of invasive wood species as energy source. Honours report, Department of Forest and Wood Science, Stellenbosch University, Stellenbosch

Smith DM (1959) Report 2014. Forest Products Laboratory, Madison

Stephen JD, Mabee WE, Saddler JN (2010) Biomass logistics as a determinant of second-generation biofuel facility scale, location and technology selection. Biofuels Bioprod Bioref 4:503–518

Sturos JB, Coyer LA, Arola RA (1983) Air drying of chunkwood and chips. Research note NC-293, North Central Forest Experiment Station, Saint Paul
Sykes R, Kodrzycki B, Tuskan G, Foutz K, Davis M (2008) Within tree variability of lignin composition in populus wood science and technology. Wood Sci Technol 42:649–661
Tsoumis G (2009) Science and technology of wood. Verlag Kessel (reprint), Remagen, Germany
Van Loo S, Koppejan J (2008) Biomass combustion and co-firing. Earthscan, London/Sterling
Vidal J, Dien BS, Ting KC, Singh JR (2011) Influence of feedstock particle size on lignocellulose conversion – a review. Appl Biochem Biotechnol 164(8):1405–1421
Walker JCF (2006) Primary wood processing. Springer, Dordrecht
Wondifraw D (2010) Air drying of wood chips and logs under shelter and outside in the Western Cape. Honours report, Department of Forest and Wood Science, Stellenbosch University, Stellenbosch
Wünning P (2001) Applications and use of lignin raw materials. In: Hofrichter M, Steinbuechel A (eds.) Biopolymer: lignin, humic acids and coal. Wiley-VCH, Weinheim

Chapter 9
Socio-Economic Aspects of Rural Bioenergy Production

Cori Ham and Theo E. Kleynhans

9.1 Introduction

Energy from biomass is one of the oldest forms of energy used by mankind due to its general availability and low technology requirements (Buchholz et al. 2007). Wood was used all over the world as the principle source of energy until about the mid nineteenth century after which it had been replaced by more efficient and convenient energy sources such as coal, gas and electricity. The move towards more convenient fuel types was especially prevalent in industrialised countries but wood has remained the dominant source of energy in developing countries where people are less able to afford alternative sources of energy and their associated technology (Arnold et al. 2003).

The move to more convenient types of energy in the industrial world set the scene for two crisis events in the late twentieth century that would refocus the world's outlook on bioenergy. The first event in the early 1970s was the so called "Energy Crisis" caused by the rise in fossil fuel prices which triggered an increased focus on fuelwood resources. Attention was drawn to the fact that in developing countries, fuelwood was the principle source of energy and that projected use levels could be above sustainable replacement levels, leading to a total depletion in wood supplies. Assumptions were made that fuelwood use was a major contributor to deforestation and forest degradation, and subsequent solutions were based on providing sustainable sources of fuelwood through forestry programmes (Ham and Theron 1999).

C. Ham (✉)
Department of Forest and Wood Science, Stellenbosch University, Stellenbosch, South Africa
e-mail: cori@sun.ac.za

T.E. Kleynhans
Department of Agricultural Economics, Stellenbosch University, Stellenbosch, South Africa

T. Seifert (ed.), *Bioenergy from Wood: Sustainable Production in the Tropics*, Managing Forest Ecosystems 26, DOI 10.1007/978-94-007-7448-3_9,

The second crisis event, in the 1990s, was the realisation that the reliance on convenient fossil fuels has an effect on the CO_2 levels in the atmosphere and could lead to a change in the world's climate. The so called "Climate Change Debate" prompted the search for more sustainable energy sources that would reduce CO_2 outputs. Energy from biomass is seen as one of the more sustainable sources of energy and it is expected that future development of bioenergy will take place in two directions. The first being an increase in bioenergy production in developed countries to try and reduce CO_2 levels, and secondly an increase in total bioenergy production in developing countries to cope with population growth and a move towards modern bioenergy conversion technologies (Buchholz et al. 2007).

The events of the 1970s and 1990s highlighted the potential positive effect on carbon balances of substituting fossil fuels with bioenergy and also emphasised the negative environmental costs associated with unsustainable fuelwood harvesting in terms of forest loss and degradation (FAO 2005). Accelerated tree planting programmes were promoted as a technical supply side solution to the "Fuelwood Gap" of the 1970s and modern bioenergy technology can be seen as a way of moving people, especially in developing countries, away from primary firewood use to more efficient bioenergy conversion technologies. The positive relationship between access to energy and human well-being has led many to conclude that improving access to modern energy, such as electricity, is a key component of poverty reduction and development (Buchholz et al. 2007).

It can be assumed that bioenergy will continue to play an important role in global energy supply as a source of renewable energy. Solid biomass contributes 45 % of the primary renewable energy in member countries of the Organisation for Economic Co-operation and Development (OECD) and provides more than 90 % of the energy needs for many developing countries (Sims 2003, ex Buchholz et al. 2007). One of the important questions related to the future of bioenergy production is how the developed and developing world will balance the positive and negative externalities associated with bioenergy use. Pure technical solutions might not be enough as it is important to consider that human well-being depends on more than just energy need and should consider every aspect of socio-economic development (Ham et al. 2008). This chapter will explore the current role that bioenergy plays in socio-economic development. It will begin with a summary of biomass as a primary source of energy in developing countries and progress to a review of the role of the rural poor in modern day bioenergy enterprises.

9.2 Use of and Dependency on Biomass for Energy in Developing Countries

Fuelwood is still the predominant source of primary energy in most of the developing world. It is estimated that only 10 % of wood removals in Africa are used for industrial roundwood, while the rest is used as fuelwood. Africa accounts

Table 9.1 Number of people relying on biomass for cooking and heating in developing countries (million) (IEA 2002, ex Arnold et al. 2003)

	Year	
Region	2000	2030
China	706	645
India	585	632
Other Asia	420	456
Africa	583	823
Latin America	96	72

for 33 % of the global fuelwood removals and only 5 % of the global industrial wood removals (FAO 2011). In Ethiopia for example, 93 % of households use open fires for cooking, while burning firewood, charcoal, crop residue and animal dung (Friends of the Earth 2010).

It is predicted that by 2030 biomass energy will still account for more than 75 % of total residential energy in Africa and that the region will have surpassed China and South Asia in terms of quantities of fuelwood used. More than 500 million people in Africa were relying on biomass as a primary source of energy in 2000 and it is projected that this number will rise to more than 800 million by 2030 as seen in Table 9.1 (IEA 2002, ex Arnold et al. 2003).

The high rate of fuelwood use in developing countries as described above can be linked to poverty and the cost of alternative fuels. In many countries fuelwood collection is the only affordable energy option and can also serve as a source of income for the poor. Despite the link between fuelwood use and income, it is interesting to note that fuelwood shortages and associated increases in collection efforts are not seen as priority problems amongst rural people and that they are more concerned with aspects related to income, health and food security (Arnold et al. 2003). It is therefore necessary to look at how people combine fuelwood and other energy sources in areas where there is a transition from traditional fuelwood to alternative energy, to understand new energy adoption patters and processes.

9.3 Rural Energy Use and Alternatives to Fuelwood

South Africa presents an interesting case study of a society in transition from utilising fuelwood for energy generation to using electricity. The country produces and consumes more than 60 % of Africa's electricity but a large percentage of the population is still dependent on fuelwood and paraffin for household energy production. It is estimated that in the Limpopo and Eastern Cape Provinces more than 50 % of households rely on fuelwood, while also using other energy sources for food preparation including paraffin, coal, liquid petroleum gas (LPG), cow dung and other biomass fuels such as crop residue (Chirwa et al. 2008).

Despite the implementation of a substantial household electrification programme most newly electrified households continue to use firewood and other energy sources because of the cost of appliances and the monthly cost associated with electricity (Shackleton et al. 2007). This multiple fuel usage is reflected in the wide variety of cooking devices available to households. Households often use a combination of braziers, paraffin stoves, wood/coal stoves, open fires with three-legged pots, LPG stoves and electric stoves (SPARKNET 2004). This is illustrated by rural households in the Cata Village in the Eastern Cape Province where 93 % of respondents in a survey indicated that they use electricity for lighting but less than 20 % use electricity for cooking and heating purposes. More than three quarters of the respondents use firewood as a primary energy source and combine it with paraffin and electricity (Chirwa et al. 2010).

Even in urban areas shifts away from fuelwood use are not taking place at a large scale. The cost of buying new stoves, adoption of new cooking methods required by unfamiliar cooking devices and the cost of alternative fuels often hinder the adoption of new and improved stoves (Arnold et al. 2003). Ham and Theron (2001) found that in towns in the Eastern Cape Province where electricity is used for lighting, a combination of fuelwood and paraffin is used for heating and cooking. Chirwa et al. (2010) found that people will use electric stoves for foods that require short cooking times or to re-heat food but still prefer to use firewood for foods that require long preparation times. The residents of Kentani, a small town in the Eastern Cape Province, used an estimated 3.7 tonnes of fuelwood per year per household for cooking and heating purposes despite the fact that the town has access to electricity (Ham 2000).

It is important to remember that people use fuelwood out of necessity and would switch to another source of energy higher up the energy ladder if it becomes available at a price point within reach and is considered as good value in terms of utility for the extra money. The energy ladder represent the continuum between the most basic use of fuelwood in an open fire to the most flexible, clean and convenient source of energy such as electricity. The major barriers between different fuels on the energy ladder (such as charcoal, paraffin, gas and electricity) are related to the costs of appliances and the costs of fuels (Practical Action Consulting 2009).

Converting people from using the most basic bioenergy resources to bioenergy resources higher on the energy ladder could be one of the main objectives of bioenergy initiatives. It is, however, important to consider that bioenergy presents a complex combination of feedstock supply, conversion technology and energy allocation. These three factors are influenced simultaneously by social, economic and ecological factors. Understanding these factors and their interdependency is essential in ensuring effective and efficient bioenergy solutions when moving up the energy ladder (Buchholz et al. 2007).

9.4 Progress Up the Energy Ladder and Implications for Rural Users of Energy

The argument that fuelwood is not necessarily a preferred energy source but one forced upon the rural poor due to economic considerations, was mentioned briefly in the previous section. The main economic factors that drive the use of fuelwood in developing countries are the low costs of procuring fuelwood and the low income of consumers. Often the main cost associated with fuelwood collection is the opportunity cost of the time (which could be substantial) taken to collect fuelwood (Cushion et al. 2010) (Fig. 9.1). In this case, rural households can allocate scarce cash resources to other needs such as education of children, investment in agricultural tools and capital for income generation activities. Such cost savings would best be reflected by the replacement value of the energy sources that fuelwood substitutes, rather than direct cost of fuelwood (Shackleton 2004).

The commercial role of fuelwood can also be significant when community members supplement their incomes by selling fuelwood. Sometimes this activity even becomes their main source of cash. Notably, this includes also the poorest of the poor where many rural landless people are among those specialising in fuelwood production (Vedeld et al. 2004).

While it seems that when household income increases, people prefer to move up the energy ladder to other energy sources (typical scenario where fuelwood is seen as an "inferior" good), some researchers also report that fuelwood use increases with increased income for very poor households where fuelwood is seen as a "normal" good (Arnold et al. 2003; FAO 2005). Other social and environmental considerations

Fig. 9.1 Rural women often spend large amounts of time per week collecting fuelwood

that might have an effect on household fuelwood consumption include climate (People might prefer to cook on fuelwood as it also provides heat for their house in winter (Chirwa et al. 2010)), access to markets and forest resources as well as health considerations (Exposure to indoor air pollution from biomass fuels is linked to many respiratory conditions and diseases (Cushion et al. 2010)).

The choice of energy source should thus be seen in the wider socio-economic developmental context and has various positive and negative implications for the user (mostly related to cost and convenience) as well as for the energy and CO_2 balance of the country. Table 9.2 summarises the key implication related to each of the energy sources along the energy ladder.

When the cost of emissions reduction is brought into the equation, progress up the energy ladder becomes less clear cut. In cases where biomass is the least expensive source of fuel and is produced sustainably, traditional biomass energy use could result in much lower emissions than fossil fuel alternatives. In such cases it might not be beneficial from a national emissions point of view to promote coal derived electrification schemes but rather to focus on more efficient traditional use of fuelwood and sustainability of fuelwood supply (Cushion et al. 2010).

It is, however, also important to consider that the long term success of bioenergy programmes are linked to the socio-economic impact that it will have on rural producers of biomass. The socio-economic implication of bioenergy programmes has often been neglected in academic studies on the sustainability of biofuels. Most work focussed on the environmental implications and the interaction between the environment and economy while the social dimension remained weakly defined (Lehtonen 2011). The following sections of this chapter will consider the costs and benefits of bioenergy programmes to rural producers of biomass and users of bioenergy.

9.5 Costs and Benefits of Rural Energy Production

The previous sections focussed on the use of biomass in simple energy production systems such as open fires made from gathered fuelwood. Such systems produce a large percentage of the cooking and heating energy required by the rural poor in developing countries. The "Climate Change Debate" has, however, brought with it an increasing demand for biofuels and the need for carbon sinks plus the system of carbon emissions trading, especially in developed countries (Tauli-Corpuz and Tamang 2007). "Modern bioenergy chains" encompass advanced concepts related to biomass feedstock production, supply chain logistics and conversion technologies such as combustion, gasification, fermentation, etc. The end uses of biomass derived energy can range from household heating with modern fuel pellets stoves to international distribution chains for liquid biofuels (Buchholz et al. 2007).

In developing countries the impact of these "modern bioenergy chains" on poverty alleviation will depend on the opportunities that are presented for agricultural development and the potential to increase poor peoples' access to

Table 9.2 Energy sources along the energy ladder and key implication for use at household level

Energy source	Key implications
Traditional open fire	Traditional uses include firewood/charcoal, dung, and crop residue. Relatively cheap and available. Potentially unsustainable production/collection. Wasteful energy conversion. Emissions from renewable resource. Inconvenient and potential health impacts.
Energy efficient stove	Fuels include firewood/charcoal, crop residue. More expensive than open fire (capital investment required) and requires adjustment of cooking techniques. Potentially unsustainable production/ collection. Higher levels of energy conversion than traditional open fire. Requires less biomass. Could use biomass pellets and reduce health impacts. Emissions from renewable resource. More convenient than open fire and potentially less health impacts.
Alternative fuels for household use (paraffin, LPG gas, etc.)	Fuels such as paraffin, and LPG gas can be purchased at village level. More expensive than firewood/charcoal. Do not affect local sources of biomass. Higher levels of energy conversion than firewood. Require capital investment in special stoves and cooking equipment. Emissions from fossil fuel. More convenient than fuelwood based energy sources. Can be dangerous to use.
Small scale energy generation facilities at rural level (could be fuelled by biomass)	Could include smaller power stations that use only biomass; small to medium-scale facilities that provide power or heat in the forestry and agricultural processing industries and at village level. Will require electrification of village houses with associated costs. Electricity will have to be purchased by consumer. Requires capital investment in electric stoves, heaters, lights. Emissions from renewable resources if fuelled by biomass. Higher production cost (in most cases) and lower energy content of biomass makes it more expensive than coal. Conversion efficiency of biomass is slightly lower than that of fossil fuels. More convenient than fuelwood or alternative household fuels. Safer with less health impacts. Could allow for sustainable community based biomass supply projects. Co-generated heat can be used locally for drying or heating.
National grid electricity derived from coal burning power plants	Village electrification as part of national electricity grid. Capital cost in supplying electricity. Electricity will have to be purchased by consumer, most likely at lower price than from small scale generator. Requires capital investment in electric stoves, heaters, lights. Emissions from fossil fuel. Lower production cost and higher conversion efficiency than biomass. More convenient than fuelwood or alternative household fuels. Safer with less health impacts. Limited options for community involvement in electrification projects.

Sources: Adapted from Cushion et al. (2010); Arnold et al. (2003) and Practical Action Consulting (2009)

improved types of bioenergy. There are significant concerns about the impact that these systems could have on agriculture, food security and sustainable forest management and the social impacts of bioenergy development, particularly in relation to land use and security of tenure (Cushion et al. 2009). This section will review the role of bioenergy production for modern bioenergy chains on job creation; land availability and rights; food security; and the sustainability of rural livelihoods in developing countries.

9.5.1 Job Creation

The main measurable economic impact of biomass production is likely to be income and employment generation. Modern fuels provide for formal employment opportunities where traditional fuels provide informal employment for the poorest members of communities (Cushion et al. 2010).

With respect to informal employment it is recognised that fuelwood production is labour intensive and require between 100 and 170 person-days per terrajoule (TJ) of energy and between 200 and 350 person days per TJ for charcoal production (Fig. 9.2). The benefit of this type of employment generation depends, however, on the value of the labour used for production and could be considered positive if unemployment is high but low or negative when there are alternative uses for this labour (FAO 2005). Fuelwood gathering and trading helps to bridge seasonal income gaps when there is no other employment available and serves as a "safety net" in time of hardship such as during droughts and associated crop failures (Arnold et al. 2003).

The ease of entry into fuelwood trading means that it is usually characterised by strong competition and low returns. Reliance on an income from fuelwood selling is often seen as a livelihood of last resort and many fuelwood collectors best be assisted by helping them to move into more rewarding, alternative employment activities as these become available (Arnold et al. 2003).

Formal employment in biomass production to supply biofuels could serve as an alternative to informal fuelwood collection activities. Current trends indicate that biomass production for liquid biofuels could substantially increase employment

Fig. 9.2 Charcoal transported from rural producers to urban consumers in Lusaka, Zambia

opportunities in developing countries. In Brazil for instance, formal employment in the sugar sector rose by 53 % between 2000 and 2005 from about 643,000 to 983,000 people as a result of ethanol production. It is projected that the minimum number of people employed by liquid biofuel production in the world by 2030 would be 2 million with the majority in sugarcane and ethanol production. This projection assumes that most production will occur in large scale mechanised operations (Cushion et al. 2010).

The promise of jobs is, however, not always fulfilled. When jobs are created, levels of pay are sometimes so low that employees are not actually better off. In Mozambique employees of a biofuels company, with rights to plant 60,000 ha of *Jatropha* on previous communal farming and grazing land, were paid the national minimum wage but there was little improvement in their standard of living. Some employees earned less than what they could during a good farming year (Friends of the Earth 2010).

Labour requirements for liquid biofuels are also mostly restricted to short-term work for land clearing and planting and some work at harvest time. Some studies estimate that one permanent job is created for every 100 ha of biofuel planted, with greater potential for job creation in the processing and production industry. When mechanised farming methods are used, employment levels are even lower (Friends of the Earth 2010).

It is possible to deduct potential employment rates for bioenergy plantations from existing timber plantation operations. Silvicultural activities in bioenergy plantations are ideally suited to the creation of low skilled job opportunities and the formation of small scale forestry enterprises in rural areas. Most silvicultural activities require basic education and low levels of basic training (2 weeks to 2 months), enabling people from marginalised rural areas to participate in economic activities (Forestry Solutions 2007).

Silvicultural activities are spread over the life cycle of a hectare of trees that could range from as short as 6 years in the case of short rotation eucalypt plantations to 25 years in the case of pine saw timber in South Africa. Table 9.3 illustrates the typical life cycle of a pine sawtimber hectare, that could yield biomass residue for bioenergy production, with the timing of activities and labour (person days/per hectare) required to perform these activities (Established at 1,372 stems per hectare on fairly level land).

It is, important to observe that 50 % of the labour activities happen during the first 2 years after plantation establishment. Unthinned pulpwood and biomass working circles share the first operations until first thinning and would thus not differ substantially in person working days. As a result of this any job creation programme based on silvicultural activities will have to rely on a sustained programme of afforestation and/or reforestation to have a substantial economic effect in rural areas.

De Beer (2012) translates these man day values into labour per hectare and estimates that 0.15 labourers are directly employed per hectare of plantation forestry in the Western Cape of South Africa. Shackleton (2004) estimates that activities along the forestry value chain can create up to three jobs per primary forestry job.

Table 9.3 Typical life cycle of a pine plantation, excluding harvesting and transport of logs

Year	Activity	Labour Person days[a]/ha
0	Preparation of planting pits	4.93
0	Planting of seedlings	3.43
0	Fertiliser application	1.03
0	Manual & chemical weed control	5.2
1	Chemical weed control × 2	3.2
2	Chemical weed control × 2	3.2
3	1st Pruning to 1.5 m & chemical weed control	5.6
5	2nd Pruning to 3 m	5
7	3rd Pruning to 5 m	7
8	Marking for thinning	1.1
12	Marking for thinning	1.1
18	Marking for thinning	1.1
	Total man days	41.89

[a]Man day values derived from Forestry Solutions (2007) Best Operating Practice estimates

Over and above direct employment benefits, Ofoegbu (2010) found in a study related to the social benefits of plantations in South Africa that rural communities adjacent to plantations also have access to harvest residues as fuelwood and source of building material, could utilize non-timber resources from plantations and had access to free accommodation, free farmland and free grazing land.

9.5.2 *Land Availability*

There is a common perception that developing countries have vast areas of available land. It is for instance estimated that there are about 800 million hectares of cultivable land across Africa of which less than a quarter appears to be used. Land prices in developing countries are in many places very low compared to developed countries and where the host country is supportive, land can be acquired on favourable terms. Such cheap access to land and cheap labour for biofuel development can be seen as a good business opportunity for foreign companies and has given rise to so called "land grabs" (Friends of the Earth 2010).

"Land grabs," where land traditionally used by local communities is leased or sold to outside investors, are becoming increasingly common. The International Food Policy Research Institute (Headey et al. 2009, ex Friends of the Earth 2010) estimated that globally 20 million hectares of land have been sold since 2006 of which 9 million hectares were in Africa. Almost 5 million hectares are reportedly intended for biofuels, including *Jatropha*, oil palm and sweet sorghum.

International companies are keen to emphasise the benefits of these land deals to local communities in terms of job creation and economic development. A study by the Food and Agricultural Organisation (FAO) (Cotula et al. 2009, ex Friends of the Earth 2010) has, however, found that in Ethiopia, Ghana, Madagascar and Mali all biomass grown for biofuels will be exported with minimal direct benefit to the countries involved.

Rising demand for bioenergy has led to rapid expansion of large scale biofuel plantations (Cushion et al. 2010). Oil palm plantations have become the fastest growing monoculture in the tropics and have increased from 6.5 million hectares in 1997 to 14 million hectares in 2007 (Gerber 2011). The largest oil palm plantations are found in Malaysia and Indonesia with 4 and 6 million hectares, respectively, under oil palm cultivation (Tauli-Corpuz and Tamang 2007). The majority of oil palm plantations are located on land that was once tropical forest. The relationship between oil palm plantations and deforestation is debatable as it is unclear how much deforestation was caused by direct clearing for oil palm plantations or how much oil palm expansion occurred on land already deforested and degraded as a result of other factors (Gerber 2011).

The growth of biofuel from oil palm has resulted in economic benefits to national governments and companies involved but they come with serious social and environmental costs which adversely affect local people dependent on tropical forests for their livelihoods (Tauli-Corpuz and Tamang 2007). In Indonesia for instance, there is a lack of clarity of ownership over forested land, leading to widespread disagreements over land tenure. Land disputes with local communities were reported by more than 80 palm oil plantation companies in Sumatra in 2000. Large plantation areas have been cleared without adequate resettlement provisions for displaced communities (Cushion et al. 2010) and communities are deprived of common areas used for biomass collection and subsistence agricultural activities. These displaced community members become landless peasants, experience the decay of indigenous culture and are forced to engage in seasonal or long term migration to urban areas in search of employment (Gerber 2011).

Supporters of biofuel plantations often argue that bioenergy plantations are established on marginal, unused or degraded lands. Land that appears degraded or "idle" to outsiders often serves a vital function for communities as common grazing land or land to collect firewood (Friends of the Earth 2010). These areas are often managed under communal traditional laws for subsistence as well as for cultural and religious practices and local livelihoods evolved around the use of products from these areas (Friends of the Earth 2005). As a rule the more marginal their livelihoods are, the more likely rural people will depend on common, open areas for their day-to-day struggle for survival. The land will yield fuel, food, medicine and building materials to people who do not have the means to obtain similar products or services in the formal economy (Van der Horst and Vermeylen 2011). The question remains, why do local communities protest when this unused or degraded land is converted to what some perceive to be more productive land use? (Friends of the Earth 2005).

Given ambitious global targets for biofuel production it can be questioned if biofuel crops can be grown only on unused or degraded land or if it will take land out of agriculture and forestry (Cushion et al. 2009). Large changes in land use may occur as a result of biofuel production. Global estimates for the amount of land required for future biofuel production range from 118 to 508 million hectares or 36 % of the current arable land by 2030 (Ravindranath et al. 2009, ex UNEP 2012).

9.5.3 Food Versus Fuel

The rise in biofuel production has had large demand effects on agricultural markets, especially grains and oilseeds. In 2007 it was estimated that biofuel production used 5 % of world cereal production, 9 % of world oilseed production and 10 % of sugar cane production. About half of the global increase in world grain consumption (about 80 million tonnes) was used for biofuels (FAO 2009).

This increase in biofuel production has become one of the major issues in the debate over climate change and global food security. On the one hand, biofuels are criticized for promoting food shortages, utilise much needed agricultural subsidies, offer little or no greenhouse gas mitigation and drive deforestation in developing countries. On the other hand they are promoted as a tool for economic development that could increase production and revitalise many countries' agricultural sectors. The cultivation of biofuels as cash crops may lead to increased investment in infrastructure that is needed to support thriving or emergent agricultural markets (Gamborg et al. 2011).

The substantial increases in agricultural commodity prices between 2005 and 2008 have partly been blamed on the conversion of food crops to biofuel. These increases have had negative implications for food security in the short term but also for net food buying countries and particularly for low income food deficit countries (FAO 2009) which raised fundamental questions about food sovereignty and government priorities. It could for instance be asked if countries dependent on food aid such as Kenya and Ethiopia should be selling fertile land to developers for biofuel production (Friends of the Earth 2010). In the long run, however, growing demand for biofuels and the rise in agricultural commodity prices may present an opportunity for promoting agricultural growth and rural development in developing countries. Furthermore the development of environmentally friendly biofuels could promote access to cheaper and safer energy supplies in rural areas which could further stimulate economic growth (FAO 2009).

Food security may be compromised if high yield agricultural lands are used for bioenergy production, pushing agriculture into more vulnerable, lower quality lands (Cushion et al. 2010). Such situations could occur when food growing farmers are forced off their land to make way for biofuel plantations. In Ghana for instance, where 50 % of the population work on the land and grow food for local consumption, *Jatropha* plantations have forced small farmers and particularly women farmers off their land. Valuable food sources such as shea nut trees have been cleared for bioenergy plantations. Small farmers in Ghana have expressed fears that they will not be able to afford to farm the land or buy food for their families. In Tanzania a similar situation occurred when thousands of rice and maize farmers were forced off their land in 2009 to make way for sugarcane plantations (Friends of the Earth 2010).

Converting forests into bioenergy plantations could also increase the food insecurity of forest-dependent communities. These impacts are, however, often short term and there is potential for biofuel production to have less impact on food security over the longer term. Biofuel production can for instance be beneficial to small producers when they are located far from markets and cannot sell their produce at competitive prices due to transport costs. Food production could be uncompetitive under such conditions making biofuels a better option (Cushion et al. 2010). It is impossible to make broad scale policy decisions about food security and biofuel production that will favour either the one or the other. Such decisions have to be made on a case by case, crop by crop, region by region and even location by location basis (Practical Action Consulting 2009).

9.5.4 Impact on Environmental and Social Sustainability

While biofuel production offers socio-economic opportunities it also presents environmental and social dilemmas. It has been claimed that if biofuel production leads to land degradation and deforestation any potential carbon savings benefits will be compromised (Gamborg et al. 2011). This compromise is illustrated by the clearing of forests to make way for oil palm plantations where it will take up to 150 years for the carbon savings from palm oil harvesting to replace the carbon lost from forests (Friends of the Earth 2010).

Biofuel production has the potential to lead to soil degradation and nutrient depletion (Chap. 10). Chemical inputs including fertilisers and pesticides can contaminate soils and lead to soil erosion. The removal of crop residues for co-firing may cause further declines in soil fertility (Cushion et al. 2010). Land clearing for biofuel plantations can cause considerable topsoil run-off and increase sediment loads in rivers. Soil erosion is five to seven times greater during clearing while sediment loads in rivers increase with a factor of four. Soil erosion is especially problematic when oil palms are planted on steep slopes and at high altitude (Friends of the Earth 2005).

Biofuel has also been described as one of the thirstiest products on the planet because of the amount of water needed to produce fuel. One litre of biodiesel from soya is estimated to require 9,100 L of water. A litre of bioethanol from maize requires 4,000 L of water and a litre of bioethanol from sugarcane can also use as much as 4,000 L of water (Friends of the Earth 2010).

Many bioenergy crops must be produced in large monocultures to be profitable. One of the risks of large scale bioenergy development is that land will be concentrated in the hands of a few wealthy companies or individuals and that small scale farmers will lose their land. Small scale farmers might have limited or no accesses to the capital required for large biofuel operations and as such miss out on investment opportunities and associated profits. The only benefit for many small scale farmers in biofuel production is to lease their land to producers. In Indonesia where 44 % of productive oil palm plantations are managed by small scale farmers there have been persistent reports that such farmers receive minimal compensation for their produce and remain in debt to the palm oil companies (Cushion et al. 2010).

In South Africa a survey amongst communities who could become involved in a potential bioenergy project in the Eastern Cape Province revealed that community members were concerned about the potential environmental, social and ecological risks associated with such a project. Some of the specific concerns included:

- fear of losing control over land seen as their inheritance;
- breakdown or distortion of the social fabric/character of the community;
- enslaving by big business and falling into poverty;
- water pollution and environmental damage;
- migration of other people to the communities; and
- unequal share of benefits from the project (Amigun et al. 2011).

The involvement of rural communities in the production of biofuels cannot be evaluated fully through simplistic proxies such as the number of jobs created and average wages paid to plantation workers. Long established and indigenous rural communities are experienced in living in a highly variable and seasonal environment and their traditional methods of survival are based on managing risk rather than optimising income. A switch from subsistence farming to farming of a cash crop brings with it an increased dependency on outside agents. This may be fine as long as there is a steady stream of income but when the prices of cash crops drop it might lead to food insecurity and increased poverty. It is therefore necessary to consider in much more detail how the livelihood strategies and outcomes of rural communities will change with a change in land ownership, land management and land use associated with a switch from the production of subsistence food crops to biofuel cash crops (Van der Horst and Vermeylen 2011). When there are uncertainties regarding the impacts of biofuels on environmental and social sustainability it is important to consider that changes in land use could significantly outweigh any carbon benefits that may result from planting biofuels (Cushion et al. 2009). The following section will consider ways of improving the benefits that biofuel production can play in the livelihoods of the rural poor.

9.6 Role of Rural Communities in BioEnergy Strategies

Despite some of the problems associated with large scale bioenergy production such as security of tenure, food security and environmental concerns, bioenergy has the potential to generate more income and employment than any other type of fuel. The higher levels of employment generated by bioenergy are a result of the relatively small size of production facilities and the large volumes of material used to produce each unit of energy output (Cushion et al. 2010). Human well-being depends, however, on more than income but includes availability of food, access to energy for cooking, shelter and heating, health care and cultural components such as political rights, education communication, transport and material comfort (Buchholz et al. 2007).

Maintaining and improving human well-being is the moral foundation of most societies. The creation of wealth in terms of products and services has a nearly one to one relationship with the use of energy per capita but when energy generation impacts negatively on human well-being the role of energy supply for wealth creation and sustainable development can be questioned (Buchholz et al. 2007). The risks of negatively affecting human well-being in bioenergy production systems can be addressed through innovative business ventures between rural producers and biofuel companies that empower rural producers of biomass and include them as equal business partners in bioenergy ventures (Ham and Thomas 2008). Such a strategy will depend on the risks associated with the production of biofuel crops, role of rural producers in relation to bioenergy companies and institutional support.

9.6.1 Biofuel Production Risk

Rural producers can play a key role in growing biomass for biofuel production. It is, however, important to consider their exposure to risks brought on by aspects such as crop failures and delays between planting and harvesting of the crop. In the case where the small scale producer bears all the risk it can be seen as a direct threat to security of livelihoods. *Jatropha* (Fig. 9.3) is for instance a labour intensive crop that is harvested by hand and is regarded as being ideally suited to small scale producers. Farmers that sign up to grow *Jatropha* can, however, face years of investment before they can harvest seeds. Studies in Mozambique and Swaziland found that many subsistence farmers gave up growing *Jatropha* after the first year due to difficulty with growing the plant. Without a harvest to sell they have no income from the land and if food crops were replaced by *Jatropha* it can leave farmers without food and income (Friends of the Earth 2010). Risk reducing strategies based on a production portfolio of biomass and food crops on a larger scale, as discussed in Chap. 5 for commercial plantations, are often not a viable option for small scale producers, since their manpower and access to land is often limited. Therefore innovative options for risk reduction must be found.

Initiatives to spread the risk between biofuel companies and small scale farmers could include diversification of crops where biofuels are intercropped with food crops and the use of unproductive land for biofuel crops. In the cases where there is a time delay in biofuel production, farmers could be encouraged to grow short rotation food crops (Practical Action Consulting 2009).

Risks for small scale farmers also arise from market isolation and lack of awareness of market trends and prices. Initiatives that encourage cooperatives

Fig. 9.3 *Jatropha* trees in Mozambique grown by small scale farmers

and producer groups enable joint bargaining and the pooling of resources for mechanisms such as bridging loans could help to reduce the risk for each producer (Practical Action Consulting 2009). Linkages between small scale farmers and bioenergy companies can reduce risk by functioning as a market-based ecosystem that allows companies and farmers to act together and create wealth in a symbiotic relationship. These actors depend on each other, the system adapts and evolves and will often be both resilient and flexible (Prahalad 2006).

9.6.2 Biofuel Producers and BioEnergy Companies

A review of a switch grass bioenergy project in the United States (Rossi and Hinrichs 2011) highlighted participating farmers' generally strong scepticism regarding the role of large scale agribusiness and energy companies. Most project participants felt that for small scale farmers and local communities to receive substantial economic benefits from bioenergy, corporate dominance in the bioenergy industry should be avoided or restricted (Rossi and Hinrichs 2011).

These views were mirrored by communities in the Eastern Cape Province of South Africa where 54 % of participants in a household survey regarding bioenergy development were not willing to produce crops for biodiesel production. One of the main objections was that local people were excluded from the project development and were asked to accept industrial scale development that could lead to further poverty (Amigun et al. 2011).

Disagreement between communities and bioenergy companies increase when the development is involuntarily imposed on the community's locality, when technology is unfamiliar, the community has no decision making power and the development is for corporate profit rather than local benefit. The lack of community support for bioenergy projects in developing countries has completely prevented development of some bioenergy projects and caused significant delays in others (Amigun et al. 2011).

Distrust and scepticism on behalf of small scale producers can be overcome when bioenergy companies co-develop solutions to problems through two-way information flow. The focus should be on working with local partners to co-design every aspect of the bioenergy value chain instead of imposing pre-existing solutions from above (Hart 2005). In managing diverse linkages between small scale producers and bioenergy companies, techniques such as participatory rural appraisal and rapid assessment processes open up valuable ways of communicating with grassroots partners and helping with mutual learning and the creation of responsive strategies (Ham and Thomas 2008).

Tenure arrangements and assets of the small scale producers are not always clearly defined nor can they be easily turned into capital or used for collateral for loans or investments (De Soto 2000). This increases the vulnerability of the arrangements between the producer and the bioenergy company since conventional legal arrangements would be impossible. In this regard trust and social capital, rather than legal contracts, could form the basis for sustainable and fair business arrangements between small scale producers and bioenergy companies (Hart 2005).

9.6.3 Institutional Support

Institutional support for small scale producers of biofuel could include financing and loans, factors of production such as fertiliser, transport, legal and contract assistance, technology research and development, bargaining support, training and capacity building and marketing and market information. Various actors in the biofuel value chain could provide such support from cooperative small scale producer entities to government and large bioenergy companies (Practical Action Consulting 2009).

Switch grass producers in the United States advocated the formation of cooperative organisations as a way of countering corporate dominance in bioenergy projects. It was seen as a way of addressing some of the risks of becoming involved in a new sector. Some producers also saw government involvement as a way of avoiding corporate dominance in bioenergy but expressed reservations about the feasibility of such an approach (Rossi and Hinrichs 2011).

Through co-operatives, small scale producers can collectively provide the "critical mass" that allows them to facilitate the harvesting, transport and marketing of their products. Cash flow regulation to individual members of the cooperative is a further opportunity to be realised by following a collaborative approach (Keyworth 2000).

Co-operatives are an ideal vehicle for emerging business people to establish themselves in a competitive market. Due to the democratic nature of representation, in simple terms one member one vote, it is difficult for a co-operative to be dominated by a minority of members. This can also be a drawback in that it can slow down decision-making, which can be a major disadvantage in competitive markets (Keyworth 2000).

Another type of institutional arrangement is that of outgrower schemes. South African timber companies have achieved success with such outgrower arrangements where they increase the supply of timber to mills by entering into partnership arrangements with small scale growers who have access to land where timber can be grown. The timber companies provide technical inputs (extension advice, seedlings, etc.), capital in the form of soft loans against the proceeds from the sale of the harvested timber and a market for the timber. In return, the grower provides land and labour as the means of growing the trees and undertakes to sell the timber to the processing companies at a market related price (Howard et al. 2005).

It is estimated that there are 24,000 small growers in formal schemes, and between 5,000 and 10,000 independent small growers in South Africa (Chamberlain et al. 2005). One of the more significant economic impacts of outgrower schemes in South Africa is that revenue is circulated within the community when growers employ the labour of both family and community members to work in their plantations. Small-scale growing has also created entrepreneurial opportunities for others, which has contributed to growth and diversification of economic opportunities of the broader community. In many cases growers do not have their own means of transporting timber from their land to the mill, creating an opportunity for the development of local transport contracting businesses where community members who have vehicles sell their services these growers (Howard et al. 2005).

Outgrower schemes not only contribute directly to the economies of rural communities but also empower some of the poorest of the poor. The effect of poverty on women and female-headed households is compounded by many cultural norms, for example lack of rights and independence (Lewis et al. 2003). Timber outgrower schemes have, however, provided women with an economic alternative. It is reported that within the Sappi Project Grow outgrower scheme, 80 % of the growers are women (van Loggerenberg 2004).

9.7 Conclusion

Energy from biomass is used by millions of poor people in developing countries as a primary source of energy. It is also becoming increasingly popular in developed countries as a way of mitigating greenhouse gas emissions. Energy pricing programmes in developed countries have enabled consumers to choose renewable energy options while corporate consumers are starting to purchase renewable energy to improve their environmental image and as part of corporate social responsibility programmes (FAO 2005). This developed country focus on renewable energy has led to an increase in demand for bioenergy, often produced in developing countries.

Bioenergy companies that produce biofuels in developing countries often use the arguments of job creation and income generation as a way of justifying large scale biofuel operations. Questions can, however, be raised about the net gain to rural communities from these biofuel plantations. Often these plantations lead to displacement of the rural poor and environmental degradation. While it may seem that rural people are willing to engage in these biofuel production activities it cannot be assumed that these operations have positive social impacts (Van der Horst and Vermeylen 2011).

Several issues should be considered when bioenergy projects are planned in developing countries. These include:

- The potential to improve the sustainability of traditional biomass use (or even to substitute it with more sustainable forms of energy).
- The appropriate level and scale of bioenergy development.
- Definition of land tenure issues and economic opportunities for the rural poor.
- Feedstock choice in relation to land and environmental suitability.
- Sustainable land use planning to reduce conflicts and environmental degradation (Cushion et al. 2010).

Biofuel development driven by agreements between small scale producers and bioenergy companies could be seen as a potential way of ensuring social and environmental benefits, especially when it addresses local development. Carefully controlled development of biofuels that recognises the dangers of rapid unregulated expansion is more likely to result in a bioenergy industry that minimises its social and environmental impact and that ensures equitable benefits sharing (Phalan 2009). Sustainable long term business ventures could be established between small scale producers and bioenergy companies if care is take to define the relationship between the various actors, spread and share risks amongst the actors and define institutional structures. Timber outgrower schemes could serve as an example for such bioenergy business arrangements between small scale producers and bioenergy companies.

References

Amigun B, Musango JK, Brent AC (2011) Community perspectives on the introduction of biodiesel production in the Eastern Cape Province of South Africa. Energy 36:2502–2508

Arnold M, Köhlin G, Persson R, Shepherd G (2003) Fuelwood revisited: what has changed in the last decade? CIFOR occasional paper no. 39. CIFOR, Jakarta, 47pp

Buchholz TS, Volk TA, Luzadis VA (2007) A participatory systems approach to modeling social, economic, and ecological components of bioenergy. Energy Policy 35:6084–6094

Chamberlain D, Essop H, Hougaard C, Malherbe S, Walker R (2005) Part I: the contribution, costs and development opportunities of the Forestry, Timber, Pulp and Paper industries in South Africa, Genesis Analytics (Pty) Ltd. Johannesburg

Chirwa PW, Ham C, Maphiri S (2008) Baseline study determining consumer behaviour with regard to kitchen management and efficient cooking habits in South Africa. Report prepared for The Programme for Basic Energy and Conservation (ProBEC) by Stellenbosch University, Stellenbosch

Chirwa PW, Ham C, Maphiri S, Balmer M (2010) Bioenergy use and food preparation practices of two communities in the Eastern Cape Province of South Africa. J Energy South Afr 21(4): 26–31

Cotula L, Vermeulen S, Leonard R, Keeley J (2009) Land grab or development opportunity? Agricultural investment and international land deals in Africa. IIED/FAO/IFAD, London/Rome

Cushion E, Whiteman A, Dieterle G (2009) Bioenergy development: issues and impacts for poverty and natural resource management. Agricultural and rural development notes 49. World Bank, Washington, DC, 4pp

Cushion E, Whiteman A, Dieterle G (2010) Bioenergy development: issues and impacts for poverty and natural resource management. World Bank, Washington, DC, 272pp

De Beer MC (2012) The Socio-economic impact of the phasing out of plantations in the Western and Southern Cape regions of South Africa – a case study of three plantations. Unpublished M.Sc. thesis, University of Stellenbosch, 92pp

De Soto H (2000) The mystery of capital: why capitalism triumphs in the West and fails everywhere else. Black Swan, London

FAO (2005) State of the world's forests. FAO, Rome

FAO (2009) The market and food security implications of the development of biofuel production. Committee on commodity problems. Sixty-seventh Session, Rome, 20–22 Apr 2009, 10pp

FAO (2011) State of the world's forests. FAO, Rome

Forestry Solutions (2007) Best operating practices for silviculture. http://www.forestrysolutions.net/. Accessed 12 Feb 2012

Friends of the Earth (2005) Greasy palm. The social and ecological impacts of large-scale oil palm plantation development in Southeast Asia. Friends of the Earth, London, 54pp

Friends of the Earth (2010) Africa: up for grabs: the scale and impact of land grabbing for agrofuels. Friends of the Earth, London, 36pp

Gamborg C, Millar K, Shortall O, Sandøe P (2011) Bioenergy and land use: framing the ethical debate. J Agr Environ Ethic. 17pp. doi:10.1007/s10806-011-9351-1

Gerber J-F (2011) Conflicts over industrial tree plantations in the South: who, how and why? Glob Environ Chang 21:165–176

Ham C (2000) The importance of woodlots to local communities, small scale entrepreneurs and indigenous forest conservation – a case study. Instruments for sustainable private sector forestry, South Africa series. International Institute for Environment and Development and CSIR-Environmentek, London/Pretoria

Ham C, Theron JM (1999) Community forestry and woodlot development in South Africa: the past, present and future. South Afr For J 184:71–79

Ham C, Theron F (2001) Community forestry resources: a case study of selected woodlots in the Eastern Cape Province. South Afr For J 191:65–74

Ham C, Thomas W (2008) Pro-poor enterprises and the base of the pyramid concept: learning from natural plant product ventures in South Africa. In: Kandachar P, Halme M (eds) Sustainability challenges and solutions at the base-of-the-pyramid: business, technology and the poor. Greenleaf Publishing, Sheffield

Ham C, Chirwa PW, Theron F (2008) The forester as a change agent: from trees between the people to people between the trees. In: Theron F (ed) The development change agent. A micro-level approach to development. JL van Schaik Publishers, Pretoria

Hart SL (2005) Capitalism at the crossroads: the unlimited business opportunities in solving the world's most difficult problems. Wharton School Publishing, Philadelphia

Headey D, Malaiyandi S, Shenggen F (2009) Navigating the perfect storm reflections on the food, energy, and financial crises, IFPRI. http://www.ifpri.org/publication/navigatingperfect-storm

Howard M, Matikinca P, Mitchell D, Brown B, Lewis F, Mahlangu I, Msimang A, Nixon P, Radebe T (2005) Case studies of partnerships for timber production: a contribution to an initiative aiming to answer the question: what role does forestry play in reducing poverty in South Africa, and how can that role be improved? Forestry and land use programme. International Institute for Environment and Development, London

International Energy Agency (IEA) (2002) Chapter 13: Energy and poverty. In: World energy outlook 2002. OECD, Paris

Keyworth P (2000) Organisation of marketing. In: Owen D (ed) South African forestry handbook 2000. Southern African Institute of Forestry, Menlo Park, pp 485–487

Lehtonen M (2011) Social sustainability of the Brazilian bioethanol: power relations in a centre-periphery perspective. Biomass Bioenergy 35:2425–2435

Lewis F, Blanché C, Todd M (2003) A review of poverty in South Africa in relation to forest based opportunities. Prepared by the INR for DWAF WFSP Forestry Programme, HTS Development Ltd., Making Forestry Markets Work for the Poor Project

Ofoegbu C (2010) An evaluation of the socio-economic impact of timber production with and without the inclusion of biomass energy production. Unpublished M.Sc. thesis, Stellenbosch University, Stellenbosch, 168pp

Phalan B (2009) The social and environmental impacts of biofuels in Asia: an overview. Appl Energ 86:S21–S29

Practical Action Consulting (2009) Small-scale bioenergy initiatives: brief description and preliminary lessons on livelihood impacts from case studies in Asia, Latin America and Africa. Prepared for PISCES and FAO by Practical Action Consulting, Jan 2009. http://www.fao.org/docrep/011/aj991e/aj991e00.HTM

Prahalad CK (2006) The fortune at the bottom of the pyramid: eradicating poverty through profits. Wharton School Publishing, Upper Saddle River

Ravindranath, NH, Manuvie R, Fargione J, Canadell JG, Berndes G, Woods J, Watson H, Sathaye J (2009) Greenhouse gas implications of land use and land conversion to biofuel crops. In: Howarth RW, Bringezu S (eds) Biofuels: environmental consequences and interactions with changing land use. Proceedings of the scientific committee on problems of the environment (SCOPE) International biofuels project rapid assessment, 22–25 September 2008, Gummersbach Germany. Cornell University, Ithaca, pp 111–125. http://cip.cornell.edu/biofuels/

Rossi AM, Hinrichs CC (2011) Hope and skepticism: farmer and local community views on the socio-economic benefits of agricultural bioenergy. Biomass Bioenergy 35:1418–1428

Shackleton CM (2004) Assessment of the livelihoods importance of forestry, forests and forest products in South Africa. Rhodes University, Grahamstown

Shackleton CM, Buiten E, Annecke W, Banks DA, Bester J, Everson T, Fabricius C, Ham C, Kees M, Modise M, Phago M, Prasad G, Smit W, Twine W, Underwood M, Von Maltitz G, Wenzel P (2007) Exploring the options for fuelwood policies to support poverty alleviation policies: evolving policy dimensions in South Africa. For Tree Livelihood 17:269–292

Sims REH (2003) Climate change solutions from biomass, bioenergy and biomaterials. Agric Eng Int CIGR E-J V:28

SPARKNET (2004) South Africa country report synthesis. http://db.sparknet.info/goto.php/SouthAfricaCountrySynthesis. Accessed 12 Aug 2008

Tauli-Corpuz V, Tamang P (2007) Oil palm and other commercial tree plantations, monocropping: impacts on indigenous peoples' land tenure and resource management systems and livelihoods. Permanent Forum on Indigenous Issues. Sixth session, New York, 14–25 May 2007, 19pp

UNEP (2012) Land use, land use change and bioenergy. Bioenergy issue paper series no. 1. http://www.unep.fr/energy/bioenergy/issues/pdf/No%201_FINAL.pdf. Accessed 15 Mar 2012

Van der Horst D, Vermeylen S (2011) Spatial scale and social impacts of biofuel production. Biomass Bioenergy 35:2435–2443

Van Loggerenberg C (2004) Towards community forestry: a South African perspective. Unpublished paper presented to B.Tech. Saasveld students, Nelson Mandela Metropolitan University, George

Vedeld P, Angelsen A, Sjaastad E, Berg GE (2004) Counting on the environment: forest incomes and the rural poor. Environmental economics series, Paper no. 98, The World Bank Environment Department. Washington

Chapter 10
Ecological Impacts of Biomass Production at Stand and Landscape Levels

Ben du Toit, Mark Gush, James S. Pryke, Michael J. Samways, and Steven B. Dovey

10.1 Introduction

In Chaps. 4, 5 and 6 of this book, we discussed the production and procurement of biomass from various sources, including extensively managed systems such as woodlands, and much more intensively managed systems such as short-rotation bio-energy plantations. It is generally accepted that intensive, production orientated land uses will have an impact on carbon stocks, biodiversity, growth resource use or resource quality (Achten and Verchot 2011; German et al. 2011). Some of these impacts may be exacerbated if exotic species are used in the system. Conversely, the ecological impacts on areas of low management intensity or near natural vegetation are usually lower. This chapter will focus more specifically on the ecological impacts of intensively managed tree-based biomass production systems, and it is structured according to the scale of impacts, i.e. the landscape level and site level. At both levels, we have attempted to highlight the major threats to sustained productivity and the provision of ecosystem services. However, even intensive biomass production systems can arguably be managed in ways that mitigate the

B. du Toit (✉)
Department of Forest and Wood Science, Stellenbosch University, Private Bag X1, Matieland, Stellenbosch 7602, South Africa
e-mail: ben@sun.ac.za

M. Gush
CSIR, Natural resources and Environment, P.O. Box 320, Stellenbosch 7599, South Africa

J.S. Pryke • M.J. Samways
Department of Conservation Ecology and Entomology, Stellenbosch University, Private Bag X1, Matieland, Stellenbosch 7602, South Africa

S.B. Dovey
Institute for Commercial Forestry Research, P.O. Box 100281, Scottsville 3209, South Africa

T. Seifert (ed.), *Bioenergy from Wood: Sustainable Production in the Tropics*, Managing Forest Ecosystems 26, DOI 10.1007/978-94-007-7448-3_10,

ecological impacts of such systems. This chapter will therefore also focus on some case studies where ecological impacts could be limited or mitigated through the adoption of specific management strategies.

10.2 Impacts at the Landscape Level

The conservation of biodiversity and the use of scarce resources such as water are issues that belong at the landscape scale. The challenge is then to balance these impacts at the landscape scale and to ensure biological corridors/linkages in the indigenous vegetation that can facilitate the movement of individuals and genetic material. Evaluation of landscape scale impacts will also consider the effects of biomass production systems on water and air quality.

10.2.1 Water Use

Large-scale changes in land-use, such as those proposed for intensive biomass production, constitute a change in the structure, functioning, species composition and management of the vegetation growing on the land. This, in turn, signifies a change in how water is intercepted, infiltrated, transpired and evaporated from the land surface. The resultant impacts on the availability and quality of water in rivers is of great importance to the downstream users of that water. Consequently a good understanding and quantification of land-use driven water resources impacts is required when land-use changes are proposed. Stream-flow changes associated with vegetative land-use changes may be described using a simplified water balance equation, namely:

$$\mathrm{Q} = \mathrm{P} - \mathrm{Et} \pm \Delta\mathrm{S}$$

where Q = streamflow, P = precipitation, Et = evapotranspiration and ΔS = changes in soil water storage.

This equation is best applied over a suitably long time period (e.g. several years), where changes in soil water storage are assumed to balance out, and rainfall is representative of the long-term mean for the area. In this case, changes in Et caused by vegetative land-use change equate to changes in streamflow at the landscape level, if the water use of the replacement vegetation is significantly different to that of the existing land-use. In South Africa, plantation forestry with introduced (exotic) species is an extensive and profitable land use in many of the high-rainfall regions of the country (Chamberlain et al. 2005) with an area of approximately 1.25 million hectares currently under commercial plantations (FSA 2010). Growth in the industry

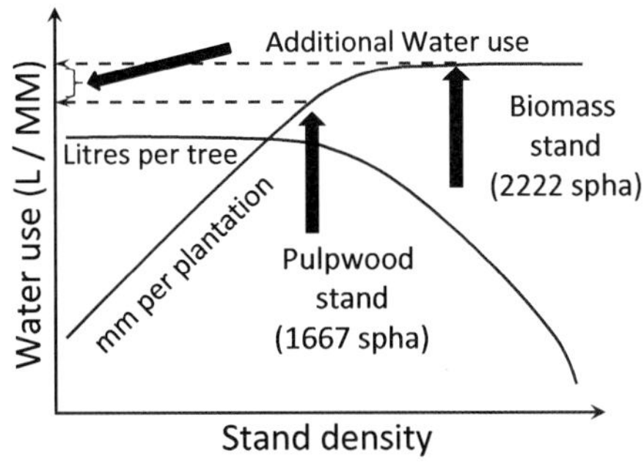

Fig. 10.1 Hypothetical changes in the water-use of individual trees (L·tree^{-1}) compared to a commercial plantation (mm), relative to increases in stand density. Current and future scenarios under intensive biomass production are illustrated

is restricted by legislation (National Water Act of 1998, Section 36), which, amongst other aspects, declares commercial plantation forestry to be a Streamflow Reduction Activity (SFRA) due to the high water-use of forest plantations and their impact on catchment water yields (Dye and Versfeld 2007). However, with the resultant efforts to maximise biomass production in intensively managed commercial forests, it is important to consider the associated hydrological impacts. This section considers the potential hydrological impacts, particularly streamflow changes, likely to be associated with intensive woody biomass production at a landscape scale. The primary land-use changes predicted to occur under intensified biomass production, and for which hydrological impacts need to be considered, include increased stand densities, shorter rotation lengths, earlier canopy closure and changes in site/species preferences.

10.2.1.1 Stand Density

Increases in tree densities are usually associated with increases in water-use, due to greater competition for resources driving increases in leaf area per unit of land, higher root intensities and colonisation of greater soil volumes by roots. While it is true that the water-use of individual trees of a given age within a plantation will decrease as tree densities increase (due to competition), over-all water use for the plantation is likely to increase until a threshold is reached where water availability is the limiting factor and the water use levels off. This threshold tree density will vary depending on the site and species, but in general for intensively managed tree plantations of a given age, where effective understorey/weed control is practiced, moving from a typical pulpwood stand density of 3 m × 2 m (1,667 spha) to a spacing of 3 m × 1.5 m (2,222 spha), envisaged for intensive biomass production, will increase overall water-use of the stand (Fig. 10.1). A potential exception to this is when understorey vegetation with a particularly high water-use is suppressed by an increase in trees with relatively lower water-use rates (e.g. certain indigenous tree species).

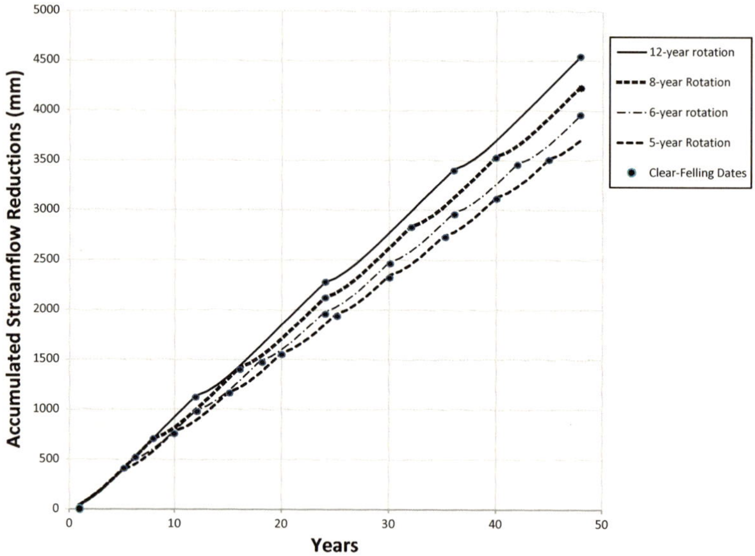

Fig. 10.2 Accumulated streamflow reductions for Eucalypts simulated under a range of rotation lengths (Data from Gush et al. 2002)

10.2.1.2 Rotation Length

Tree water-use also changes relative to the age of the tree, typically following a sigmoid curve, with a low initial water use followed by rapid increases flattening out to a plateau once canopy cover is achieved (Scott and Smith 1997; Dye and Bosch 2000). In certain long-rotation sawtimber stands (>30 years for pines, and >15 years for eucalypts), tree water-use (streamflow reduction) has even been observed to tail off with age (Scott and Prinsloo 2008). However, with the advancement of tree improvement programmes and genetic selection for faster growth rates, rotation lengths have generally decreased, particularly in eucalyptus plantations (Verryn 2000). Rotation lengths are predicted to shorten even further with the advent of intensive biomass production plantations, typically reducing from 6 to 5 years on good quality pulpwood sites, and from 12 to 8 years on sawtimber sites. Combined with higher stand densities, these plantations will produce trees of smaller individual volumes at harvesting, but greater biomass production per land unit overall. Due to the shape of the tree water-use curve relative to tree age, reductions in rotation length are likely to decrease the overall water-use of the plantation, resulting in lower streamflow reductions (Fig. 10.2).

A challenge in evaluating the wider scale hydrological impacts of rotation length changes associated with biomass plantations is that individual compartments within a commercial forest plantation go through a growth cycle from planting to clearfelling and back to planting again, with constantly changing water use impacts. However, at a landscape scale it may be helpful to consider the planted area as

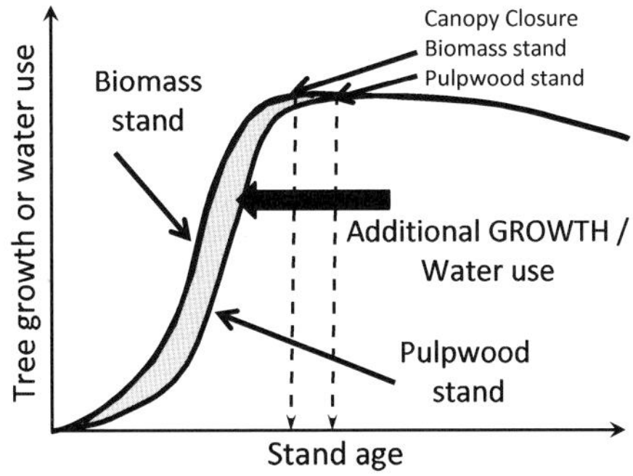

Fig. 10.3 Hypothetical changes in tree growth and water-use relative to stand age, before canopy closure within a pulpwood stand compared to an intensive biomass production stand

a mosaic of compartments representing all ages from seedlings to mature trees, growing simultaneously, and cycling through the various stages of growth and water use. For areas growing just one species under one rotation length, the net water use impact will be that of plantations at the "water-use mid point" of their rotation. However for areas growing multiple species at different rotation lengths water use impacts may need to be weighted relative to species and rotation length predominance.

10.2.1.3 Canopy Closure

Tree improvement programmes and advances in site management have increased the growth rates of commercial forestry tree species. This has also lead to earlier canopy closure in plantations. Under increased stand densities and shorter rotation lengths associated with intensive biomass production, canopy closure, with an associated Leaf Area Index (LAI) of 5 is now likely to occur after just 2 years on a favourable site. There are well established links between growth and water-use in trees, so an increase in the gradient of the water-use curve during the first few years of the rotation may be expected under faster growing trees. However, at canopy closure, competition for light, as opposed to water, may become the limiting factor to further increases in leaf area and hence water-use. Consequently, under intensive biomass production, the water-use curve is likely to peak and plateau earlier than before, resulting in an overall increase in water-use relative to tree age (Fig. 10.3).

10.2.1.4 Site and Species Choices

The location within the landscape of commercial forestry plantations; be they existing pulpwood/saw-timber stands or proposed future intensive biomass production stands, undoubtedly has an impact on their water-use. This is most pronounced in the distinction between riparian and non-riparian sites. In a study quantifying the effect of changes in riparian zone vegetation on catchment water yield (streamflow),

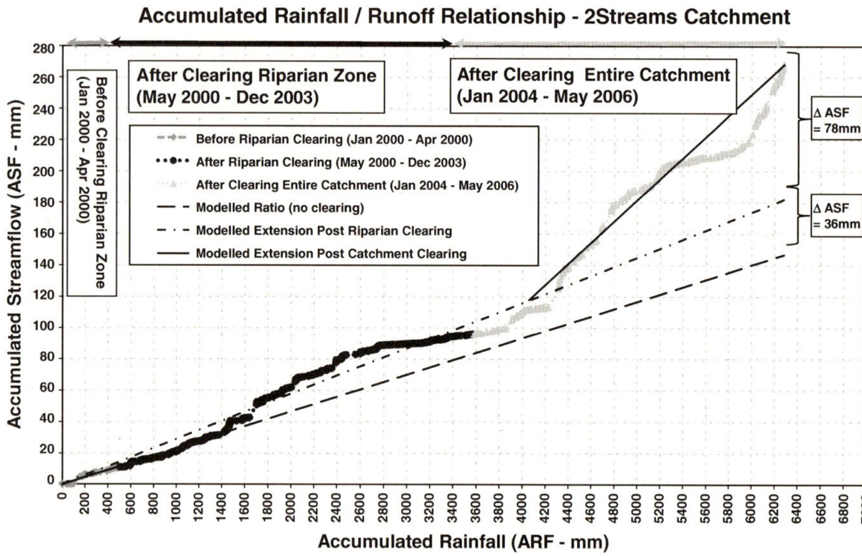

Fig. 10.4 Changing relationships between accumulated streamflow and rainfall in response to progressive clearing of *Acacia mearnsii* trees from riparian and upslope sites respectively (Everson et al. 2007)

Table 10.1 Calculation of relative contributions of riparian and upslope areas to streamflow following clearing of *Acacia mearnsii* (Black Wattle) stands

Zone	Area (ha)	% of total	Streamflow gain (mm)	% of total	Streamflow gain (mm ha^{-1})
Riparian Zone (RZ)	7.5	11.5	36	31.5	4.8
Non-RZ	58	88.5	78	68.5	1.34
Total cleared area	65.5	100	114	100	1.74

Based on data from Everson et al. (2007)

Everson et al. (2007) showed significant responses in streamflow following clearing of *Acacia mearnsii* (Black Wattle) trees from riparian and upland areas in a small catchment in KwaZulu-Natal (Fig. 10.4).

During the 6 year period of the study (May 2000 to May 2006), increases in streamflow associated with the clearing of the *A. mearnsii* trees, which had initially been planted throughout the catchment including the stream channel, were monitored. Based on the areas cleared and the resultant streamflow changes observed, these results indicate that streamflow gains following clearing operations were 4.8 mm for every hectare of riparian area cleared, and 1.34 mm for every hectare of upslope area cleared (Table 10.1). A unit of land in the riparian area under *A. mearnsii* consequently represented the hydrological equivalent of 3.58 times the upslope area (4.8/1.34 = 3.58) when cleared.

While legislation currently prohibits the establishment of commercial forestry plantations in riparian areas (FIEC 1995), the above findings illustrate the importance of focusing on riparian areas when clearing invasive exotics through activities such as the Working for Water programme (Turpie et al. 2008). They also help to quantify the water released by such activities, particularly when harvesting for biomass/bio-energy production. A further site-related hydrological consideration is the utilization of more marginal forestry areas, particularly in terms of rainfall. Intensive biomass production in such areas is likely to have lower *absolute* water-use impacts (mm), but significantly greater *relative* water-use impacts (%), compared to optimum sites. Linked to this is the need to make distinction between impacts on total flows and impacts on low flows (e.g. driest 3 months of the year). Scott and Smith (1997) argued that low flows may be of greater relevance to decision makers than reductions in total flows, and several South African studies have focused on this aspect, most recently Jewitt et al. (2009). The significance of low flows is also attributable to the emphasis placed by the National Water Act (NWA 1998) on the human and ecological "reserve", both of which are critical during periods of low flow.

As far as species selection is concerned, intensive biomass production is likely to favour coppicing Eucalyptus species due to their rapid growth, despite a relatively high ash content after combustion, However, pines will still be considered due to better pellet quality producing less ash, particularly where multiple-use of tree biomass is practiced (e.g. quality saw timber used for conventional sales, with off-cuts and branches used for biomass production). The implications of this in terms of water-use are that Eucalyptus species use more water than pines and wattle in turn. Allocation is made for these differences in the current SFRA water-use licensing system, however, changes from one species to another (e.g. Pine to Eucalyptus) will constitute an increase in water-use (greater streamflow reduction) and hence will be subject to species exchange adjustments to existing water-use licenses.

Given the scenarios discussed in this section, the evidence suggests that higher stand densities and faster growth rates (earlier canopy closure) will increase water-use while shorter rotations will reduce water-use. In general, however, a move from conventional pulpwood and saw-timber plantations to intensive biomass production plantations is likely to result in increases in water-use per unit area. If large-scale changes to this form of land-use are to be approved innovative solutions will be required to offset the increased water resource impacts. Options include accelerated clearing of high water-using invasive exotic trees from riparian areas, and possible replacement with low water-using indigenous tree species of high economic and ecological value (see Gush et al. 2011; Wise et al. 2011).

10.2.2 The Biodiversity Threat of Commercial Timber Plantations

With the growth of the global human population, the demand for food, wood and fuel will increase, so more areas of the world will turn to intensive agriculture and timber cropping systems like plantation forestry (Cubbage et al. 2010). Plantation forestry is a serious risk to global biodiversity, as the plantations themselves are often non-native and contribute little to biodiversity (Samways and Moore 1991; Pryke and Samways 2009; Bremer and Farley 2010). Biodiversity can be directly impacted by the plantations themselves, especially as large amounts of natural habitat are often transformed into plantations. Even so there are often natural areas left between plantation blocks in planted areas. These are left aside as areas of high conservation value such as protected grasslands, wetlands or indigenous forests, or for management requirements such as firebreaks, power lines and vehicle tracks (Samways et al. 2010). While not directly affected by the plantations these natural areas are often indirectly impacted through landscape fragmentation. This fragmentation isolates populations and leads to ecological relaxation, which is the loss of species from these fragments due to stochastic events or loss of ecological interactions leading to further losses of overall biodiversity. Abiotic disturbances caused by timber plantations such as water loss and soil nutrient depletion also complicate the conservation of biodiversity in and around plantations. Ecological networks (ENs) are a way to mitigate many of the adverse effects of plantation forestry on the local biodiversity (Samways et al. 2010). ENs work by connecting as many of the fragments together, although they need to be designed properly and be well managed to optimise their conservation value.

10.2.2.1 The Role of Ecological Networks in Conserving Biodiversity in Highly Transformed Landscapes

ENs are systems of natural or semi-natural landscape elements that are configured to **best** conserve and maintain biodiversity and ecological function (Fig. 10.5) (Bennett and Witt 2001). ENs consist of core natural patches that comprise of either existing nature reserves, areas of high conservation value or even areas within plantations that for various reasons remain unplanted, and are then connected by natural linkages (Jongman 2004). These linkages usually are either stepping stone patches or corridors of natural vegetation (Jongman 1995). Corridors are often simply defined as movement corridors for focal species (Hilty et al. 2006), but they also can function as habitats per se, especially when connected among themselves to form ENs. As the aim of ENs is to conserve biodiversity, they also need to include the inherent abiotic and biological complexity of the whole ecosystem (Fig. 10.1) (Jongman 1995). Conceptually much work has gone into the biodiversity value of ENs, although only a few areas of the world have actually implemented them, most notably the Pan European Ecological Networks (Jongman et al. 2011), the greenways in China (Yu et al. 2006) and the South African timber industry (Samways et al. 2010).

These ENs reduce the isolation of populations or even individuals, allowing better gene flow and reduce founder effects (loss of genetic variation that occurs when a new population is established by a very small number of individuals). ENs also allow species to recolonize areas after localized extinctions. This reduction of

Fig. 10.5 Ecological networks need to conserve ecological processes and services such as hydrology

isolation and fragmentation helps prevent ecological relaxation (the loss species and their interactions) and so prevent further biodiversity loss. This means that when ENs are designed and managed correctly, with large-scale interconnecting corridors and reserve areas, they can play an important role in ensuring connectivity between habitat patches for organism dispersal on evolutionary as well as on ecological time scales (Beier and Noss 1998; Samways et al. 2010) (Fig. 10.5).

10.2.2.2 Designing Ecological Networks

The design of ENs is a major determinant of their effectiveness. For example, stepping-stone patches within ENs, provided they are large enough, are useful for some of the larger animals using the ENs (particularly large mammals and birds). However, continuous corridors of good quality habitat in ENs are still preferable, as they allow smaller, less mobile, animals, such as frogs, insects and spiders, to use the linkages. In fact, these small animals use these corridors as habitats in their own right, giving the corridors themselves their own inherent biological value (Pryke and Samways 2011).

ENs function at the landscape scale. So, to be effective, they need to incorporate as many landscape features as possible. The inclusion of as many habitat types, as well as landscape features, is fundamental to good EN design. For example, grassland ENs with indigenous forests embedded within them have high biological value, not only because of the additional species in the forests themselves, but also because there are more species in the associated grasslands (Pryke and Samways 2011). This is due to grassland-indigenous forest interface, which is seemingly so essential for some species.

Recently, there has been much interest in the edge effect between the transformed plantation blocks and the corridors of the ENs. One reason for this is that ENs have more edge than occurs naturally because of the linear nature of corridors (Koh et al. 2010). Understanding these EN edge effects is important for conservation planning, in that it determines minimal width of corridors. Edge effects are caused by structural changes along the edge boundary (Cadenasso et al. 2003; Harper et al. 2005), as well as through changes in soil moisture and nutrients (Li et al. 2007). Over time, secondary effects, such as roads and invasion by exotic plants and animals, can further deteriorate the habitat along the edge.

The influence that transformed areas have on the ENs is often a two-zoned effect: the edge zone, which is influenced by the interface between a transformed area and a natural one, and the interior zone, where species richness, abundance and assemblage composition are no longer influenced by the distance to the edge (Cadenasso et al. 2003; Ries et al. 2004). Disturbance on the edge allows generalist species to disrupt natural systems (Pinheiro et al. 2010; Ivanov and Keiper 2010), although given enough space, it gives way to a more valuable interior zone (Slawski and Slawska 2000; Hochkirch et al. 2008).

Edge effects of exotic plantation blocks on indigenous grasslands are larger in size than that between natural forest patches and grasslands (Wilson et al. 2010), while there seems to be no general edge effects between natural Afromontane forest and its associated grassland (Kotze and Samways 2001). The type of transformed landscape also contributes to the extent of the edge, and determines those species found in it, as has been shown with changes in edge zones in rural versus urban contexts (Vallet et al. 2010) and in edges between different age classes of timber plantation blocks (Armstrong and van Hensbergen 1994).

Although some biodiversity responds positively to the edge, and many species have their habitat at the edge (van Halder et al. 2011), it is the interior zone which is of most concern. The reason for this is that the interior is harder to conserve, as it requires enough space for edge zones to completely surround it. When corridors are too small, they consist entirely of edge zone, without the important interior zone. When edge effects for a variety of arthropods are tested between plantation blocks and adjacent grasslands, there are many different responses, but overall edge effects for all arthropod groups are absent beyond 32 m into the grassland corridor (Pryke and Samways 2011).

Although this 32 m edge zone is a conservative estimate of grassland edge effects around timber plantation blocks, this result suggests that corridors of less than 64 m will be mainly edge and have specific conservation value only as disturbed sites. In fact, interiors of the corridors have similar biodiversity to reserves, suggesting that corridors with widths over 64 m have a biodiversity profile similar to that of nearby reserves (Pryke and Samways 2011). Provided corridor width is wide enough, these corridors have considerable biodiversity value. The 250 m suggested by Pryke and Samways (2001) is appropriate, as this incorporates a great deal of interior space for more sensitive species. Furthermore, a network of larger habitat corridors, as suggested by Samways et al. (2010), will reduce the area of edge zones across the entire network. When planning agroforestry landscapes with conservation in mind, we need to consider the edge zone around intensive land-use areas as a transitional area from a transformed to a natural ecosystem.

The concept of corridor and ENs is based on connectivity to enable organisms to move through the fragmented landscape (Hilty et al. 2006). For arthropods, these concepts need to be put into perspective, especially as dispersal in most species is strongly linked to resource-searching behaviours (foraging, mate or lek location etc.) (Baguette and Van Dyck 2007). Corridors need to be habitats that allow less mobile arthropods to use them as pathways for dispersal. This means that these corridors need to be of high enough habitat quality to encourage resource use (e.g. feeding, breeding etc.). The best way to ensure this high quality habitat is to manage the ENs optimally.

10.2.2.3 Managing Ecological Networks

Along with design, management considerations are vital for ensuring that the ENs function optimally. Such management includes using correct fire regimes, grazing densities and controlling invasive plants, to ensure availability of high quality habitat. Using grasshoppers as sensitive indicators, management of the ENs was found to be three to five times more important than EN design (Bazelet and Samways 2011). This means that all the expense and time put into designing and setting up ENs can be undone if ENs are not managed correctly.

Using the correct fire regimes, grazing densities and the clearing of invasive plants are the three most important management considerations for the ENs to work correctly. Fire regimes are critically important for grasslands and need to simulate natural fire regimes as closely as possible (O'Connor et al. 2004). The simulation of natural fire regimes is also important to savanna and fynbos, although fire intervals are usually longer in these biomes. This management technique is often problematic in timber production areas, as fire is a major risk to plantations (Kirkman and Pott 2002). Because of this, managers in fire prone areas tend to burn entire ENs too early, so essentially using the ENs as fire breaks. The consequences of continuous annual burning are not fully understood, but generally it seems that it is not ideal for biodiversity and it would be better to rest the land occasionally (Chambers and Samways 1998; Uys et al. 2004). This would require fire protection zones to be set up next to the plantations, requiring bigger corridors to accommodate them. In fact, bigger corridors would allow half the corridors to be alternately burned between years and so have fire protection value allowing more space to bring fires under control, while being easier to manage, as well as being more accommodating for

Fig. 10.6 Ecological network with an unburned corridor in the foreground, a burned corridor in the middle and a reserve in the background

biodiversity (Fig. 10.6). An alternative to annual burning of ecological corridors would be to alternate it with prescribed burning operations within plantation blocks where this is possible (e.g. semi-mature plantations).

A way to reduce the fuel load is to allow grazing on the ENs. Ideally native fauna should be used (Fig. 10.7). However, in areas where this is not an option, domesticated animals could be used, as long as their densities are controlled (Fynn and O'Connor 2000). Where fire and grazing are not available as realistic options, then mowing has been shown to have some success for at least grasshopper diversity (Chambers and Samways 1998).

Invasive plants are to some degree controlled in grasslands by fire and grazing. If invasive plants are allowed to take over corridors then they lose their ecological value and essentially become transformed areas themselves (Magoba and Samways 2011). This can lead to the breakdown of connectivity and of the optimal functionality of the EN.

The effectiveness of ENs within the South African timber industry has been assessed for a wide range of different organisms from the habitat base of plants, through to large mammals and birds (Joubert 2011). Much of this research has been based on arthropods, as they are small, hyperdiverse, habitat sensitive, resource dependent, ecologically important and can be sampled in large numbers (Bazelet 2011; Pryke and Samways 2011). Although there is much variation among these groups, they all benefit from ENs, provided that the ENs are well designed (corridors wide enough and all landscape feature are considered) and good quality habitat is maintained (Figs. 10.6 and 10.7).

Fig. 10.7 Native animals grazing in an ecological network

10.3 Impacts at the Site Level

Management of the ecological impacts at the site level has the goal to ensure sustainable productivity over the long term. Site-level impacts such as nutrient depletion, soil loss or degradation, pest and disease outbreaks or uncontrolled fires are likely to impact negatively on the sustainable production capacity of the system. Careful monitoring of intensively managed systems is thus needed, along with strategies to mitigate against negative site impacts that may take place. It is in this regard that agro-forestry systems and mixed cropping systems with nitrogen fixing trees can contribute to mitigate against nutrient depletion. Silvicultural systems have to be so designed to offer a degree of buffer against biotic or abiotic risks. The discussion on mitigation strategies and ecological buffering is supported by information from case studies in plantation and agro-forestry systems.

10.3.1 Nutritional Sustainability

The biogeochemical cycling of nutrients in forest ecosystems has been well described for several forest ecosystems (Jorgenson et al. 1975; Likens and Bormann 1995; Ranger and Turpault; du Toit and Scholes 2002; Laclau et al. 2005, 2010a; Dovey 2012). Nutrients reside within forest and plantation systems in several nutrient pools that differ in size and in the form in which a particular plant nutrient is held. Figure 10.8 is a simplified representation of a number of basic nutrient pools as well the movement of nutrients (usually termed nutrient fluxes) (a) into

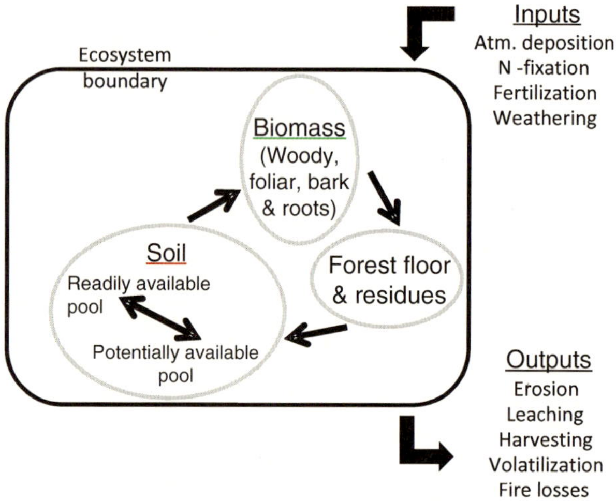

Fig. 10.8 Schematic representation of a simplified forest biogeochemical cycle (nutrient pools and fluxes within a forest ecosystem) as well as inputs/outputs from such a system (After Ackerman et al. 2013)

and out of the system (as part of the so-called input–output budget), and (b) among pools within the system (Ranger and Turpault 1999; du Toit and Scholes 2002). Forests are at risk of malnutrition and subsequent decline in productivity if the biogeochemical cycles of nutrients are decoupled in time or space. For example, an ecosystem can systematically be depleted of nutrients if outputs exceed inputs by a large margin. Several short rotation commercial forestry systems where only stemwood is harvested, are not at risk of nutrient depletion (du Toit and Scholes 2002; Ackerman et al. 2013). However, in the context of this chapter, increased nutrient removals in bio-energy plantations resulting from the harvesting of tree crowns, bark (and even roots) in addition the conventional stem wood harvest is probably the biggest concern (see examples in Fig. 10.9 below). Forest productivity can also be negatively affected if nutrients are not necessarily lost from the system, but end up in pools that are (temporarily) decoupled from their usual cycle and thus unavailable to the stand during a particular phase of growth. An example would be the lock-up of nutrients in the forest floor (Morris 1986) or the precipitation of a micronutrient in a plant-unavailable form following measures to raise soil pH, such as liming. In practice, management operations that affect nutrient pool sizes usually have a direct or indirect effect on nutrient fluxes too. For example, slash removal for bio-fuels will affect (for example) the content of nitrogen on the site, but also the rate of mineralisation which transforms N in a form that can readily be taken up by trees (du Toit and Dovey 2005; Deleporte et al. 2008; Gonçalves et al. 2008b; Mendham et al. 2008). Similarly, fire will affect the quantity of N and P oxidised during burning, but it may also strongly influence the availability of soil N through changes in mineralisation rates after the fire, as well as P availability (through changes in soil pH caused by the increase in pH (the ash-bed effect).

10.3.1.1 Gauging Nutritional Sustainability

The only way in which a comprehensive understanding of nutritional sustainability of a particular forest system can be gained, is by studying the majority of the more important nutrient fluxes in the system as brought about by specific management regimes or operations, (e.g. Ranger and Turpault 1999; du Toit and Scholes 2002; Laclau et al. 2005, 2010a; Dovey 2012). However, it has been known for a long time that nutrient dynamics may change significantly with the stage of stand development (Miller 1995). It follows that such studies has to be repeated in time, (or perhaps be done in a chronosequence approach) to paint the full picture. This fact, coupled to the reality that many different site types (and potentially even more than one management regime per site type) will have to be studied, makes it a very daunting task.

Nutrient fluxes consist of some processes in the biogeochemical cycle that are comparatively easy to monitor, e.g. uptake from the soil to the plant or return to the slash/soil layer following clear felling, thinning or through litterfall and fine root turnover. However, it also includes more complex processes that are difficult to quantify, such as transformations in the soil that are specific to each nutrient and the prevailing microclimatic and edaphic conditions in a specific soil, which are largely responsible for plant nutrient availability (e.g. nitrogen mineralisation, sorption and desorption processes on soil phosphate, oxidation/reduction processes on soil sulphur, etc.) This chapter deals with these fluxes in a simplistic way (see Fig. 10.8), and the reader is referred to soil chemistry texts for more detailed descriptions of these processes.

10.3.1.2 Nutrient Input–Output Budgets and an Index of Nutritional Sustainability

A simple index of nutritional sustainability has been proposed by du Toit and Scholes (2002) to gauge the nutritional sustainability of a variety of management regimes across different site types. While this is a fairly coarse indicator (it does not take transformations within the system into account) it is comparatively easy to use because it requires estimates of only (a) the larger input–output fluxes and (b) the major system nutrient pools sizes of the macronutrients. These can be estimated to an acceptable degree of accuracy in many regions of the world. Minor nutrient fluxes (such as weathering rates in very old soils) does not have to be gauged to high degrees of accuracy as they will not materially influence the system. Du Toit and Scholes (2002) proposed to express the net nutrient output from a system as a fraction of either (1) readily available or (2) potentially available nutrient pools in the system, to judge potential short- and long term effects. The index of nutritional sustainability thus developed carries the acronym pINS, where:

$$p(INS) = -\log_{10}\left(\frac{Net\ annual\ nutrient\ loss}{Nutrient\ pool}\right)$$

Table 10.2 Scenario's for biomass harvesting intensity per genus and per silvicultural regime in the case study of Ackerman et al. (2013)

Scenario	Genus and silvicultural regime	Harvesting intensity
A	Eucalypt pulpwood	Regular (75 % of stem wood only)
B		As above plus slash burning
C		Whole tree harvest with 75 % efficiency
D	Pine pulpwood	Regular (75 % of both stem wood and bark)
E		As above plus slash burning
F		Whole tree harvest

In its original form, du Toit and Scholes (2002) made provision for the nutrient pool to be either can be calculated as the readily plant available fraction or the long term (potentially) plant available fraction. We have used the fraction of the nutrient pool that is likely to be available to trees on a time scale of months to several years, because estimation of the long term potentially available pool sizes requires more developmental work.

Notes:

- In most intensively managed forestry and agricultural systems, there is a net loss of nutrients over time until such time as ameliorative action is taken.
- If the net input–output budget does not constitute a loss, it is simply reported as a gain and the pINS index is not calculated.
- A value of 1 (log scale) has been tentatively chosen as a value that should raise a red flag (i.e. if the net nutrient loss is more than 1/10th of the readily available nutrient pool) as defined by du Toit and Scholes 2002, the site may be at risk of nutrient depletion if the current management regime continues to be implemented.
- A feature of the pINS index is that different scenario's can be developed, for example where the portion of biomass harvested is increased or the rotation length is shortened (as is likely to happen in bio-energy crops), and the new scenario's can be compared with conventional systems.

This approach was developed further by Dovey and du Toit (2006) and by du Toit and Dovey as part of a more comprehensive study reported by Ackerman et al. (2013) dealing with nutrient fluxes and nutrient pools, respectively, in South African short-rotation plantation systems. Ackerman et al. (2013) chose three scenario's each for short-rotation pine and eucalypt systems as shown in Table 10.2. These scenario's were applied to 28 short-rotation pine sites and 21 short-rotation eucalypt sites in Southern Africa for which adequate data was available.

The pine sites in the study of Ackerman et al. (2013) virtually all showed net gains in N and P with relatively higher pINS values for K, Ca and Mg than the eucalypt sites. The main reason for this is twofold: Firstly, the longer average rotations of pine pulpwood (18.0 years as opposed to 7.1 years for eucalypt sites tested) has the effect that harvesting outputs are offset by a larger number of years' worth of atmospheric deposition. The regions where most of the pine test sites are located receives higher loads of atmospheric deposition than other remote rural locations, due to its proximity to a large number of coal-fired power plants (Olbrich 1993; Lowman 2004). Secondly, the pine sites are mostly located on more fertile sites in the region (clays and loams with high organic matter contents in the topsoils) whereas many of the eucalypt test sites in the test battery are located on sandy soils with low levels of organic matter. Nonetheless, the case study does illustrate the relative resilience from the most vulnerable to the more resilient sites in the region.

A summary of the results for the eucalypt sites is presented in Fig. 10.9, where it can be seen that the pINS value frequency curves all shift to the left (lower pINS values) when moving from Scenario A via B to C. This means that under scenario C (whole tree harvesting with 75 % efficiency), a large number of stands are coming close to a situation where nutrients may be depleted over the scope of several rotations unless corrective action is taken. Many tropical soils under short-rotation plantations have undergone more intensive leaching and are substantially poorer in nutrient pools and organic carbon, than Southern Africa's eucalypt sites presented in the case study in Fig. 10.9 (Gonçalves et al. 1997; Deleporte et al. 2008; Tiarks and Ranger 2008). It follows that intensively harvested bio-energy plantations on infertile sites are at much higher risk of nutrient depletion than is the case for the Southern African eucalypt data set.

It is important to keep in mind that increasingly intensive harvesting regimes and shortened rotations may result in a net loss of nutrients in many plantations, yet this may not have any immediate effect of decreasing the subsequent rotation's productivity. This may happen because the new rotation may still have access to sizable pools of readily available nutrient reserves on the site. Furthermore, there may not always be a positive growth response when the net loss of nutrients are replaced (e.g. by fertilization). Most intensively managed, short-rotation plantation forests respond mainly to macronutrient additions of N and P (Gonçalves et al. 1997). Indeed, it is only after several rotations of intensive biomass harvesting in plantations and/or plantations grown on poor soils that widespread responses to the addition of base cations started to become common (Gonçalves et al. 2008a). In short rotation tree stands where fertilization regimes are very basic or non-existent, there may be a net loss of several nutrient elements, and although there may not be an immediate growth response to (say) replacing Ca lost during harvesting, there will still be a constraint on the ability of the site to supply Ca in successive rotations. Furthermore, Laclau et al. (2010a) have presented evidence to show that short-rotation plantations of eucalypts may be capable of extremely efficient nutrient conservation and cycling, but that many such plantation systems in the tropics apparently depend on soils having been pre-enriched with nutrients by the natural vegetation before the plantations were established. For these reasons, and because bio-energy plantations are usually grown on very short rotations with large percentages of the biomass harvested, it would be wise to monitor nutrient exports in bio-energy plantations very intensively, and to upgrade the fertilization regime where necessary.

One of the most important measures to ensure sustained productivity on infertile sites, is the conservation of organic matter in the system (Laclau et al. 2010a, b).

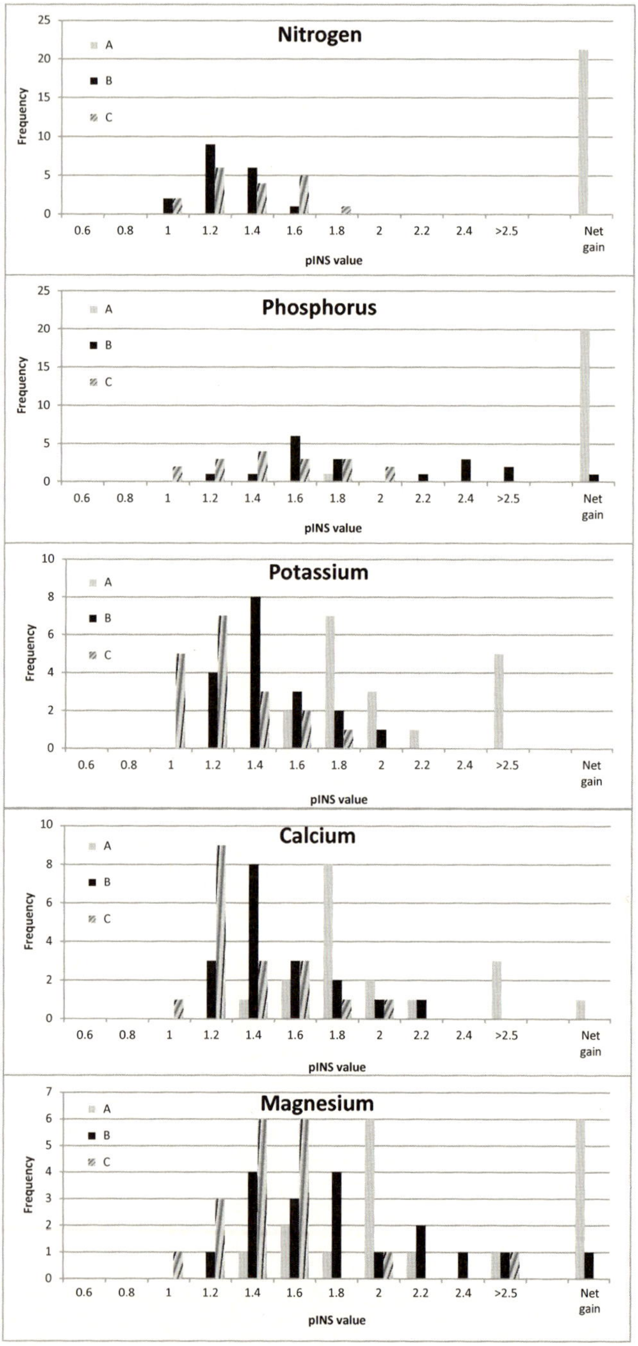

Fig. 10.9 pINS indices for five macronutrients (after Ackerman et al. 2013) calculated for 21 short-rotation eucalypt crops under scenarios A, B and C (Refer to Table 10.2 for scenario explanation)

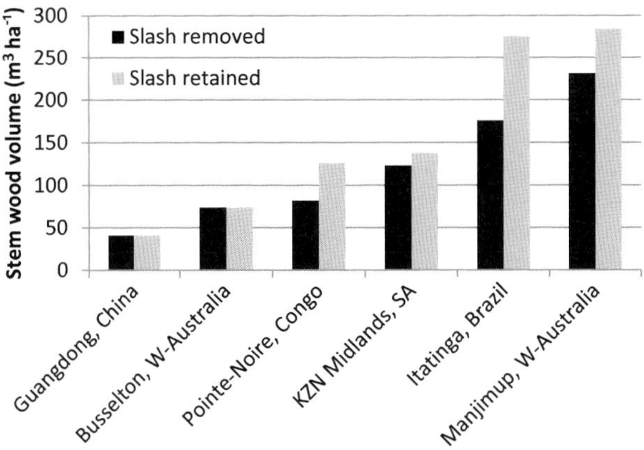

Fig. 10.10 Stemwood volume production in short rotation eucalypt case studies under slash retention and slash removal scenario's (After Nambiar and Kallio 2008)

Several experiments in a tropical network study (reviewed by Nambiar and Kallio 2008) tested the effects of slash management (and in particular, slash removal) on short rotation stand productivity. The slash removal treatments in this trial series constituted the removal of the slash plus the un-decomposed portion of the forest floor, which is a more intensive treatment than whole aboveground tree harvesting (which effectively only excludes the return of harvesting residue to the soil). However, it does give an indication of what can potentially be the result after successive rotations of either whole tree harvesting or some form of intensified biomass harvesting. The stand productivities (stem wood volume production at rotation end) of eucalypt case studies in this network of trials under slash retained and slash removed scenario's are given in Fig. 10.10.

The case studies by (Deleporte et al. 2008 (Congo); du Toit et al. 2008 (South Africa); Gonçalves et al. 2008b (Brazil); Mendham et al. 2008 (Manjumup, Australia)) all showed decreases in forest productivity following removal of harvesting residue and un-decomposed material in litter layers. The largest decline in stand productivity due to slash removal occurred in sandy soils (arenosols) and dystrophic loams (oxisols), both with low topsoil organic matter contents. This result underscores the point that a combination of poor soils, short rotations and intensified biomass harvesting means that many bio-energy plantation systems in the warm climate countries will not be nutritionally sustainable in the long run unless significant additional nutrient inputs are made. Inputs may be in the form of fertilizers, and/or ash replacement (from biomass burners), and/or incorporation of N fixation, either through mixed cropping (Binkley and Giardina 1997; Bouillet et al. 2012), or crop rotations with symbiotic N-fixers in the broader silvicultural management system.

10.3.2 Soil Conservation and Protection

The major threats to the ability of soils to sustain highly productive forest, other than nutrition-related issues discussed in Sect. 10.3.1, is soil displacement, soil erosion and soil compaction.

When viewed simplistically, soil compaction sensitivity by mechanical equipment is strongly related to machine mass, soil texture and soil water content at the time of impact. Coarse textured soils such as sands can be trafficed by fairly heavy loads with low risk of compaction in both wet and dry conditions (Smith et al. 1997a, b; du Toit et al. 2010; Ponder et al. 2012). Soils with very sandy textures seldom suffer from compaction problems and may even experience improved water holding capacity and sometimes improved growth following moderate compaction (Smith et al. 1997b; du Toit et al. 2010; Ponder et al. 2012). Conversely, soils with a fairly even mixture of particle size classes such as sandy loams, loams and sandy clay loams are moderately compactable when dry but strongly compactable when moist (Smith et al. 1997a, b; du Toit et al. 2010). Harvesting operations should where possible be scheduled to avoid soils with an even particle size distributions during wet conditions. Furthermore, machines travelling on plantation soils could be matched with the soils load bearing capacity, which is strongly related to texture, organic matter and initial bulk density (Ampoorter et al. 2012). Most compacting occurs in soils with the first few passes of machines over a specific area. This fact, combined with efforts to limit the spatial extent of compaction, calls for controlled vehicular movement on designated skid trails (Ampoorter et al. 2010). The effects of compaction in short-rotation bio-energy plantations can be thus minimised (a) by using designated skid trials, (b) by matching machine mass with soil texture and thus with load bearing ability, (c) by limiting harvesting operations on fine textured soils during wet conditions and (d) by retaining as much of the harvesting residue as is possible, given the harvesting system chosen.

The sustainability of bio-energy plantations will be severely compromised if erosion rates significantly exceed soil formation rates. Soil erosion is generally affected by rainfall & runoff, slope gradient and length, soil erodibility, vegetation cover and soil surface cover, soil tillage, and any other man-made support practices to limit erosion, e.g. contour banks or windbreaks.

On a site with a give soil, slope and climate, forest management practices can play a major role in limiting erosion. The single factor giving the most effective protection against wind, rain-splash and water erosion is the presence of absence of a soil cover layer, e.g. a mulch layer in agricultural fields or, the forest floor/slash layer in plantations and forests (Morgan 1995). This is echoed by several case studies where forest floor layers have been removed or destroyed by intensive fires (du Toit 2002; Miura et al. 2003; Fernández et al. 2004). Soil loss through erosion in plantation-based systems therefore depends very strongly on management of the slash and the forest floor. If destruction of the slash/forest floor is combined with other factors that favour soil erosion, it often results in an increase in erosion by orders of magnitude. Sherry (1953, 1954, 1961, 1964, 1971) documented the effects

Table 10.3 Average soil loss over two crop cycles of short-rotation Acacia mearnsii plantations under varying management regimes

Number of erosion enhancing factors present	Description	Soil loss (tons ha^{-1})
One	Steep slopes	Nil
	Slash burning	0–0.8
	Hoeing to control weeds	0–0.4
Two	Burning on slopes	11.4
	Burning and hoeing	10.1–17.6
	Hoeing on slopes	4.6
Three	Burn + Hoe on slopes	113.7

From du Toit (2002); after the work of Sherry (1953, 1954, 1961, 1964, 1971)

of three factors: slash burning, slope steepness and soil tillage (and combinations thereof) on soil erosion. The results have been re-analysed by du Toit (2002) and show an order of magnitude increase in soil erosion with the number of factors present (Table 10.3).

In non-planted areas, road design, construction and maintenance is critically important to minimize erosion because cuttings and road construction will lay the soil bare and poor road drain maintenance may cause the water flow to be concentrated in certain areas.

References

Achten WMJ, Verchot LJ (2011) Implications of biodiesel-induced land-use changes for CO_2 emissions: case studies in tropical America, Africa, and Southeast Asia. Ecol Soc 16(4):14

Ackerman P, Ham C, Dovey S, du Toit B, de Wet J, Kunneke A, Seifert T, Meincken M, von Doderer C (2013) The use of forest residue for bioenergy in Southern Africa. ICFR Bulletin Series (No. 03/2013). Institute for Commercial Forestry Research, Pietermaritzburg, 173pp

Ampoorter E, Van Nevel L, De Vos B, Hermy M, Verheyen K (2010) Assessing the effects of initial soil characteristics, machine mass and traffic intensity on forest soil compaction. For Ecol Manage 260(10):1664–1676

Ampoorter E, Schrijver A, Nevel L, Hermy M, Verheyen K (2012) Impact of mechanized harvesting on compaction of sandy and clayey forest soils: results of a meta-analysis. Ann For Sci 69(5):533–542

Armstrong AJ, van Hensbergen HJ (1994) Comparison of avifaunas in *Pinus radiata* habitats and indigenous riparian habitat at Jonkershoek, Stellenbosch. S Afr J Wildl Res 24:48–55

Baguette M, Van Dyck H (2007) Landscape connectivity and animal behavior: functional grain as a key determinant for dispersal. Landsc Ecol 22:1117–1129

Bazelet CS (2011) Grasshopper bioindicators of effective large-scale ecological networks. Ph.D. dissertation, Stellenbosch University, Stellenbosch

Bazelet CS, Samways MJ (2011) Relative importance of management vs. design for implementation of large-scale ecological networks. Landsc Ecol 26:341–353

Beier P, Noss RF (1998) Do habitat corridors provide connectivity? Conserv Biol 12:1241–1252

Bennett G, Witt P (2001) The development and application of ecological networks: a review of proposals plans and programmes. AIDEnvironment/IUCN, Gland

Binkley D, Giardina C (1997) Nitrogen fixation in Tropical Forest Plantations. In: Nambiar EKS, Brown AG (eds) Management of Soil, vol 43, Nutrients and Water in Tropical Plantation Forests. ACIAR Monograph No., pp 297–337

Bouillet J-P, Laclau J-P, Gonçalves JLM, Voigtlaender M, Gava JL, Leite FP, Hakamada R, Mareschal L, Mabiala A, Tardy F, Levillain J, Deleporte P, Epron D, Nouvellon Y (2012) Eucalyptus and acacia tree growth over entire rotation in single- and mixed-species plantations across five sites in Brazil and Congo. For Ecol Manage. http://dx.doi.org/10.1016/j.foreco.2012.09.019

Bremer LL, Farley KA (2010) Does plantation forestry restore biodiversity or create green deserts? A synthesis of the effects of land-use transitions on plant species richness. Biodivers Conserv 19:3893–3915

Cadenasso ML, Pickett STA, Weathers KC, Bell SS, Benning TL, Carreiro MM, Dawson TE (2003) An interdisciplinary and synthetic approach to ecological boundaries. BioScience 53:717–722

Chamberlain D, Essop H, Hougaard C, Malherbe S, Walker R (2005) The Genesis Report, Part I: the contribution, costs and development opportunities of the forestry, timber, pulp and paper industries in South Africa. Report to the Department of Trade and Industry and the Pulp and Paper Manufacturers Association, South Africa. Genesis Analytics

Chambers BQ, Samways MJ (1998) Grasshopper response to a 40-year experimental burning and mowing regime, with recommendations for invertebrate conservation management. Biodivers Conserv 7:985–1012

Cubbage F, Koesbandana S, Mac Donagh P, Rubilar R, Balmelli G, Olmos VM, De La Torre R, Murara M, Hoeflich VA, Kotze H, Gonzalez R, Carrero O, Frey G, Adams T, Turner J, Lord R, Huang J, MacIntyre C, McGinley K, Abt R, Phillips R (2010) Global timber investments, wood costs, regulation, and risk. Biomass Bioenerg 34:1667–1678

Deleporte P, Laclau JP, Nzila JD, Kazotti JG, Marien JN, Bouillet J-P, Szwarc M, D'Annunzio R, Ranger J (2008) Effects of slash and litter management practices on soil chemical properties and growth in second rotation eucalypts in the Congo. In: Nambiar EKS (ed) Site management and productivity in tropical plantation forests: proceedings of workshops in Piracicaba (Brazil), 22–26 Nov 2004 and Bogor (Indonesia), 6–9 Nov 2006. CIFOR, Bogor, pp 5–22

Dovey SB, du Toit B (2006) A review of nutrient fluxes across South African plantation forestry areas. ICFR Bulletin Series. 11/2006. Institute for Commercial Forestry Research, Pietermaritzburg, 15pp

Dovey SB (2012) Effects of clear felling and residue management on nutrient pools, productivity and sustainability in a clonal eucalypt stand in South Africa. Ph.D. thesis, Stellenbosch University, Stellenbosch, 218pp

du Toit B (2002) Chapter 14: Long-term ecological sustainability of wattle plantations. In: Dunlop RW, MacLennan L (eds) Black Wattle: the South African research experience. Institute for Commercial Forestry Research, Pietermaritzburg, pp 135–144

du Toit B, Dovey SB (2005) Effect of site management on leaf area, biomass development and stand growth efficiency of a young stand of Eucalyptus grandis in South Africa. Can J For Res 35(4):891–900

du Toit B, Scholes MC (2002) Nutritional sustainability of Eucalyptus plantations: a case study at Karkloof, South Africa. South Afr For J 195:63–72

du Toit B, Dovey SB, Smith CW (2008) Effects of slash and site management treatments on soil properties, nutrition and growth of a Eucalyptus grandis plantation in South Africa. In: Nambiar EKS (ed) Proceedings of workshops in Piracicaba (Brazil) 22–26 November 2004 and Bogor (Indonesia) 6–9 November 2006. Center for International Forestry Research (CIFOR), Bogor, pp 63–77

du Toit B, Smith CW, Little KM, Boreham G, Pallett RN (2010) Intensive, site-specific silviculture: manipulating resource availability at establishment for improved stand productivity. For Ecol Manage 259:1836–1845

Dye PJ, Bosch JM (2000) Sustained water yield in afforested catchments – the South African experience. In: von Gadow K, Pukkala T, Tomé M (eds) Sustainable forest management. Kluwer Academic Publishers, Dordrecht, pp 99–120

Dye PJ, Versfeld D (2007) Managing the hydrological impacts of South African plantation forests: an overview. For Ecol Manage 251:121–158

Everson CS, Gush MB, Moodley M, Jarmain C, Govender M, Dye PJ (2007) Effective management of the riparian zone vegetation to significantly reduce the cost of catchment management and enable greater productivity of land resources. Report to the Water Research Commission, Pretoria. WRC Report No. 1284/1/07 (ISBN: 978-1-77005-613-8)

Fernández C, Vega JA, Gras JM, Fonturbel T, Cuiñas P, Dambrine E, Alonso M (2004) Soil erosion after *Eucalyptus globulus* clearcutting: differences between logging slash disposal treatments. For Ecol Manage 195(1–2):85–95

FIEC (1995) Guidelines for environmental conservation management in commercial forests in South Africa. Forestry Industry Environmental Committee, Forest Owners Association, Johannesburg. FSA, 2010. Abstract of South African Forestry Facts for the year 2007/2008. Forestry South Africa, report to Department Water Affairs and Forestry. Available from http://www.forestry.co.za

FSA (2010) Forestry South Africa annual report 2010. FSA, Pietermaritzburg

Fynn RWS, O'Connor TG (2000) Effect of stocking rate and rainfall on rangeland dynamics and cattle performance in a semi-arid savanna, South Africa. J Appl Ecol 37:491–507

German L, Schoneveld GC, Gumbo D (2011) The local social and environmental impacts of smallholder-based biofuel investments in Zambia. Ecol Soc 16(4):12

Gonçalves JLM, Barros NF, Nambiar EKS, Novais RF (1997) Soil and stand management for short-rotation plantations. In: Nambiar EKS, Brown AG (eds) Management of soil, nutrients and water in tropical plantation forests, ACIAR (Australian Centre for International Agricultural Research) Monograph No. 43. ACIAR, Canberra, pp 379–417

Gonçalves JLM, Wichert MCP, Gava JL, Masetto AV, Arthur JC Jr, Serrano MIP, Mello SLM (2007) Soil fertility and growth of *Eucalyptus grandis* in Brazil under different residue management practices. South Hemisph For J 69(2):95–102

Gonçalves JLM, Stape JL, Laclau JP, Bouillet JP, Ranger J (2008a) Assessing the effects of early silvicultural management on long-term site productivity of fast-growing eucalypt plantations: the Brazilian experience. South For 70(2):105–118

Gonçalves JLM, Wichert MCP, Gava JL, Serrano MIP (2008b) Soil fertility and growth of *Eucalyptus grandis* in Brazil under different residue management practices. In Nambiar EKS (ed) Site management and productivity in tropical plantation forests: proceedings of workshops in Piracicaba (Brazil), 22–26 Nov 2004 and Bogor (Indonesia), 6–9 Nov 2006. CIFOR, Bogor, pp 51–62

Gush MB, Scott DF, Jewitt GPW, Schulze RE, Lumsden TG, Hallowes LA, Görges AHM (2002) Estimation of streamflow reductions resulting from commercial afforestation in South Africa. Water Research Commission, Pretoria, Report No. TT173/02 (ISBN: 1-86845-845-8)

Gush MB, Dye PJ, Geldenhuys CJ, Bulcock HH (2011) Volumes and efficiencies of water-use within selected indigenous and introduced tree species in South Africa: current results and potential applications. In: Proceedings of the 5th natural forests and woodlands symposium, Richards Bay, 11–14 Apr 2011

Harper KA, Macdonald SE, Burton PJ, Chen J, Brosofske KD, Saunders SC, Euskirchen ES, Roberts D, Jaiteh MS, Esseen PA (2005) Edge influence on forest structure and composition in fragmented landscapes. Conserv Biol 19:768–782

Hilty JA, Lidicker WZ, Merenlender AM (2006) Corridor ecology: the science and practice of linking landscapes for biodiversity conservation. Island Press, Washington, DC

Hochkirch A, Gartner AC, Brandt T (2008) Effects of forest-dune ecotone management on the endangered heath grasshopper, *Chorthippus vagans* (Orthoptera: Acrididae). Bull Entomol Res 98:449–456

Ivanov K, Keiper J (2010) Ant (Hymenoptera: Formicidae) diversity and community composition along sharp urban forest edges. Biodivers Conserv 19:3917–3933

Jewitt GPW, Lorentz SA, Gush MB, Thornton-Dibb S, Kongo V, Blight J, Stuart-Hill SI, Versfeld D, Wiles L (2009) Methods and guidelines for the licencing of SFRAs with particular reference to low flows. Water Research Commission Report No.1428-1-09, WRC, Pretoria. (ISBN 978-1-77005-877-4)

Jongman RHG (1995) Nature conservation planning in Europe – developing ecological networks. Landsc Urban Plan 32:169–183

Jongman RHG (2004) The context and concept of ecological networks. In: Jongman RHG, Pungetti G (eds) Ecological networks and greenways – concept, design and implementation. Cambridge University Press, Cambridge, pp 7–33

Jongman RHG, Bouwma IM, Griffioen A, Jones-Walters L, Doorn AM (2011) The Pan European Ecological Network: PEEN. Landsc Ecol 26:311–326

Jorgenson JR, Wells CG, Metz LJ (1975) The nutrient cycle: key to continuous forest production. J Forest 73:400–403

Joubert L (2011) Biodiversity value of grassland ecological networks in afforested areas, KwaZulu-Natal, South Africa. M.Sc. thesis, Stellenbosch University, Stellenbosch

Kirkman KE, Pott RM (2002) Biodiversity conservation in plantation forestry. In: Pierce SM, Cowling RM, Sandwith T, MacKinnon K (eds) Mainstreaming biodiversity in development – case studies from South Africa. The World Bank Environmental Department, Washington, DC, pp 33–42

Koh LP, Lee TM, Sodhi NS, Ghazoul J (2010) An overhaul of the species-area approach for predicting biodiversity loss: incorporating matrix and edge effects. J Appl Ecol 47:1063–1070

Kotze DJ, Samways MJ (2001) No general edge effects for invertebrates at Afromontane forest/grassland ecotones. Biodivers Conserv 10:443–466

Laclau J-P, Ranger J, Deleporte P, Nouvellon Y, Saint-Andre L, Marlet S, Bouillet JP (2005) Nutrient cycling in a clonal stand of *Eucalyptus* and an adjacent savanna ecosystem in Congo 3. Input–output budgets and consequences for the sustainability of the plantations. For Ecol Manage 210:375–391

Laclau J-P, Ranger J, de Moraes Gonçalves JL, Maquère V, Krusche AV, M'Bou AT, Nouvellon Y, Saint-André L, Bouillet J-P, de Cassia Piccolo M, Deleporte P (2010a) Biogeochemical cycles of nutrients in tropical *Eucalyptus* plantations: main features shown by intensive monitoring in Congo and Brazil. For Ecol Manage 259(9):1771–1785

Laclau J-P, Levillain J, Deleporte P, Nzila JDD, Bouillet J-P, Saint André L, Versini A, Mareschal L, Nouvellon Y, M'Bou AT, Ranger J (2010b) Organic residue mass at planting is an excellent predictor of tree growth in *Eucalyptus* plantations established on a sandy tropical soil. For Ecol Manage 260(12):2148–2159

Li L, He X, Li X, Wen Q, He HS (2007) Depth of edge influence of the agricultural-forest landscape boundary, southwestern China. Ecol Res 22:774–783

Likens GE, Bormann FH (1995) Biogeochemistry of a forested ecosystem, 2nd edn. Springer, New York

Lowman GRP (2004) Deposition of nitrogen to grassland versus forested areas in the vicinity of Sabie, Mpumalanga. M.Sc. thesis, University of the Witwatersrand, Johannesburg

Magoba RN, Samways MJ (2011) Comparative footprint of alien, agricultural and restored vegetation on surface-active arthropods. Biol Invasions 14:165–177. doi:10.1007/s10530-011-9994-x

Mendham DS, Grove TS, O'Connell AM, Rance SJ (2008) Impacts of inter-rotation site management on soil nutrients and plantation productivity in *Eucalyptus globulus* plantations in South-western Australia. In: Nambiar EKS (ed) Site management and productivity in tropical plantation forests: proceedings of workshops in Piracicaba (Brazil), 22–26 Nov 2004 and Bogor (Indonesia), 6–9 Nov 2006. CIFOR, Bogor, pp 79–92

Miller HG (1995) The influence of stand development on nutrient demand, growth, and allocation. Plant and Soil 168–169:225–232

Miura S, Yoshinaga S, Yamada T (2003) Protective effect of floor cover against soil erosion on steep slopes forested with *Chamaecyparis obtusa* (hinoki) and other species. J For Res 8(1):27–35

Morris AR (1986) Soil fertility and long term productivity of *Pinus patula* in Swaziland. University of Reading: In Partial fulfilment of the Degree of Doctor of Philosophy United Kingdom. 388pp
NWA (1998) National Water Act of the Republic of South Africa (No. 36 of 1998): RSA Government Gazette No. 19182. Cape Town
Nambiar EKS, Kallio MH (2008) Increasing and sustaining productivity in tropical forest plantations: making a difference through cooperative research partnership. In: Nambiar EKS (ed) Proceedings of workshops in Piracicaba (Brazil) 22–26 November 2004 and Bogor (Indonesia) 6–9 November 2006. Center for International Forestry Research (CIFOR), Bogor, pp 205–227
O'Connor TG, Uys RG, Mills AJ (2004) Ecological effects of fire-breaks in the montane grasslands of the southern Drakensberg, South Africa. Afr J Range Forage Sci 21:1–9
Olbrich K (1993) Final report: the chemistry and frequency of mist events on the Eastern Transvaal escarpment. DWAF Report FOR-DEA 658, CSIR, Pretoria
Pinheiro ERS, Duarte LD, Diehl E, Hartz SM (2010) Edge effects on epigeic ant assemblages in a grassland-forest mosaic in southern Brazil. Acta Oecol Int J Ecol 36:365–371
Ponder F, Fleming RL, Berch S, Busse MD, Elioff JD, Hazlett PW, Berch S, Busse MD, Elioff JD, Hazlett PW, Kabzems RD, Kranabetter JM, Morris DM, Page-Dumroese D, Palik BJ, Powers RF, Sanchez FG, Scott DA, Stagg RH, Stone DM, Young DH, Zhang J, Ludovici KH, McKenney DW, Mossa DS, Sanborn PT, Voldseth R (2012) Effects of organic matter removal, soil compaction and vegetation control on 10th year biomass and foliar nutrition: LTSP continent-wide comparisons. For Ecol Manage 278:35–54
Pryke SR, Samways MJ (2001) Width of grassland linkages for the conservation of butterflies in South African afforested areas. Biol Conserv 101:85–96
Pryke JS, Samways MJ (2009) Recovery of invertebrate diversity in a rehabilitated city landscape mosaic in the heart of a biodiversity hotspot. Landsc Urban Plan 93:54–62
Pryke JS, Samways MJ (2011) Conservation management of complex natural forest and plantation edge effects. Landsc Ecol 27:73–85. doi:10.1007/s10980-011-9668-1
Ranger J, Turpault MP (1999) Input–output nutrient budgets as a diagnostic tool for sustainable forest management. For Ecol Manage 122:139–154
Ries L, Fletcher RJ, Battin J, Sisk TD (2004) Ecological responses to habitat edges: mechanisms, models, and variability explained. Ann Rev Ecol Evol Syst 35:491–522
Samways MJ, Moore SD (1991) Influence of exotic conifer patches on grasshopper (Orthoptera) assemblages in a grassland matrix at a recreational resort, Natal, South-Africa. Biol Conserv 57:117–137
Samways MJ, Bazelet CS, Pryke JS (2010) Provision of ecosystem services by large scale corridors and ecological networks. Biodivers Conserv 19:2949–2962
Scott DF, Prinsloo FW (2008) Longer-term effects of pine and eucalypt plantations on streamflow. Water Resour Res 44:W00A08. doi:10.1029/2007WR006781
Scott DF, Smith RE (1997) Preliminary empirical models to predict reductions in total and low-flows resulting from afforestation. Water SA 23:135–140
Sherry SP (1953) The effect of different methods of brushwood disposal upon site conditions in wattle plantations. I. Experimental layout and apparatus designed for the measurement of run-off and soil-loss. In: Wattle Research Institute Report for 1952–1953 (sixth year), pp 33–41, Institute for Commercial Forestry Research; Pietermaritzburg, South Africa
Sherry SP (1954) The effect of different methods of brushwood disposal upon site conditions in wattle plantations. II. A study of run-off and erosion during the first two years after clear-felling and regeneration. In: Wattle Research Institute Report for 1953–1954 (seventh year), pp 27–36, Institute for Commercial Forestry Research; Pietermaritzburg, South Africa
Sherry SP (1961) The effect of different methods of brushwood disposal upon site conditions in wattle plantations. III. Run-off behaviour over an nine-year rotation. In: Wattle Research Institute Report for 1960–1961 (fourteenth year), pp 32–40, Institute for Commercial Forestry Research; Pietermaritzburg, South Africa
Sherry SP (1964) The effect of different methods of brushwood disposal upon site conditions in wattle plantations. IV. A study of run-off and erosion during the first two rainy seasons of

the second crop cycle of the experiment. In: Wattle Research Institute Report for 1963–1964 (seventeenth year), pp 41–50, Institute for Commercial Forestry Research; Pietermaritzburg, South Africa

Sherry SP (1971) The black wattle. University of Natal Press, Pietermaritzburg, South Africa

Slawski M, Slawska M (2000) The forest edge as a border between forest and meadow: vegetation and Collembola communities. Pedobiologia 44:442–450

Smith CW, Johnston MA, Lorentz S (1997a) Assessing the compaction susceptibility of South African forestry soils. I. The effect of soil type, water content and applied pressure on uni-axial compaction. Soil Till Res 41(1–2):53–73

Smith CW, Johnston MA, Lorentz S (1997b) Assessing the compaction susceptibility of South African forestry soils. II. Soil properties affecting compactibility and compressibility. Soil Till Res 43(3–4):335–354

Tiarks A, Ranger J (2008) Soil properties in tropical plantation forests: evaluation and effects of site management: a summary. In: Nambiar EKS (ed) Site management and productivity in tropical plantation forests: proceedings of workshops in Piracicaba (Brazil), 22–26 Nov 2004 and Bogor (Indonesia), 2–9 Nov 2006. Center for International Forestry Research (CIFOR), Bogor, pp 63–78

Turpie JK, Marais C, Blignaut JN (2008) The working for water programme: evolution of a payments for ecosystem services mechanism that addresses both poverty and ecosystem service delivery in South Africa. Ecol Econ 65:788–798

Uys RG, Bond WJ, Everson TM (2004) The effect of different fire regimes on plant diversity in Southern African grasslands. Biol Conserv 118:489–499

Vallet J, Beaujouan V, Pithon J, Rozé F, Daniel H (2010) The effects of urban or rural landscape context and distance from the edge on native woodland plant communities. Biodivers Conserv 19:3375–3392

van Halder I, Barbaro L, Jactel H (2011) Conserving butterflies in fragmented plantation forests: are edge and interior habitats equally important? J Insect Conserv 15:591–601

Verryn SD (2000) Eucalypt hybrid breeding in South Africa. In: Dungey HS, Dieters MJ, Nikles DG (eds) Hybrid breeding and genetics of forest trees. QFRI/CRC-SPF, Noosa, pp 191–199

Wilson JW, Stirnemann RL, Shaikh ZS, Scantlebury M (2010) The response of small mammals to natural and human-altered edges associated with Afromontane forests of South Africa. For Ecol Manage 259:926–931

Wise RM, Dye PJ, Gush MB (2011) A comparison of the biophysical and economic water use efficiencies of indigenous and introduced forests in South Africa. For Ecol Manage 262:906–915

Yu K, Li D, Li N (2006) The evolution of greenways in China. Landsc Urban Plan 76:223–239

Chapter 11
Determination of the Environmental Implications of Bio-energy Production Using a Life-Cycle Assessment Approach

Clemens C.C. von Doderer and Theo E. Kleynhans

11.1 Introduction

A variety of reasons have led to the promotion of indigenous renewable energy sources and to an entirely new energy paradigm from fossil to renewable energy resources. These include, amongst others, the need for security and diversification of energy supplies as well as for less reliance on fossil fuels, the uncertainty surrounding oil prices, and increasing concerns over environmental degradation and climate effects.

An implied aim of renewable energy production is the degree it can reduce an eventual impact on the environment associated with the use of the fossil energy that it will replace. However, as the bioenergy crop is growing, production of ancillary materials, conversion to an energy product and the use of the energy product are not necessarily free of environmental impacts. Thus it is essential that the benefits of bioenergy schemes be investigated from a life-cycle perspective. This has led to the development of a variety of methods to better comprehend the investigated system and, eventually, to reduce those environmental impacts.

There is broad agreement in the scientific community that life-cycle assessment (LCA) is one of the best methodologies for the evaluation of environmental burdens associated with the production of bioenergy and related products (Consoli et al. 1993; Davis et al. 2009; Cherubini and Strømman 2011). It allows an identification of opportunities for environmental improvement by identifying energy and materials used as well as waste and emissions released to the environment.

The energy- and greenhouse gas balances of bioenergy systems can differ significantly. A variety of factors need to be considered, such as the type of feedstock, type of procurement system or the type of conversion technology.

C.C.C. von Doderer (✉) • T.E. Kleynhans
Department of Agricultural Economics, University of Stellenbosch, Stellenbosch, South Africa
e-mail: cvd@sun.ac.za

T. Seifert (ed.), *Bioenergy from Wood: Sustainable Production in the Tropics*, Managing Forest Ecosystems 26, DOI 10.1007/978-94-007-7448-3_11,

Regional differences can also be significant, especially with respect to land use and biomass production patterns, as well as the reference system with which the bioenergy system is compared. In addition, bioenergy production usually results in the generation of co-products, which can replace conventional products providing further environmental benefits to the biofuel process chain (Cherubini et al. 2009).

The aim of this chapter is to provide an introduction into the LCA methodology. The usefulness and the functionality of an LCA process will be illustrated through a recent case study of a lignocellulosic bioenergy production systems.

11.2 Life-Cycle Assessment

The heightened awareness of environmental protection, and the possible impacts associated with products manufactured and consumed on it, has led to the development of methods to better comprehend and reduce those impacts (ISO 14040 1997). Life-cycle assessment (LCA) has been postulated as an important and comprehensive technique. In an LCA study, the whole system involved in the production, use and waste management of a product or a service is described. Intuitively, one can understand LCA as a structured and comprehensive technique to assess the potential environmental impacts and resources used throughout a product's life-cycle, i.e. from raw material acquisition via production and use phases, to waste management (Baumann and Tillman 2004). LCA studies are done in various contexts (ISO 14040 1997), assisting in

- identifying opportunities for improving the environmental aspects of products at various points in their life-cycle;
- decision-making in industry, governmental or non-governmental organisations (e.g. strategic planning, priority setting, product or process design or redesign);
- the selection of relevant indicators of environmental performance, including measurement techniques; and
- marketing (e.g. environmental claim, eco-labelling or environmental product declaration.

The LCA method has its origins mainly in the packaging and waste management, as well as the oil crisis and the energy debate at the beginning of the 1970s. Pioneers in the field of LCA were industrialised countries such as the USA, the UK, Germany and Sweden. It is generally accepted the first LCA was a study on the consequences of packaging and manufacturing beverage containers by the Midwest Research Institute on behalf of Coca Cola (Baumann and Tillman 2004). Subsequent to that other studies were initiated in Europe in both the private (Tetra Pak) and public (German Federal Ministry of Education and Science) sectors. Although public interest waned due to the ending of the first energy crisis in the late 1970s, private businesses and certain industries (e.g. product design) remained interested in the LCA approach. An increased environmental awareness in the 1980s saw once again the need in focussing on a systematic approach for capturing the environmental

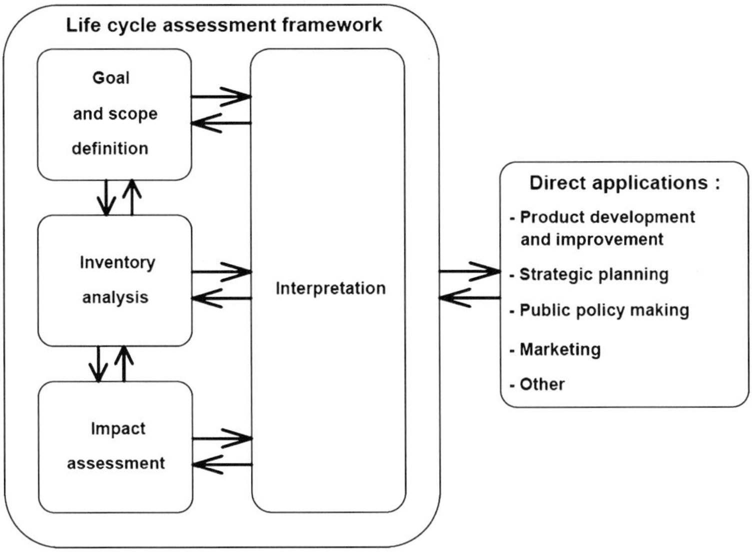

Fig. 11.1 Phases of a life-cycle assessment (Source: ISO 14040 1997)

impacts of product or service systems. The 1990s were characterised by the harmonisation and standardisation of the LCA methodology (e.g. ISO standards 14040–14044). Today LCA represents a common environmental assessment tool which is predominantly applied in the primary and secondary production sectors. The importance and relevance of LCA can be identified by a steadily growing number of LCA studies. Various software suppliers, such as SimaPro, GaBi and Umberto, have developed user-friendly LCA interfaces.

An LCA is structured into of four phases, namely (1) goal and scope definition, (2) inventory analysis, where an inventory of the relevant inputs and outputs of a product, process or service system is compiled, (3) impact assessment, where the potential environmental impacts associated with those inputs and outputs are evaluated and (4) interpreting the results of the inventory analysis and impact assessment phases in relation to the objectives of the study (refer to Fig. 11.1). Each phase is elaborated on below.

11.2.1 Goal and Scope Definition

The definition of goal and scope represents the foundation of life-cycle assessment. As stated in the ISO standard 14041 (1998), the definition of goal refers to; 'unambiguously state the intended application, the reason for carrying out the study and the intended audience', i.e. to whom the results of the study are intended to be communicated. The definition of the scope of the LCA sets the boundaries of the assessment, which includes the functional unit to be used, the product system to be studied and the product system's boundaries.

Since LCA is an iterative technique, the scope of the study may need to be modified while conducting the study as additional information is collected.

11.2.1.1 Functional Unit

After the goal, the product(s) and the system have been decided on, the functional unit needs to be defined. The functional unit corresponds with a reference flow to which all other modelled flows of the system are related (Cherubini et al. 2009). This is why the functional unit needs to be quantitative. The functional unit provides a reference to which the input and output process data are normalised, and the basis on which the final results are presented. Generally four types of functional units can be found in the bioenergy-LCA literature (Cherubini and Strømman 2011):

1 Input related – the functional unit is the unit of input biomass, measured in either mass or energy. With this type of functional unit, results are independent of conversion processes and types of end-products. This unit can be selected by referring to studies that aim at comparing the best uses for a given biomass feedstock.
2 Output unit related – here the functional unit is the unit of output, e.g., units of heat or power produced, or kilometres of transportation provided. This type of functional unit is usually selected by referring to studies aimed at comparing the provision of a given service using different feedstocks.
3 Unit of agricultural land – this functional unit refers, for instance, to hectare of land used to produce the biomass feedstock. This unit should be the first parameter to take into account when biomass is produced from dedicated bioenergy crops.
4 Time – results of the assessment are reported on a time basis. This type of functional unit is used in studies characterised by multiple final products, since it allows the avoidance of an allocation step.

Typical functional units in a bioenergy context are emissions/sequestrations per unit of energy produced, emissions/sequestrations per unit of land required.

11.2.1.2 Systems Boundaries

The system boundaries of a product, process or service system need to be specified in terms of several dimensions (Tillman et al. 1994), namely:

- Boundaries in relation to the natural system
- Geographical boundaries
- Time boundaries
- Boundaries within the technical system

In general, activities included in the flow model of a technical system (the inventory model) are activities under human control. However, when a flow enters (or leaves) human control, it also enters (or leaves) the technical system. While it is relatively easy for non-renewable resources such as oil and minerals to be defined in the 'cradle', i.e., during the extraction thereof, the boundaries for renewable resources, between the technical and the natural system, are less easy to draw. Renewable resources may be divided into fund resources (e.g. forests and agricultural land) and flowing resources (e.g. solar radiation and fresh water streams). In many cases, the boundary between the technical and the environmental system is obvious. However, when the life cycle includes forestry, agriculture, emissions to external wastewater systems and landfills, the system boundary needs to be explicitly defined (Finnveden et al. 2009).

Assessing bioenergy systems, geographical boundaries are an important consideration. Certain types of biomass feedstock may be limited to certain areas, and productivities may differ from area to area because they are limited by the availability of water, climate, soil or terrain conditions. Furthermore, infrastructure such as electricity production, waste management and transport systems may vary between regions (Von Doderer 2012).

Time boundaries, when defining the goal and scope of the study, are an important aspect of the LCA. Time defines the type of LCA study to be used. Change-oriented LCAs are future bound. They look forward in time, since they are about alternative choices of action. Accounting LCAs ask what environmental impact a product may be made responsible for; hence, they are retrospective (Baumann and Tillman 2004).

Boundaries within the technical system relate, to production capital and personnel. Whether the environmental impact from production and maintenance of capital goods should be included in an LCA has been debated (Baumann and Tillman 2004). For accounting LCAs, the guiding principle is that the study should be as complete as possible, and the production and maintenance of capital goods should thus be included. For change-oriented LCAs, whether or not capital goods will be affected by change has to be considered. A topic that is similar to that of capital goods is that of personnel. Processes require personnel, and personnel need food, transportation and so on. However personnel-related environmental impacts are usually not included in an LCA (Baumann and Tillman 2004).

Boundaries within the technical system include those in relation to other products' life cycles. Sometimes several products (or functions) share the same process(es). If the environmental load of these processes is to be expressed in relation to a single function, then there is an allocation problem. A detailed discussion of the types of allocation problems, the principles pertaining to allocation, and specific operational allocation methods can be found in Baumann and Tillman (2004).

11.2.1.3 Land-Use Change

Especially when assessing the environmental burdens associated with the production of bioenergy and related products, particular focus should be given to the impact due to land-use change.

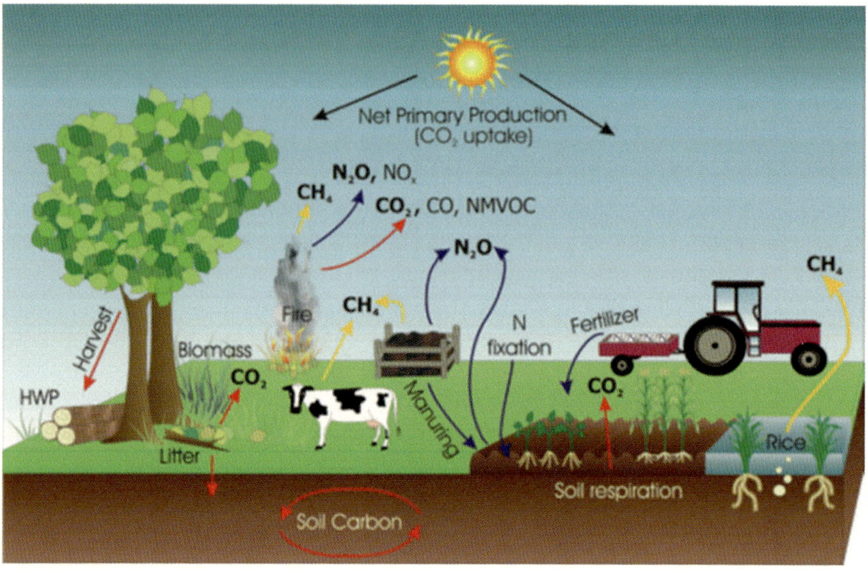

Fig. 11.2 The main greenhouse gas emission sources/removals and processes in managed ecosystems (Source: Paustian et al. 2006)

Land-use, depending on the land management practices, can be accepted as the main cause of soil degradation. Perennial crop systems, such as SRC plantation systems, tend to accumulate soil organic carbon and can serve to remediate contaminated soil (Brandão et al. 2010). Land-use change to bioenergy production can occur in two ways: (i) directly, when uncultivated land and pasture are converted to produce energy crops, or (ii) indirectly, through displacing food and feed crop production to new land areas previously not used for cultivation. From an LCA perspective, direct land-use change is often straightforward and easy to include in the assessment (Reijnders and Huijbregts 2008), although there are often uncertainties in the levels of carbon stock changes due to variations in local conditions and a lack of reliable field trial data (Fig. 11.2).

Mills et al. (2005) define six factors affecting the accumulation of carbon within an ecosystem:

- Carbon (C) storage is a function of mean annual precipitation (MAP) and temperature. Soil carbon and tends to increase with an increase in mean annual precipitation (Dalal and Mayer 1987; Hontoria et al. 1999). This is most likely due to primary productivition being a function of rainfall (Knapp and Smith 2001) and organic matter inputs into the soil tend to be greater in mesic than in arid regions.
- Carbon storage will increase with an increase in woody biomass.

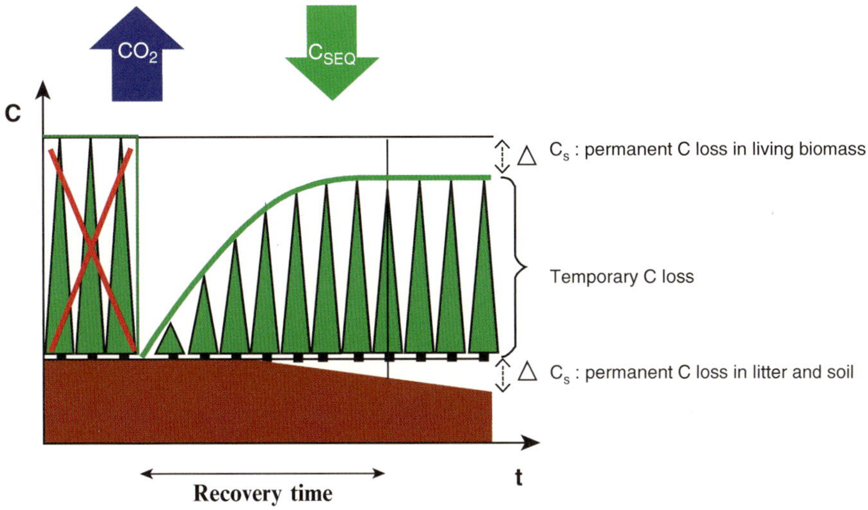

Fig. 11.3 Temporary and permanent carbon losses due to increased biomass use (Source: Bird et al. 2010)

- Frequent fires will lead to a decrease in Carbon storage in both biomass (Tilman et al. 2000) and soils (Bird et al. 2000).
- Tillage will reduce Carbon storage in biomass and soils (Tiessen et al. 1992; Gregorich et al. 1994; Aslam et al. 2000; Francis et al. 2001).
- The establishment or maintenance of a permanent cover of vegetation (e.g. pasture, thicket) will maintain or increase soil Carbon (Dalal and Chan 2001; Dominy and Haynes 2002). The effect of pasture establishment on the organic carbon storage capacity depends on the structure of the natural vegetation. Pastures may accumulate more carbon than natural grassland if a dense grass sward is established, but will have less carbon than woody systems.
- Any of the above effects will be dependent on changes to, and the inherent chemical and physical properties of the soil (Oades 1993; Zech et al. 1997; Percifal et al. 2000). The establishment of plantations on former grassland, for example, may be expected to reduce soil water content, improve soil aeration, and therefore reduce soil carbon storage (Birch 1958).

The possible change of carbon storage pools in the forest (i.e. trees, soil and litter) brought about by removing wood from forests should be considered, at least as a qualitative description (Schlamadinger et al. 1997). The most important carbon source in forest ecosystems are living vegetation (trees and other vegetation), dead organic matter and the forest soil (Jungmeier et al. 2003). In interpreting the carbon cycle, it is important to consider the following aspects: assumed rotation period of the forest ecosystem, changes to carbon storage pools, landfill by wood-based waste, and recycling (Fig. 11.3).

11.2.1.4 Critical Review Process

In some instances, for e.g. where LCA is used for certifying a system or a process, it is important to ensure the quality of the life-cycle assessment. Various factors need to be considered by the reviewer, namely that the methods used to carry out the LCA are consistent with the relevant ISO standards, and are scientifically and technically sound, that the data used are appropriate and reasonable in relation to the goal of the study, that the interpretations reflect the limitations identified and the goal of the study, and that transparency and consistency of the study report is ensured (Astrup Jensen et al. 1997).

11.2.2 Life-Cycle Inventory (LCI)

Inventory analysis involves data collection and calculation procedures to quantify the relevant inputs and outputs of a product system. The data collection can be a resource intensive process. For each unit process that is included within the systems boundaries the relevant inputs and outputs, such as use of resources and releases to air, water and land associated with the system, need to be considered. Interpretation may be drawn from these data, depending on the goal and scope of the LCA. These data also constitute the input to the LCIA (ISO 14040 1997) (Fig. 11.4).

A system model thus needs to be built according to the requirements of the goal and scope definition. The systems model is the flow model for a technical system with certain types of system boundaries (e.g. 'cradle-to-grave' or 'cradle-to-gate').

The process of conducting an inventory analysis is iterative. As data are collected and more is learned about the system, new data requirements or limitations may be identified that require a change in the data-collection procedures so that the goals of the study are still met. Sometimes, issues may be identified that require revisions to the goal or scope of the study.

The result is an incomplete mass and energy balance for the system. It is incomplete in the sense that only the environmental relevant flows are considered, which more or less include the use of scarce resources and the emissions of substances considered harmful. Environmentally indifferent flows such as water vapour emissions from combustion or industrial surplus heat are disregarded. Figure 11.4, below, is an illustration of the main steps and flows involved in an LCA.

When dealing with systems involving multiple products, allocation methods are needed. The materials and energy flows as well as associated emissions need to be allocated to the different products according to justifiable, clearly stated and well documented procedures. The allocation method can be, based on a unit of mass or energy, or in some instances on the financial value of the products.

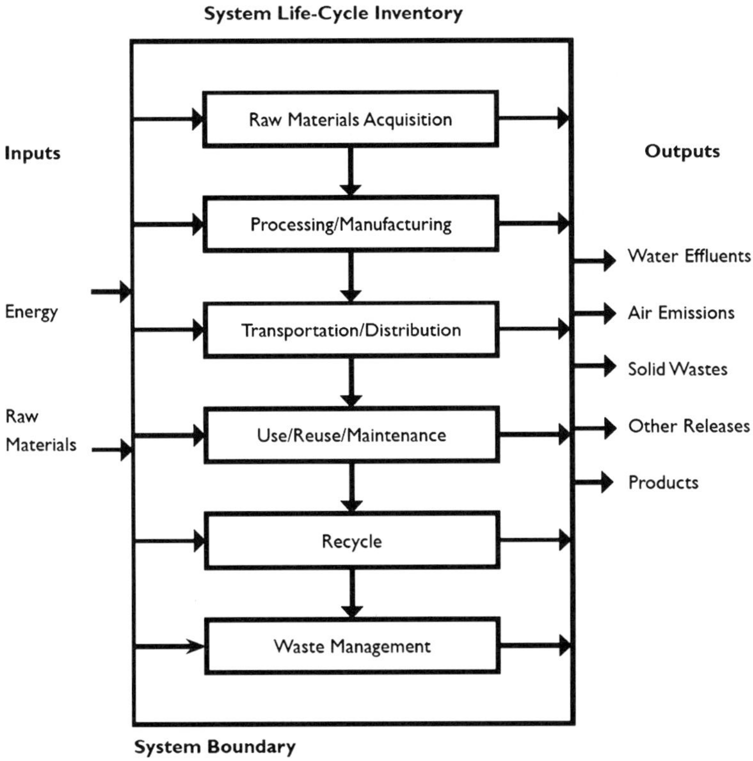

Fig. 11.4 Scheme of the main steps and flows involved in an LCA (Source: Bird et al. 2010)

11.2.3 Life-Cycle Impact Assessment (LCIA)

The impact assessment phase (LCIA) follows the LCI and involves an assessment of all relevant environmental impacts associated with the input and emissions mapped in the LCI. The LCIA also covers other chemically related impacts like global warming and tropospheric ozone formation, as well as the physical impacts on the land and input-related impacts or the availability of resources (Birkved and Hauschild 2006; Wenzel et al. 1997; Hauschild and Wenzel 1998). The level of detail, choice of impacts evaluated and methodologies used depend on the goal and the scope of the study (ISO 14040 1997).

The purpose of the LCIA is to provide additional information to help assess the results from the LCI so as to better understand their environmental significance (ISO 14040 1997). Thus the LCIA should translate the inventory results into their potential impacts in what are referred to as the "areas of protection" of the LCIA (Consoli et al. 1993), i.e. the entities that are to be protected by using the LCA. Today there is acceptance in the LCA community that the areas of protection offered by the LCA are human health, the natural environment, natural resources, and to some extent, the man-made environment (Udo de Haes et al. 2002; Finnveden et al. 2009).

LCA software packages such as the PE International's GaBi 4.4 or PRé's SimaPro 7.3 (PRé 2012) offer a variety of LCIA methods, including CML 2001, Eco-Indicator 95, EDIP 2003, Impact 2002+ or TRACI. One of the most commonly used, the CML 2001 method, is a collection of impact assessment methods that restricts quantitative modelling to the relatively early stages in the cause-effect chain, to limit uncertainties and group LCI results in mid-points categories, according to themes.

An important consideration for the LCIA is the spatial differentiation concerned. As the impacts caused by an emission depend on the quantity of substance emitted, the properties of the substance, the characteristics of the emitting source, and the receiving environment (Finnveden et al. 2009). The site-generic approach (or global default) followed in current characterisation modelling includes only the first two aspects, inherently assuming a global set of average/standard conditions concerning the properties of the source and the receiving environment. For truly global impact categories like climate change and stratospheric ozone depletion, this is not a problem, since the impact is independent of where the emission occurs. For the other impacts modelled in the LCIA, however, the situation can be different. They are often regional or local in nature, and a global set of standard conditions can disregard large and unknown variations in the actual exposure to stimuli of the sensitive parts of the environment (Potting and Hauschild 1997; Finnveden et al. 2009). At the same time, spatial differences can be reduced in the case of sources from multiple locations, particularly when these result in uniform emission distributions.

11.2.4 Interpretation

The objectives of the life-cycle interpretation are to analyse results, reach conclusions, explain limitations, and provide recommendations based on the findings of the preceding phases of the LCA or LCI study, and to report the results of the life cycle interpretation in a transparent manner. Furthermore, the interpretation phase is intended to provide a readily understandable, complete and consistent presentation of the results of an LCA or an LCI study, in accordance with the goal and scope definition of the study (ISO 14043 2000). Figure 11.5 shows the relationship of the elements within the interpretation phase with other phases of the LCA.

11.2.5 LCA Methods

Finnveden et al. (2009) distinguish between two types of LCAs: attributional and consequential LCAs. The attributional LCA is defined by its focus on describing the

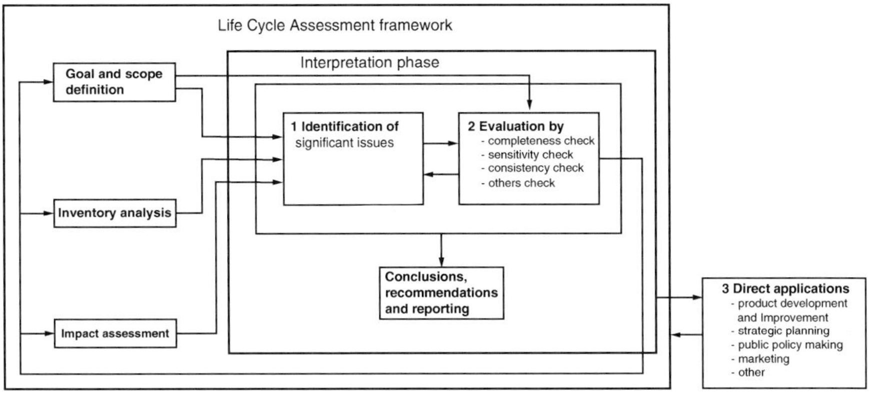

Fig. 11.5 Relationships of the elements within the interpretation phase with the other phases of LCA (Source: ISO 14043 2000)

environmentally relevant physical flows to and from a life cycle and its subsystems. The consequential LCA is defined by its aim of describing how environmentally relevant flows will change in response to possible decisions (Curran et al. 2005). Similar distinctions have been made in several other publications, but often using other terms to denote the two types of LCA, and sometimes including further distinctions of subcategories within the two main types of LCA (Guinée et al. 2002). Baumann and Tillman (2004), for instance, distinguish at least three types of LCAs, namely LCAs of the accounting type, LCAs of the change-oriented type and standalone LCAs.

LCA studies of the accounting type are comparative and retrospective. This type of LCA is well suited to different types of eco-labelling and can be used in purchasing or procurement situations, since these applications involve a comparison of existing products. LCA studies of the change-oriented type are comparative and prospective. This makes them useful in product development, building design and process choices, since decision-making involves a comparison of options that may be implemented or produced in the future. A standalone LCA is used to describe a single product, often in an exploratory way in order to get acquainted with some important environmental characteristics of that product, identifying the 'hot spots' in the life-cycle, i.e. which activities cause the greatest environmental impact (Baumann and Tillman 2004).

In general, the attributional method is the most used in LCA, but in LCA of bioenergy systems the consequential methods appears as the most broadly applied. Almost 75 % of relevant studies reviewed by Cherubini and Strømman (2011) compare the environmental impacts with those of a fossil reference system, as they are aimed at addressing the needs of policy makers, since consequential LCA is more relevant for decision-making.

11.2.6 Recent Applications in a Lignocellulosic Bioenergy Systems Context

With increasing maturity of the LCA approach, the numbers of studies using LCA have increased. A variety of LCA studies deal with forestry, forestry products, or different forestry production phases, such as harvesting, primary or secondary transport. The increasing interest in short-rotation coppice (SRC) systems is also reflected by the increasing number of LCA studies investigating the environmental impact of these bioenergy initiatives. A selection of LCA studies focussing on forestry operations and related products, as well as SRC plantings, is listed in Table 11.1, below.

11.3 Assessing Lignocellulosic Bioenergy Systems Using LCA – A Case Study

The following section provides a recent example for applying LCA in a bioenergy context: Following scheduled power cuts in 2007 due to outdated electricity infrastructure and low capacities of electricity generated, decision makers of the Cape Winelands District Municipality (CWDM), South Africa, seeked to investigate implementing local renewable bioenergy systems. By introducing alternative energy systems, the aim was to improve energy security and to reduce the dependency on the national energy supplier, ESKOM, while minimising the environmental impacts. A previous study (Von Doderer and Kleynhans 2010) identified bioenergy systems using lignocellulosic biomass grown in a short-rotation system as one of the most promising options for the CWDM. Thus, the LCA approach was used to determine the environmental performance of viable lignocellulosic bioenergy systems.

11.3.1 Goal and Scope Definition

The goal of this-study was to support the CWDM to identify the most favourable lignocellulosic bioenergy system by providing environmental performance data as decision support. Following a LCA approach, the first phase sets the foundation by defining the study's goal and scope, specifying the functional unit and then the different dimensions of system boundaries. The functional unit, which provides a reference to which the input and output process data arc normalised and the basis on which the final results are presented, was defined as 'burdens calculated for an average year's operation normalised to the electrical power produced per year', i.e. the electricity generated annually by a 5-MW system over 330 days of full production.

Table 11.1 Applications of the LCA approach in a forestry/woody biomass context

Application area	Year	Author(s)	Country of application	Nature and context of the problem
Forestry operations	1997	Berg (1997)	Sweden	Energy use and environmental impact of forestry operations
	2005	Berg and Lindholm (2005)	Sweden	General aspects of forestry operations
	2003	Klvac et al. (2003)		Energy audit of wood harvesting systems
	2007	Lawes et al. (2007)		Impact of colonial logging and recent subsistence harvesting in Afrotemperate forests
Forestry operations/ products	2001	Karjalainen et al. (2001)		Energy, carbon and other material flows of forestry and forest products
Forestry products	2003	Jungmeier et al. (2003)		Energy aspects in LCA of forest products
	2006	Nebel et al. (2006)	Germany	Wood-laminated floorings
Secondary transport	2009	González-García et al. (2009)	Sweden	Comparative environmental assessment of wood transport models: a case study of a Swedish pulp mill
	2000	Forsberg (2000)	Sweden, the Nether-lands	Comparison of biomass energy transport systems using LCA
Short-rotation coppice (SRC)	1999	Börjesson (1999a, b)	Sweden	Maximisation of environmental benefits of energy crop cultivation (SRC forest and energy grass)
	2003	Heller et al. (2003)		Willow SRC plantations for bioenergy production
	2007	Gruenewald et al. (2007)	Germany	Agroforestry systems for the production of woody biomass for energy transformation purposes
	2010	Roedl (2010)	Germany	Production and energy utilisation of wood from SRC plantations
	2011	Fiala and Bacenetti (2011)	Italy	Economic, energetic and environmental impact in SRC harvesting operations

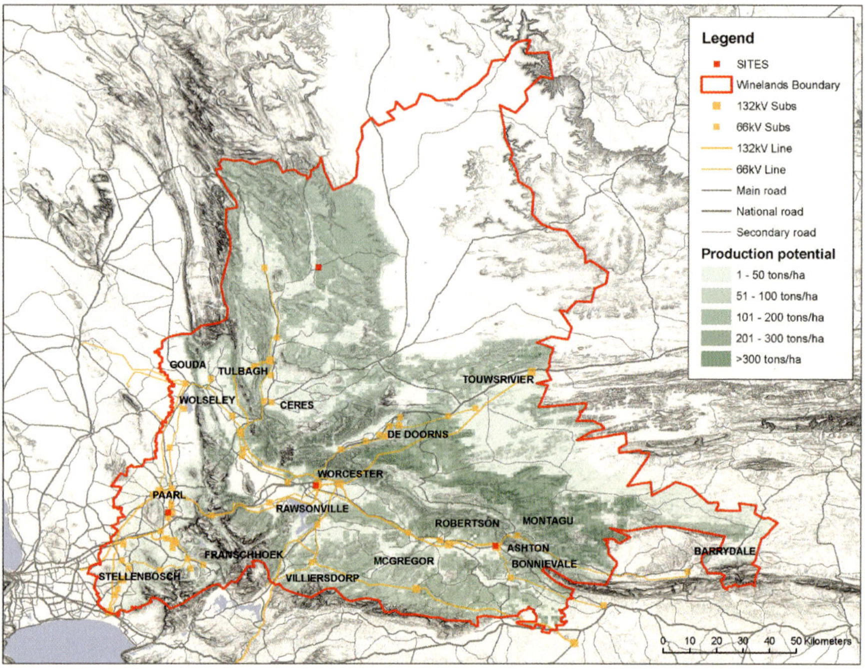

Fig. 11.6 Lignocellulosic biomass availability of the CWDM, including main electricity grid and electricity substations, as well as potential sites for bioenergy generation (Source: Van Niekerk and Von Doderer 2009)

Geographical boundaries were similar to the political boundaries of the CWDM, in the Western Cape, South Africa. Land productivity and availability for producing biomass in SRC systems was dine using GIS systems (Von Doderer 2009). Non-suitable areas, such as urban areas, areas with terrain limitations (>35 %), areas with water limitations (aridity index) and ecologically sensitive areas (protected areas, critical biodiversity areas, and water catchment areas), were excluded. About 175,000 ha was eventually determined potentially available for SRC systems. The completed biomass productivity assessment indicated that about 1.4 mt of fresh lignocellulosic biomass could be supplied annually (Von Doderer 2009).

Figure 11.6, above, illustrates the availability of potential land for producing woody biomass in the CWDM. It also shows potential sites for bioenergy conversion, based on access to infrastructure such as main electricity lines, electricity substations, road networks, and potential consumers of by-products (e.g. thermal energy) from the conversion of bioenergy.

Figure 11.7, below, gives schematic representation of biological biomass production process via photosynthesis as a simplified, static approach. CO_2, radiation from the sun and water are used to produce glucose, which is then converted in the metabolic cycle into lignin, cellulose, and hemicellulose. Additionally, as oxygen is generated; water is used as a reducing agent and is partially emitted; and the

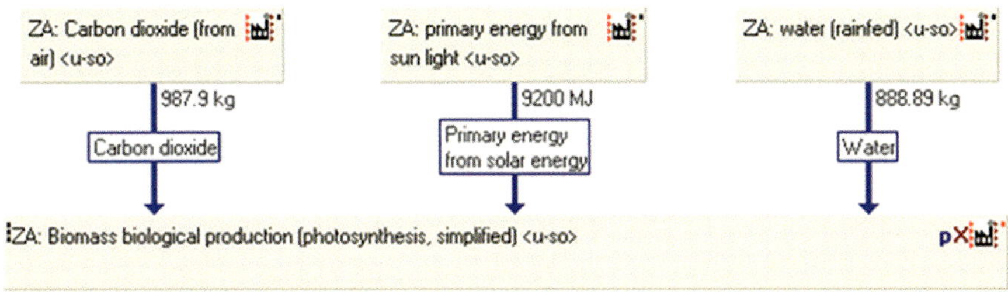

Fig. 11.7 Simplified static approach for the biological biomass production via photosynthesis (GaBi 4.4's software interface (PE International 2011))

sun's radiation is converted into chemical energy (refer also to Zimmer and Wegener 1996, or Von Doderer 2012).

GIS was also used in order to account for the carbon stock change due to the introduction of SRC plantations in the CWDM. Based on available land cover data, some generalisations were made: for instance only organic carbon taken into account; or typical production types were chosen as representatives for the land cover types available in the GIS database).

In general, the results of the case study show that when substituting current production systems (e.g. extensive dryland and improved grassland) with a biomass SRC production system, the organic carbon stock will in most cases increase. For further details, please refer to Von Doderer (2012).

11.3.1.1 Technical System Boundaries

For the case study the technical system boundaries were structured in five production phases:

1 Primary production of biomass in SRC plantations;
2 Harvesting and subsequent primary transport of the biomass from in-field to the roadside;
3 Pretreatment of the biomass, including comminution, drying and mobile fast pyrolysis;
4 Secondary transport of the bioenergy feedstock from the roadside to a central conversion plant; and
5 Further processing of the feedstock and its conversion into electricity.

This resulted in a set of 37 lignocellulosic bioenergy systems, which are characterised by different combinations of the production phases (Table 11.2). A general description of each module of the five production phases is given in the life-cycle inventory below (Von Doderer 2012). Best operating practices (BOP) are assumed for all activities/processes throughout the life-cycle.

Table 11.2 Overview of CWDM bioenergy pathways leading to a set of 37 lignocellulosic bioenergy systems

Plantation			Roadside		Road	Central conversion site		LBS
Primary Production of Biomass	Motor-Manual Harvesting (whole tree)	Extraction with forestry machinery (whole tree)	Mobile Chipping of Biomass	Transport of Comminuted Biomass		Direct Combustion of Biomass		01
						Direct Gasification of Biomass		02
						Central Pyrolysis + Conversion		03
				Mobile Pyrolysis	Transport of Mobile Pyrolysis Products	Combustion of Pyrolysis Products		04
						Bio-Oil in Gas Turbine/Bio-Char sold to Industry		05
			Transport of Uncomminuted Biomass			Stationary Chipping at Landing	Direct Combustion of Biomass	06
							Direct Gasification of Biomass	07
							Central Pyrolysis + Conversion	08
	Motor-Manual Harvesting (log)	Extraction with agricultural machinery (manual loading of logs)	Mobile Chipping of Biomass	Transport of Comminuted Biomass		Direct Combustion of Biomass		09
						Direct Gasification of Biomass		10
						Central Pyrolysis + Conversion		11
				Mobile Pyrolysis	Transport of Mobile Pyrolysis Products	Combustion of Pyrolysis Products		12
						Bio-Oil in Gas Turbine/Bio-Char sold to Industry		13
			Transport of Uncomminuted Biomass			Stationary Chipping at Landing	Direct Combustion of Biomass	14
							Direct Gasification of Biomass	15
							Central Pyrolysis + Conversion	16

Harvesting with Forestry Machinery (whole tree)	Extraction with forestry machinery (whole tree)	Mobile Chipping of Biomass	Transport of Comminuted Biomass		Direct Combustion of Biomass		17
					Direct Gasification of Biomass		18
					Central Pyrolysis + Conversion		19
			Mobile Pyrolysis	Transport of Mobile Pyrolysis Products	Combustion of Pyrolysis Products		20
					Bio-Oil in Gas Turbine/Bio-Char Sold to Industry		21
		Transport of Uncomminuted Biomass			Stationary Chipping at Landing	Direct Combustion of Biomass	22
						Direct Gasification of Biomass	23
						Central Pyrolysis + Conversion	24
	Extraction with agricultural machinery (loading of whole trees with three-wheeler)	Mobile Chipping of Biomass	Transport of Comminuted Biomass		Direct Combustion of Biomass		25
					Direct Gasification of Biomass		26
					Central Pyrolysis + Conversion		27
			Mobile Pyrolysis	Transport of Mobile Pyrolysis Products	Combustion of Pyrolysis Products		28
					Bio-Oil in Gas Turbine/Bio-Char sold to Industry		29
		Transport of Uncomminuted Biomass			Stationary Chipping at Landing	Direct Combustion of Biomass	30
						Direct Gasification of Biomass	31
						Central Pyrolysis + Conversion	32
Harvesting with Agricultural Machinery (whole tree)	Extraction with agricultural machinery (whole tree)	Transport of Comminuted Biomass			Direct Combustion of Biomass		33
					Direct Gasification of Biomass		34
					Central Pyrolysis + Conversion		35
		Mobile Pyrolysis		Transport of Mobile Pyrolysis Products	Combustion of Pyrolysis Products		36
					Bio-Oil in Gas Turbine/Bio-Char sold to Industry		37

Source: Von Doderer (2012)

11.3.2 Life-Cycle Inventory

The first production phase, primary production, is common to all the alternatives and takes all the activities and processes in the establishment and maintenance of the SRC plantations into account. It comprises, *inter alia*, mechanical and chemical land preparation, planting of fast-growing trees, weed control operations prior and after planting, and fertilising operations in order to enhance the growth rate of the trees, particularly during the first years after planting, until canopy closure is reached, after which competition from weeds is eliminated (Little et al. 1997).

The second production phase, harvesting and primary transport, comprises five harvesting system modules, including three different harvesting technologies and three types of primary transportation. The harvesting technologies modelled are motor-manual, mechanised forestry, and modified agricultural machinery. A forwarder fitted with a crane; a tractor-trailer combination loaded and unloaded, either manually or with a three-wheeler loader; and a tractor-container-trailer combination were assumed for the primary transportation.

The third production phase, pre-treatment of the biomass, entails three types of activities, namely comminution, drying and mobile fast pyrolysis. Depending on the harvesting system applied, two locations for comminution were proposed, i.e. mobile comminution at the roadside and stationary comminution at a landing of the central conversion plant. Similarly, both the location of the stored biomass and the form of the biomass (comminuted or un-comminuted) depend on the harvesting systems applied. In the case of four of the harvesting systems, uncomminuted biomass is stored in-field to air-dry for several weeks until the biomass has reached moisture content levels of about 40 % (dry basis). Once this level of moisture has been reached, the biomass is extracted to roadside for further processing. In the case of the remaining harvesting system, the trees are felled and comminuted in a single process, resulting in wood chips with moisture content levels of around 80 % (dry basis).

Irrespectively of the harvesting technology applied, additional drying of the biomass is required to reach the moisture content levels stipulated for the upgrading and conversion process. This is assumed to be achieved by using the exhaust heat from the respective conversion system. As no additional energy will be required, no additional costs and emissions arise from the active drying process.

As illustrated in Table 11.2, some of the alternatives use mobile fast pyrolysis. This is a process whereby the biomass is degraded in the absence of an oxidising agent, i.e. the volatile components of a solid carbonaceous feedstock are vaporised in primary reactions by heating, leaving a residue consisting of bio-char and ash. Pyrolysis always produces a gas vapour that can be collected as a liquid and as a solid char. Fast pyrolysis processes are designed and operated to maximise the liquid fraction by up to 75 wt% (dry basis). Thus, although fast pyrolysis can be understood as some form of pre-treatment of the biomass, it also represents one of the possible pathways for upgrading low-bulk-density biomass into densified, more homogeneous energy carriers (bio-oil and bio-char).

The fourth production phase encompasses the secondary transport of the bioenergy feedstock from the roadside to a central conversion plant. Un-comminuted biomass is assumed to be transported with a truck-trailer combination, comminuted biomass and bio-char with a truck-container-trailer combination, and bio-oil from a mobile fast pyrolysis system by a dedicated truck-tanker combination.

Five configurations of bioenergy conversion systems (BCS) were assumed for the fifth production phase. BCS I is an integrated steam-turbine system where biomass at a maximum 20 % moisture content (dry basis) is combusted to generate steam, which is then used in a steam turbine to generate electricity. The same moisture content is required for BCS II, an integrated gasifier-gas-turbine system, where the biomass is upgraded to bio-gas, which in turn, is fed into a gas turbine. BCS III consists of a stationary fast-pyrolysis plant converting biomass (10 % MC) into bio-oil and bio-char. The upgraded products are then fed into an integrated boiler-steam-turbine system to generate electricity. An integrated steam-turbine system is also assumed for BCS IV, also using bio-oil and bio-char that is produced in a mobile fast-pyrolysis system at the roadside, close to the primary biomass production sites. The last bioenergy conversion system (BCS V) also encompasses mobile fast-pyrolysis systems, but differs in the final conversion step, where only bio-oil is used to generate electricity, by directly injecting the liquid into a gas turbine. The bio-char by-product is assumed to be sold to the fertilising industry. To some extent, this effectively works as a way of capturing and storing carbon.

11.3.3 Life-Cycle Impact Assessment

The purpose of the third phase of an LCA, the life-cycle impact assessment (LCIA), is to assess a product system's life-cycle inventory results, to better understand their environmental significance (ISO 14042 2000). The impact assessment is achieved by translating the environmental loads from the inventory results into environmental impacts.

For this study the so-called CML 2001 method (normalisation factors from November 2009) was applied to translate the environmental loads of the 37 lignocellulosic bioenergy systems into environmental impacts. CML 2001 is a collection of impact assessment methods that restricts quantitative modelling to the relatively early stages in the cause-effect chain, to limit uncertainties and to group LCI results into midpoint categories, according to themes (Guinée et al. 2002).

Only one of the global impact categories, the global warming potential ($GWP_{100\ years}$), calculated as t CO_2-equivalent, will be further discussed in this chapter. The results in terms of other environmental impacts of the assessed lignocellulosic bioenergy systems, such as the abiotic depletion potential (ADP, measured in gigajoules), acidification potential (AP, t SO_2-equivalent), eutrophication potential (EP, t phosphate-equivalent) and photochemical ozone creation potential (POCP, t ethene-equivalent), can be found in Von Doderer (2012).

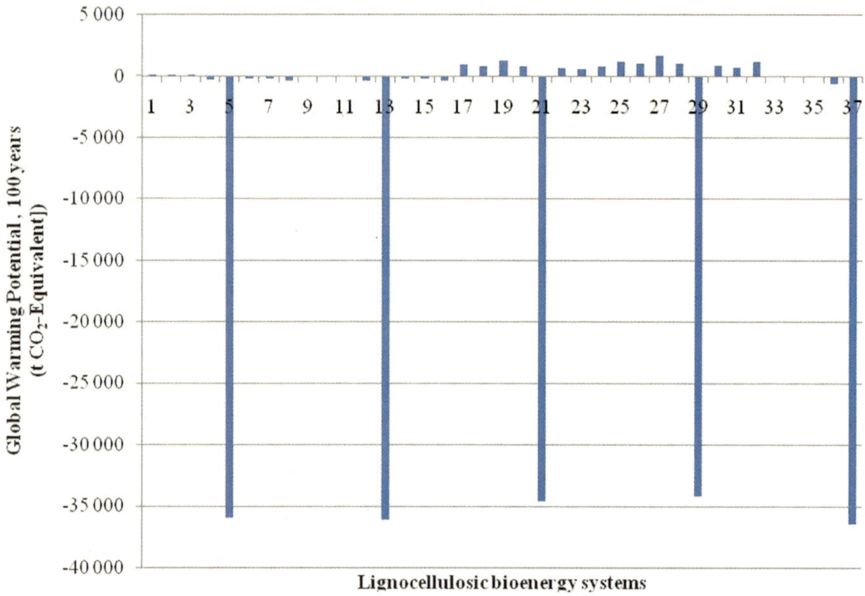

Fig. 11.8 The LBSs' global warming potential in high biomass productivity areas of the CWDM

Climate change may lead to a broad range of impacts on ecosystems and our societies, but greenhouse gases (GHG) have one property in common, which is useful for characterisation in an LCA. Characterisation of GHGs is based on the extent to which they enhance the radiative forcing in the atmosphere, i.e. their capacity to absorb infrared radiation and thereby heat in the atmosphere (Baumann and Tillman 2004: 149).

The mechanism of the greenhouse effect can be observed on a small scale, as the name suggests, in a greenhouse. These effects also occur on a global scale. Short-wave radiation from the sun reaches the earth's surface and is partially absorbed and partially reflected as infrared radiation. The reflected fraction is absorbed by greenhouse gasses (GHGs) in the troposphere and is re-radiated in all directions, including back to earth. This results in a warming effect on the earth's surface (PE International 2010).

This effect is amplified by human activities, in addition to the natural mechanism. Carbon dioxide is not the only gas that causes climate change. Methane, chlorofluorocarbons (CFCs), nitrous oxide and other trace gases also absorb infrared radiation. Compared with CO_2, they absorb much more effectively. The potential contribution of a substance to climate change is expressed as its global warming potential (GWP) (Baumann and Tillman 2004: 149).

The LBSs' overall performance in terms of global warming potential is presented in Fig. 11.8, above. Significantly, different results can be seen for LBSs 5, 13, 21, 29 and 37. These alternatives have bioenergy system V in common, where only

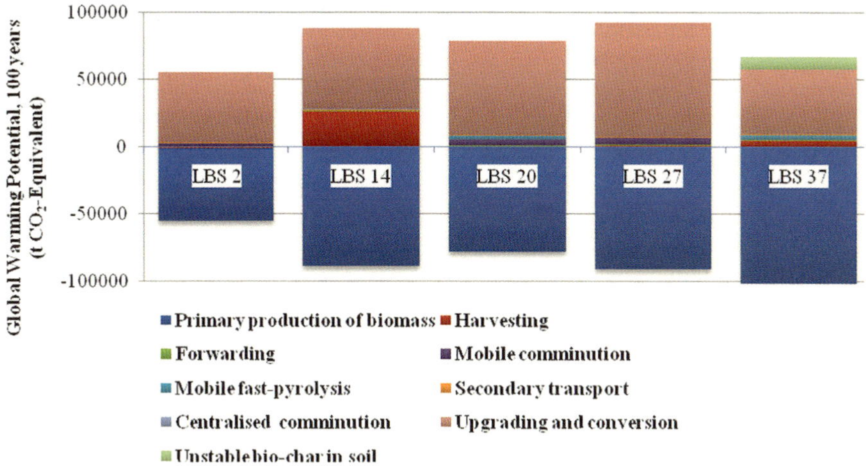

Fig. 11.9 Global warming potential of selected lignocellulosic bioenergy systems subdivided into production phases

bio-oil produced in mobile fast-pyrolysis units is used for electricity generation. The other product from the fast-pyrolysis process, bio-char, is assumed to be sold to the fertilising industry for application to soil. Eighty percent of the bio-char is assumed to be stable in the soil, resulting in negative GWP levels of more than 32,000 t CO_2-equivalent. For the other LBSs, a similar observation can be made as for the acidification and eutrophication potential impact categories: the greater the overall-conversion efficiency of the bioenergy conversion system applied, the fewer up-stream activities are required and the lower the GWP. Other LBSs also show negative GWPs, which can be explained by the positive effects of carbon stock changes when introducing SRC plantations. In these cases the increase in carbon stock compensates for the GHG emissions caused during harvesting, forwarding, pre-processing and secondary transportation. In comparison, the South African power-grid mix shows a GWP of more than 44,000 t/CO_2-equivalent, assuming the same functional unit.

Figure 11.9, above, shows the GWP performance of five selected LBSs, subdivided in production phases. LBS 14 uses bioenergy conversion system I, while LBS 2, 20, 27 and 37 deploy BCS II, IV, III and V respectively. The relatively large fraction of GWP for the harvesting phase for LBS 14 can be explained by the 30 % of unutilised biomass remaining infield. Although there is no direct relation between the harvesting and decomposition of the unutilised biomass, it is during the harvesting phase that the trees are felled, de-branched and cross-cut, leaving the tops and branches behind. LBS 37 entails an unstable carbon fraction. When using biochar as additive to soil around 20 % are assumed to be unstable, resulting in the decomposition thereof.

11.3.4 Interpretation and Outcome of Case Study

As mentioned in the goal and scope definition, the goal of this study was to support public decision-makers of the CWDM to identify the most favourable lignocellulosic bioenergy system. In the original study (Von Doderer 2012), the lignocellulosic bioenergy systems were assessed not only from an environmental perspective, but also from a financial- and socio-economic perspective. The generated performance data was further applied in a multi-criteria decision-making analysis using the analytical hierarchy process, to determine the most sustainable alternative. The study concludes that lignocellulosic bioenergy system 26 is most favourable, showing a strong financial-economic viability, a high socio-economic potential and a relatively low environmental impact.

Generally, the main driver for each criterion, whether it be of an environmental, financial-economic or socio-economic nature, is the overall conversion efficiency (OCE) of the biomass upgrading and bioenergy conversion system. The greater the OCE, the less biomass is required, resulting in fewer upstream activities and less land required for biomass production. In terms of the environmental impact of the LBSs, a greater OCE is desired, resulting in lower total emissions and, therefore, in lower impacts for each life-cycle impact category. Similarly, for the financial-economic viability of the LBSs, a greater OCE results in lower costs, both in terms of capital and operating expenditure, as well as in higher internal rates of return on the capital invested.

Another important driver is the efficiency of the harvesting system, which has an effect similar to the OCE. The greater the degree of mechanisation and automation, the lower the environmental impact and the higher the cost-effectiveness and profitability.

11.4 Conclusions

As shown in this Chapter, the life-cycle assessment (LCA) approach, originally developed as an environmental assessment tool, is a very useful tool to provide environmental performance information in a structured and comprehensive way. It can be understood intuitively as a tool to capture the environmental impacts along the entire life-cycle of a product or a service (from its 'cradle' to its 'grave'). The LCA method is structured in four phases, namely (1) goal and scope definition, (2) life-cycle inventory (LCI) analysis, (3) life-cycle impact assessment (LCIA), and (4) interpretation of the results.

The first phase sets the foundation of an LCA, by defining goal and scope and by specifying functional unit and the different dimensions of the system boundaries. All relevant inputs and outputs of the considered system are brought together during the second phase, the life-cycle inventory. In the third phase, all potential environmental impacts associated with the inputs and outputs are evaluated, by translating the environmental loads into impacts, which makes the results more environmentally relevant, comprehensible and easier to communicate. Several LCIA methods exist, and there is not always an obvious choice between them. Common areas of protection covered by LCAs are human health, natural environment, natural resources, and to some extent, the man-made environment. However, other environmental concerns, such as impact on biodiversity, water balance or land-use change, which are more difficult to specify, are not included in the LCA method.

Due to its systematic and transparent approach, LCA is well suited to being extended to measure a product's financial and social aspects along with its life-cycle. There is broad agreement in the scientific community that LCA is one of the most effective methods for evaluating the environmental burdens associated with biofuel and bioenergy production. This was confirmed in the second section of this chapter, which entails some of the results of a recent study aimed at determining the most sustainable lignocellulosic bioenergy system. Along the lines of the LCA framework, important aspects for assessing the environmental impacts of bioenergy systems are discussed. Furthermore, the from translating the environmental loads of 37 lignocellulosic bioenergy systems results in terms of their respective global warming potential were presented.

When comparing various bioenergy pathways, it can be further concluded that the overall conversion efficiency (OCE) of the biomass upgrading and bioenergy conversion system is one of the main drivers affecting the environmental performance of the assessed systems. The greater the OCE, the less biomass is required, resulting in fewer upstream activities and less land required for biomass production. In terms of the environmental impact of the LBSs, a greater OCE is desired, resulting in lower total emissions and, therefore, in lower impacts for each life-cycle impact category.

The efficiency of the harvesting system has an effect similar to the OCE. The results of this study show, that the greater the degree of mechanisation and automation, the lower the environmental impact.

References

Aslam T, Choudhary MA, Saggar S (2000) Influence of land-use management on CO_2 emissions from a silt loam soil in New Zealand. Agr Ecosyst Environ 77:257–262

Astrup Jensen A, Hoffman L, Møller B, Schmidt A, Christiansen K, Berendsen S, Elkington J, Dijk FV (1997) Life cycle assessment (LCA) – a guide to approaches, experiences, and information sources, Environmental issues series no 6. European Environment Agency, Copenhagen

Baumann H, Tillman A-M (2004) The Hitch Hiker's guide to LCA – an orientation in life cycle analysis assessment methodology and application. Studentlitteratur, Lund

Berg S (1997) Some aspects of LCA in the analysis of forestry operations. J Clean Prod 5:211–217

Berg S, Lindholm E-L (2005) Energy use and environmental impacts of forest operations in Sweden. J Clean Prod 13:33–42

Birch HF (1958) The effect of soil drying on humus decomposition and nitrogen availability. Plant Soil 10:9–31

Bird MI, Veenedaal EM, Moyo C, Llyod J, Frost P (2000) Effect of fire and soil texture on soil carbon in a sub-humid savanna (Matapos, Zimbabwe). Geoderma 94:71–90

Bird ND, Cherubini F, Pena N, Zanchi G (2010) Greenhouse gas emissions and bioenergy. In: Amezaga JM, Von Maltitz G, Boyes S (eds) Assessing the sustainability of bioenergy projects in developing countries – a framework for policy evaluation. Newcastle University, Newcastle upon Tyne

Birkved M, Hauschild MZ (2006) PestLCI – a model for estimating field emissions of pesticides in agricultural LCA. Ecol Model 198:433–451

Börjesson P (1999a) Environmental effects of energy crop cultivation on Sweden – part I: identification and quantification. Biomass Bioenergy 16:137–154

Börjesson P (1999b) Environmental effects of energy crop cultivation on Sweden – part II: economic valuation. Biomass Bioenergy 16:155–170

Brandão M, Milà I, Canals L, Clift R (2011) Soil organic carbon changes in the cultivation of energy crops: implications for GHG balances and soil quality for use in LCA. Biomass Bioenergy 35:2323–2326

Cherubini F, Strømman AH (2011) Life cycle assessment of bioenergy systems: state of the art and future challenges. Bioresour Technol 102:437–451

Cherubini F, Bird ND, Cowie A, Jungmeier G, Schlamadinger B, Woess-Gallasch S (2009) Energy- and greenhouse gas-based LCA of biofuel and bioenergy systems: key issues, ranges and recommendations. Resour Conserv Recycl 53:434–447

Consoli F, Allen D, Boustead I, Fava J, Franklin W, Jensen AA, De Oude N, Parrish R, Perriman R, Postlethwaite D, Quay B, Siéguin J, Vigon B (1993) Guidelines for life cycle assessment. A code of practice. SETAC (Society of Environmental Toxicology and Chemistry) Press, Pensacola

Curran MA, Mann M, Norris G (2005) The international workshop on electricity data for life cycle inventories. J Clean Prod 13:853–862

Dalal RC, Mayer RJ (1987) Long-term trends in fertility of soils under continuous cultivation and cereal cropping in Southern Queensland. VI. Loss of total N from different particle size and density fractions. Aust J Soil Res 25:83–93

Dalal RC, Chan KY (2001) Soil organic matter in rainfed cropping systems of the Australian cereal belt. Aust J Soil Res 39:435–464

Davis SC, Anderson-Teixeira KJ, Delucia EH (2009) Life-cycle analysis and the ecology of biofuels. Trends Plant Sci 14:140–146

Dominy CS, Haynes RJ (2002) Influence of agricultural land management on organic matter content, microbial activity and aggregate stability in the profiles of two oxisols. Biol Fertil Soils 36:298–305

Fiala M, Bacenetti J (2012) Economic, energetic and environmental impact in short rotation coppice harvesting operations. Biomass Bioenergy 42:107–113

Finnveden G, Hauschild MZ, Ekvall T, Guinée J, Heijungs R, Hellweg S, Koehler A, Pennington D, Suh S (2009) Recent developments in life cycle assessment. J Environ Manage 91:1–21

Forsberg G (2000) Biomass energy transport – analysis of bioenergy transport chains using life cycle inventory method. Biomass Bioenergy 19:17–30

Francis GS, Tabley FJ, White KM (2001) Soil degradation under cropping and its influence on wheat yield on a weakly structured New Zealand silt loam. Aust J Soil Res 39:291–305

González-García S, Berg S, Feijoo G, Moreira MT (2009) Comparative environmental assessment of wood transport models: a case study of a Swedish pulp mill. Sci Total Environ 407:3530–3539

Gregorich EG, Carter MR, Angers DA, Monreal CM, Ellert BH (1994) Towards a minimum data set to assess soil organic matter quality in agricultural soils. Can J Soil Sci 74:367–385

Gruenewald H, Brandt BKV, Schneider BU, Bens O, Kendzia G, Hüttl RF (2007) Agroforestry systems for the production of woody biomass for energy transformation purposes. Ecol Eng 29:319–328

Guinée JB, Gorrée M, Heijungs R, Huppes G, Kleijn R, De Koning A, Van Oers L, Wegener Sleeswijk A, Suh S, Udo De Haes HA, De Brujin H, Van Duin R, Huijbregts MAJ, Lindeijer E, Roorda AAH, Van Der Ven BL, Weidema B (2002) Handbook on life cycle assessment – operational guide to the ISO standards. Kluwer Academic Publishers, Dordrecht

Hauschild MZ, Wenzel H (1998) Environmental assessment of products, vol 2, Scientific background. Chapman and Hall/Kluwer Academic Publishers, London/Hingham

Heller MC, Keoleian GA, Volk TA (2003) Life cycle assessment of a willow bioenergy cropping system. Biomass Bioenergy 25:147–165

Hontoria C, Rodriguez-Murillo JC, Saa A (1999) Relationships between soil organic carbon and site characteristics in Peninsular Spain. Soil Sci Soc Am J 63:614–621

PE International (2011) LBP: GaBi 4.4 software-system and databases for life-cycle engineering. University of Stuttgart, Stuttgart/Leinfelden-Echterdingen

ISO 14040 (1997) Environmental management – life-cycle assessment – principles and framework. International Organisation for Standardisation (ISO), Geneva

ISO 14041 (1998) Environmental management – life cycle assessment – goal and scope definition and inventory analysis. International Organisation for Standardisation (ISO), Geneva

ISO 14042 (2000) Environmental management – life cycle assessment – life cycle impact assessment. International Organisation for Standardisation (ISO), Geneva

ISO 14043 (2000) Environmental management – life cycle assessment – life cycle interpretation. International Organisation for Standardisation (ISO), Geneva

Jungmeier G, Mcdarby F, Evald A, Hohenthal C, Petersen A-K, Schwaiger H-P, Zimmer B (2003) Energy aspects in LCA of forest products. Int J Life Cycle Anal 8:99–105

Karjalainen T, Zimmer B, Berg S, Welling J, Schwaiger H, Finér L, Cortijo P (2001) Energy, carbon and other material flows in the life cycle assessment of forestry and forestry products: achievements of working group 1 of the COST Action E9. In: Päivinen R (ed) Discussion paper. European Forest Institute, Joensuu

Klvac R, Ward S, Owende PMO, Lyons J (2003) Energy audit of wood harvesting systems. Scand J Forest Res 18:176–183

Knapp AK, Smith MD (2001) Variation among biomes in temporal dynamics of aboveground primary production. Science 291:481–484

Lawes MJ, Griffiths ME, Boudreau S (2007) Colonial logging and recent subsistence harvesting affect the composition and physiognomy of a podocarp dominated Afrotemperate forest. For Ecol Manage 247:48–60

Little KM, Kritzinger J, Maxfield M (1997) Some principles of vegetation management explained. ICFR Bulletin, 9/97

Mills AJ, O' Connor TG, Donaldson JS, Fey MV, Skowno AL, Sigwela AM, Lechmere-Oertel RG, Bosenberg JD (2005) Ecosystem carbon storage under different land uses in three semi-arid shrublands and a mesic grassland in South Africa. South Afr J Plant Soil 22:183–190

Nebel B, Zimmer B, Wegener G (2006) Life cycle assessment of Wood Floor Coverings – a representative study for the German flooring industry. Int J Life Cycle Anal 11:172–182

Oades JM (1993) The role of biology in the formation, stabilisation and degradation of soil structure. Geoderma 56:377–400

Paustian K, Ravindranath N, Van Amstel A, Gytarsky M, Kurz WA, Ogle S, Richards G, Somogyi Z (2006) Introduction to volume 4: agriculture, forestry and other land use. In: Eggleston S, Buendia L, Miwa K, Ngara T, Tanabe K (eds) 2006 IPCC guidelines for national greenhouse gas inventories. IGES, Hayama

Percifal HJ, Parfitt RL, Scott NA (2000) Factors controlling soil carbon levels in New Zealand grasslands: is clay content important? Soil Sci Soc Am J 64:1623–1630

Potting J, Hauschild MZ (1997) Predicted environmental impact and expected occurrence of actual environmental impact, part II: spatial differentiation in life-cycle assessment via the site-dependent characterisation of environmental impact from emissions. Int J Life Cycle Assess 2:209–216

Reijnders L, Huijbregts MAJ (2008) Biogenic greenhouse gas emissions linked to the life cycles of biodiesel derived from European rapeseed and Brazilian soybeans. J Clean Prod 16:1943–1948

Roedl A (2010) Production and energetic utilization of wood from short rotation coppice – a life cycle assessment. Int J Life Cycle Ass 15:567–578

Schlamadinger B, Apps M, Bohlin F, Gustavsson L, Jungmeier G, Marland G, Pingoud K, Savolainen I (1997) Towards a standard methodology for greenhouse gas balances of bioenergy systems in comparison with fossil energy systems. Biomass Bioenergy 13:359–375

Tiessen H, Salcedo HI, Sampaio EVSB (1992) Nutrient and soil organic matter dynamics under shifting cultivation in semi-arid north eastern Brazil. Agric Ecosyst Environ 38:139–151

Tillman A-M, Ekvall T, Baumann H, Rydberg T (1994) Choice of system boundaries in life cycle assessment. J Clean Prod 2:21–29

Tilman D, Reich P, Phillips H, Menton M, Patel A, Vos E, Peterson D, Knops J (2000) Fire suppression and ecosystem carbon storage. Ecology 81:2680–2685

Udo De Haes HA, Finnveden G, Goedkoop M, Hauschild MZ, Hertwich EG, Hofstetter P, Jolliet O, Klöpffer W, Krewitt W, Lindeijer E, Müller-Wenk R, Olsen SI, Pennington DW, Potting J, Steen B (2002) Life cycle impact assessment (LCIA): striving towards best practice. SETAC (Society of Environmental Toxicology and Chemistry) Press, Pensacola

Van Niekerk A, Von Doderer CCC (2009) Map: biomass production potential of the Cape Winelands District Municipality (personal communication). Stellenbosch

Von Doderer CCC (2009) Viability of producing lignocellulosic biomass in the Cape Winelands District Municipality for bioenergy generation. Department of Agricultural Economics, University of Stellenbosch, Stellenbosch

Von Doderer CCC (2012) Determining sustainable lignocellulosic bioenergy systems in the Cape Winelands District Municipality, South Africa. Department of Agricultural Economics, University of Stellenbosch, Stellenbosch

Von Doderer CCC, Kleynhans TE (2010) Financial viability of agroforestry for bioelectricity generation on various farm types in the Western Cape, South Africa. Agrekon Agric Econ Res Policy Pract South Afr 49:168–194

Wenzel H, Hauschild MZ, Alting L (1997) Environmental assessment of products, vol 1, Methodology, tools, techniques and case studies. Kluwer Academic Publishers, Hingham

Zech W, Senesi N, Guggenberger G, Kaiser K, Lehmann J, Miano TM, Miltner A, Schroth G (1997) Factors controlling humification and mineralisation of soil organic matter in the tropics. Geoderma 79:117–161

Zimmer B, Wegener G (1996) Stoff- und Energiefluesse vom Forst zum Saegewerk. Holz als Roh- und Werkstoff 54:217–223

Index

T. Seifert (ed.), *Bioenergy from Wood: Sustainable Production in the Tropics*, Managing Forest Ecosystems 26, DOI 10.1007/978-94-007-7448-3,

Printed by Printforce, the Netherlands